AF677011

WATER CONSERVATION

HOW TO ORDER THIS BOOK

BY PHONE: 800-233-9936 or 717-291-5609, 8AM–5PM Eastern Time

BY FAX: 717-295-4538

BY MAIL: Order Department
Technomic Publishing Company, Inc.
851 New Holland Avenue, Box 3535
Lancaster, PA 17604, U.S.A.

BY CREDIT CARD: American Express, VISA, MasterCard

BY WWW SITE: http://www.techpub.com

PERMISSION TO PHOTOCOPY—POLICY STATEMENT

Authorization to photocopy items for internal or personal use, or the internal or personal use of specific clients, is granted by Technomic Publishing Co., Inc. provided that the base fee of US $3.00 per copy, plus US $.25 per page is paid directly to Copyright Clearance Center, 222 Rosewood Drive, Danvers, MA 01923, USA. For those organizations that have been granted a photocopy license by CCC, a separate system of payment has been arranged. The fee code for users of the Transactional Reporting Service is 1-56676/00 $5.00 + $.25.

VOLUME 3

WATER CONSERVATION

Water Management, Purification & Conservation in Arid Climates

EDITED BY

Mattheus F. A. Goosen, Ph.D.

Walid H. Shayya, Ph.D.

TECHNOMIC PUBLISHING CO., INC.

LANCASTER · BASEL

Water Conservation, Volume 3

a**TECHNOMIC**® publication

Technomic Publishing Company, Inc.
851 New Holland Avenue, Box 3535
Lancaster, Pennsylvania 17604 U.S.A.

Copyright © 2000 by Technomic Publishing Company, Inc.
All rights reserved

No part of this publication may be reproduced, stored in a retrieval system, or transmitted, in any form or by any means, electronic, mechanical, photocopying, recording, or otherwise, without the prior written permission of the publisher.

Printed in the United States of America
10 9 8 7 6 5 4 3 2 1

Main entry under title:
Water Management, Purification and Conservation in Arid Climates—Water Conservation, Volume 3

A Technomic Publishing Company book
Bibliography: p.
Includes index p. 329

Library of Congress Catalog Card No. 99-68513
ISBN No. 1-56676-771-7

Table of Contents

Foreword

Water is essential for all living things. An adequate supply of good quality water is fundamental to the economic and social development of any society. However, a significant increase in population and per capita demand is imposing tremendous strains on global water resources. Water shortages have historically caused conflicts within societies. Proper water management is the key to avoiding such conflicts and to a sustainable development. Today, managers of water resources require multi-disciplinary skills and approaches to strive toward balancing the views of competing users and interest groups.

Purification of saline water and low-quality wastewater is one way of increasing the supply of usable water to meet growing demands. However, high energy costs of desalination and environmental and health concerns related to the use of treated wastewater are the main constraints that need to be addressed by society. Appropriate technology, legislation and its strict enforcement, and a high level of public awareness are also required to achieve satisfactory levels of water conservation.

Water is an especially scarce resource in arid countries. Desertification and the consequences of global warming will further strain water resources. Many arid countries are realistically facing the challenge to ensure an adequate supply of good quality water. Political leaders, decision-makers, scientists, engineers and the public need to work together to alleviate this problem. Failure to do so will have catastrophic consequences for future generations.

This three-volume book series covers various aspects of water management, purification, and conservation with up-to-date knowledge and information. The authors are from a multitude of disciplines and backgrounds. We are proud of the leading role that scientists at Sultan Qaboos University have taken to assemble such an important document.

Christopher D. Lu
Dean, College of Agriculture
Sultan Qaboos University

Preface

POPULATION GROWTH AND socioeconomic development in many regions of the world have led to growing pressures on the environment. Historically, civilizations evolved close to sources of water as this natural resource is an indispensable commodity that supported their growth and economic activity. On the other hand, surface and underground water resources are unevenly distributed in many regions of the world both in terms of quantity and quality. Economic and population growth intensifies the demand for water especially during drought seasons. This could lead to local and regional conflicts that would hamper development and affect long-term prosperity.

The demand for fresh water resources has seen marked increases in many parts of the world due to the combined effects of community expansion and increased water demand per capita. In dry climates where the availability of good quality water is limited, it is essential that this natural resource is exploited rationally and conserved at the national, regional, and local levels. As the water consumption rate in many regions of the world far exceeds the natural recharge rate, there is the need to both conserve and manage water resources more effectively. The management of these limited resources should focus on the proper exploitation of water to avoid conflicts. It should involve an integrated management approach through the participation of users. Such an approach should be based on environmentally sustainable levels as it must take into consideration the current and future surface and underground water potentials, with special emphasis on both the quantity and quality of available water. For example, the overexploitation of coastal aquifers in many regions has resulted in the problem of seawater intrusion. Consequently, restrictions on groundwater extraction have been instated to address this problem.

When renewable surface and underground water resources are not sufficient to meet the ever increasing demands from the agricultural, industrial, and domestic sectors, the focus will shift to finding alternative water sources. Examples include the use of reclaimed wastewater,

water reuse systems, and water purification. A more effective augmentation of scarce water resources can result by improving the efficiency of current desalination techniques. Alternative desalination techniques, such as utilization of solar radiation, also need to be considered.

This three-volume book series presents a broad and integrated approach to water management, purification, and conservation. Volume one includes an introductory chapter on water problems and water resources in arid climates followed by specific chapters covering various aspects of water management. Volumes two and three deal with water purification and water conservation, respectively. Many textbooks on water issues normally deal with only one of these areas. Our book series will differ from these volumes by covering all three areas with an emphasis on the problems faced by arid regions.

The intended audience for this series includes industry specialists in desalination and wastewater treatment, irrigation engineers, graduate and undergraduate students in hydrology, water management and water conservation, government personnel involved in water resources development, decision-makers, environmentalists, employees of the petrochemical industry, and individuals wishing to specialize in water management, purification, and conservation.

The views expressed in the chapters of this book series are those of the authors and not necessarily those of their respective institutions. Trade and company names mentioned do not institute an endorsement as these are provided for identification and information purposes only.

In closing, the processing of correspondence involved in the development of a book project involves a lot of work. We wish to offer a special thanks to Ms. Susan Liu for her diligent efforts and assistance in the preparation of this book series.

Mattheus F. A. Goosen
Walid H. Shayya
Sultan Qaboos University

1

Water Conservation: What Can We Do?

GORDON STANGER
32 Dutchman Drive
Hallet Cove SA 5158, Australia

INTRODUCTION

HOW SERIOUS IS the need to conserve water? Most of us have heard horror stories of unfortunate peasants, squatters, and refugees, who have to get up while it is still dark, walk many kilometers to the nearest well, line up to fill a container or two, and then walk back with the family's daily ration of water, which may or may not be safe to drink without sterilization. We regard these stories of debilitating drudgery as exceptional, while we take it for granted that our own domestic supply is virtually unlimited. As long as water emerges from our own tap, there can't really be a serious problem—or can there? With over six billion people, utilizing nearly half of the gross biological productivity of the planet, and with a strongly skewed water distribution, more than two-thirds of all countries have now experienced some degree of water shortage. Even by the most conservative estimates, the global population cannot stabilize at less than ten billion people, all of whom will probably demand a water supply of at least 250 liters per person per day, not to mention the much larger hidden water demand required to produce food and nearly all other consumer items. A water crisis is imminent, if not upon us already, and it will not be long before the principles of water conservation, at present only an optional novelty, become an unavoidable necessity. Several good references concerning the topics presented in this chapter may be used for further reading. These include Akad (1989; 1990), Brengle (1982), Charles (1991), Dandy (1987), DeCook et al. (1988), Karpiscak et al. (1990; 1991), Lynden (1990), Syme et al. (1983), Tuscon Water (1992), UNESCO (1979), and Walters (1990).

The varied spectrum of feasible water conservation measures may be considered in terms of: agricultural, industrial, infrastructural, domestic household, and domestic garden.

CONSERVATION IN AGRICULTURE AND NATIONAL WATER SUPPLIES

Examination of Figure 1 indicates the overwhelming importance of irrigation water as the major consumptive use in all but the wettest and most intensely industrialized countries. Japan's preoccupation with growing rice in paddy fields makes it somewhat anomalous in terms of water usage. Otherwise, the countries are listed in order of decreasing aridity, and clearly indicate both the correlation between the aridity and irrigation water demand and the need for more than 80% of all water to be devoted to agriculture in arid to semi-arid regions.

Measures required to conserve agricultural water in arid areas are many, varied and somewhat specialized, being beyond the scope of a general overview. However, there are three general principles that are just as applicable to the small garden as to farms on the grand scale. First, the intensity of runoff in deserts threatens soil, nutrient and

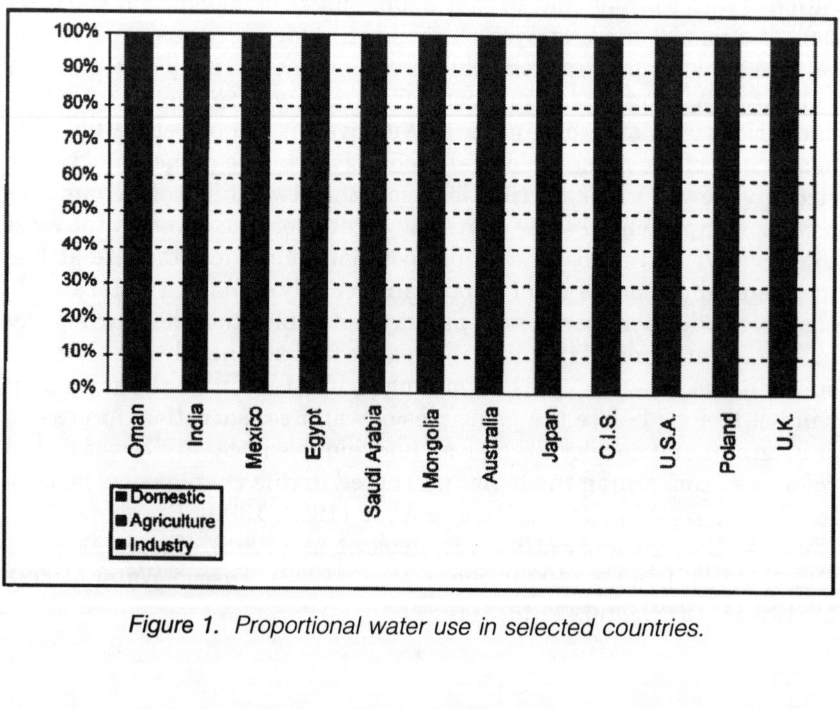

Figure 1. Proportional water use in selected countries.

water loss, and therefore, the advantages of field leveling, perhaps with steps or terraces are obvious. The modern high-tech practice of laser leveling of fields may be appropriate in mediterranean areas, where there is no serious shortage of water, but are too extravagant for semi-arid to arid environments. In the case of large-scale contour farming, there may be a conflict between wind orientation and the orientation required to maintain level strips of irrigated land. The effect of hot low-humidity winds blowing along a strip of land can be extremely desiccating, and hence cross-strip windbreaks may be useful. If high wind exposure or sloping land is inevitable, then some form of mulching, which is almost always beneficial, becomes even more important.

Second, unless large-scale inter-basin water transfer is possible, the scarcity of water in arid regions precludes very extensive cultivation, in which case the "oasis effect" becomes important. This is shown schematically in Figure 2, where the cultivated perimeter is exposed to hot drying winds which cause greatly increased transpiration. Examples of eucalypts have been studied where edge exposure increases the transpiration rate from about 2.5 to 4.5 meters per year. In the case of trees or large shrubs, the perimeter plants afford partial shelter and protection to more interior plants, but may themselves be subject to extreme desiccation and transpiration stress. There is some argument as to whether this oasis effect is due solely to greater leaf area exposure to insolation along the perimeter, as opposed to the greater advective effect of wind exposure. Whatever the cause, and there is probably an element of truth in both arguments, the oasis effect can be minimized by making oases as large as possible, and merging gardens/parks/small

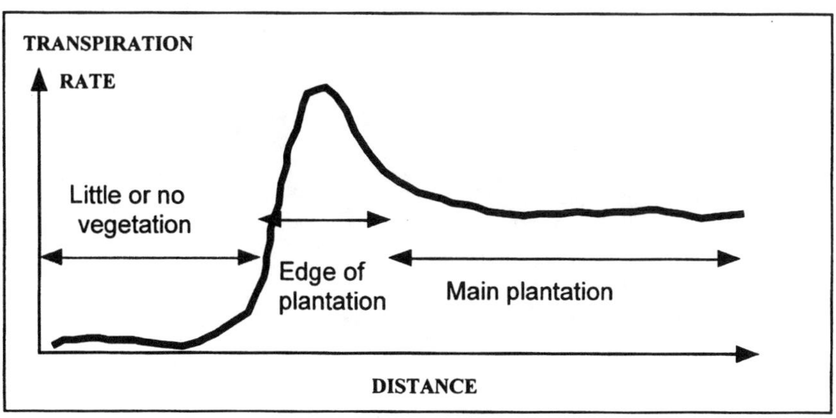

Figure 2. Schematic crop water consumption under desert conditions.

farms wherever possible. If large-scale cultivation is not possible, then at least the available shading, contours, and building positions should be taken into account to minimize wind and insolation exposure.

Third, high evapotranspiration rates, together with total dissolved salt concentrations, which tend to be higher than equivalent water supplies in wetter environments, lead to rapid build-up of sodium salts in the root zone or shallow water tables. Hence good drainage is of paramount importance.

Industrial water consumption is too varied to be treated in any generic discussion, but if the ethos of water conservation discussed below in the domestic context is successful, then a corresponding water-conscious attitude should not be difficult to transfer to industry. However, before considering water conservation in the home and garden, there is one preeminent source of water wastage that requires attention. Water authorities typically provide nearly all the domestic and municipal water, together with about half of the industrial water, all of which is supplied by distribution systems that vary enormously in their efficiency. Therefore, probably the most important water conservation measure of all consists of cutting down on distribution losses, some examples of which are shown in Figure 3.

Before accepting the above figures at face value, it should be remembered that domestic and industrial water meters are generally under-calibrated to about 95% of the true usage. This is a kind of insurance

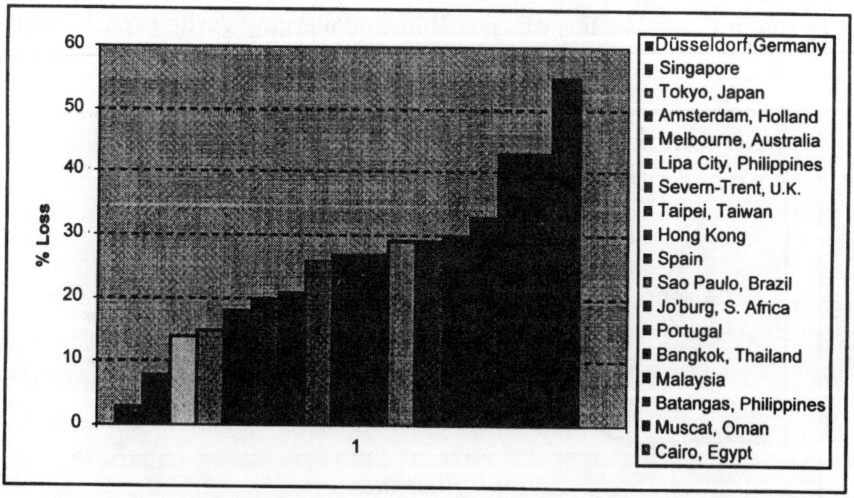

Figure 3. Selected international water system losses (also known as the "angels' share" = "non-revenue water").

on the part of water authorities to avoid legal actions arising from overcharging and, as such, shows up as an apparent 5% loss in the overall statistics. In addition, illegal connections (some of them ingeniously disguised!) cause loss of revenue, but are not strictly water losses in the sense of nonconsumption.

As shown in Figure 3, the typical water losses make up between 20% and 30% of the system. There is a marked tendency for the most prosperous and best developed countries to have the most conservative distribution systems, with the converse being inefficient water distribution in old, heavily overcrowded cities with grossly underfunded infrastructures. Cairo, with a loss rate of about 55%, is one of the most notorious and oft-cited examples, but hundreds of other cities are just as inefficient. This is not the kind of statistic that authorities are proud of, while its remedy is "out of sight, out of mind"—not the kind of high-profile solution that wins votes!

There are, of course, exceptions to the above trend: In many of the older cities of Europe, there are numerous cases of third world water distribution standards, all awaiting the allocation of the financial resources necessary to upgrade. Also, in the above histogram two adjacent cities in the Philippines are listed, Lipa and Batangas, having remarkably different loss rates of 20% and 43%, respectively. The expertise and resources available to the two cities are similar, but the political will to tackle the problem is radically different.

Referring again to Figure 1, domestic consumption seems like a minor part of overall water use, but this is no reason for complacency. Most rural water can never be available to the urban consumer, while both seasonal and annual fluctuations in irrigation water availability can be accommodated by varying the amount of fallow land, cropping frequency, or the ratio of annual to perennial crops. The urban context is less flexible and is more sensitive to water availability. Hence, it is incumbent upon of all of us to play a responsible part in water conservation.

In domestic and municipal contexts, successful water conservation requires a partnership between the consumer and the water utility management. Heightened problem-solution awareness, and motivation, is required on behalf of both parties if a significant shift in water usage is to be achieved. Moreover, a sustained commitment to conservation is required if the outcome is to become anything more than a cosmetic exercise. The objective is to promote a permanent shift in water use attitudes. This will require long-lasting effort, perhaps for as long as a generation, on the part of water authorities. Such sustained education must be maintained over perhaps seven or eight terms of political office. Any lesser commitment will not achieve optimum results.

WATER CONSERVATION IN THE DOMESTIC HOUSE

It is an unusual household that cannot cut its water consumption by at least a quarter. The first 15 to 20% is relatively easy and can be achieved with a minimum of technological aid. The next 10% is much harder, and requires real commitment and significantly changing attitudes on the part of the consumer. A 40% reduction in water consumption, in both indoor and outdoor use, is the kind of target that is currently attained only in experimental demonstration houses. It requires a great deal of research, together with recycling of grey water, water harvesting, and implementation of most of the conservation measures described below. Barring a radical shift in public attitudes and priorities, a water reduction target of about 30% seems to be about the maximum that is realistically achievable. More drastic reduction is possible but would be technologically innovative, involve a high capital expense, and would almost certainly result in an unacceptable reduction in living standards.

By far the most important factor in water conservation is the will to conserve; to develop a family ethos of water saving. Water conservation can, to a degree, be enforced by imposing draconian water charges, but this fosters resentment against the water authority and, where a base allowance is given per household, has even been known to induce deliberate wastage by people determined to receive every drop that they have paid for! Obviously, a happier and more successful aim is to foster a sense of cooperation between the consumer and the water authority for the common good. Sadly, such an attitude is missing in much of modern society, but it can be achieved, to the benefit of all.

It is well known that the primary influence upon water consumption is the mean degree of affluence in the society. Rising affluence is inevitably accompanied by higher expectations of water supply, irrespective of whether this is realistic in terms of locally available water resources. This is a deeply ingrained social attitude that is not likely to change without great effort on behalf of the water authorities. While education in water conservation is necessary to decrease the average household water consumption, or at least to stabilize it during a period of economic growth, an easier target is to reduce the highest bracket of water consumption in each economic class. Studies in Australia have shown that for any given style of house and garden, the relative water usage varies by at least a factor of two. That this is not due to peculiarities of particular housing infrastructure can be shown by comparing the water usage of successive occupants of the same house, in which the same order of variation in water usage (about twofold) still occurs. An obvious objective must therefore be to demonstrate that for a given standard of living, a lower water usage is normally feasible.

Leakage can occur in a multitude of ways. An easy periodic check is to note the water meter setting at night after all the household water use has occurred, and to compare this with the reading first thing in the morning before any water is used. Any significant difference should be investigated.

Quite apart from the moral imperative, the financial benefits of stopping leaks and adopting conservation measures should be a significant incentive. For example, a study in Tucson showed that the simple introduction of low-flow taps and low-flush toilets saved the average household U.S.$100 annually (1993 prices). Whether motivation takes the form of economic or social benefits, the incentive for water conservation must be blended with know-how. The following are some of the more common suggestions appropriate to arid regions that can and should be implemented, albeit with local modifications where appropriate.

General Points

A significant fraction of tap water is never actually used, but is run to waste while the user waits for the water to attain the required temperature. In particular, both water and energy is wasted as hot water displaces ambient water in the domestic pipe system, and as it is cooled in unlagged pipes. Obviously, insulation of hot water pipes will cut down on this wastage, but the main saving cannot be achieved easily by the user once the house is built. Water saving is achieved by minimizing the distance between end-user points and the source of water heating (or cooling). Therefore, the bathroom(s), kitchen and laundry should be designed to be in close proximity to the immersion heater, gas heater, solar panel, or whatever other heating systems might be in use. In a hot climate, this also has the advantage of minimizing the dissipation of unwanted heat throughout the house. Water-/energy-efficient plumbing is a specialized design consideration which, historically, has not received much attention. In this respect public education is ineffective. Rather, it requires the water authority to bring this conservation point to the attention of architects.

Even in nonarid regions, many water authorities have introduced mandatory water-efficient measures, such as low-flush or dual-flush toilets, low-flow/aerator taps, and low-flow showerheads as part of the building codes. One might have expected that arid regions would have been in the forefront of such legislation, but this is frequently not the case. Hence the educational ethos of water conservation as a way of life needs to be directed to the highest levels of government as well as to the general public.

Rainwater Tanks

Collection of rainwater from roofs is one of the oldest water sources in history. In many parts of Australia, where surface water is nonexistent, and where ground-water is brackish, rainwater tanks are still the only source of potable water—a matter of critical importance when significant replenishment occurs only once every few years. In many semi-arid to arid regions, this rural practice has also been adopted in suburban and even urban areas, but not without some misgivings. The problem is not the conservation ethos (which is admirable) but the water quality. Bird droppings, wind-blown bacteria and fungi, and airborne pollution (particulate traffic exhaust and other industrial emissions), all contribute to a potentially unhealthy "soup" that gets washed into the water tank after each rainstorm. The popular image of rainwater as "pure" is a myth. In addition to the microbial cocktail in the rain tank, urban/suburban air pollution is rich in nutrients (organic, nitrogenous, and phosphatic). Consider the rainwater in an outside tank, warmed by the sun, but shielded from the sun's sanitizing (ultraviolet) radiation, and inoculated with microbiota and nutrients: This is close to the ideal bacterial culture conditions! Even if the rainwater is filtered, microbial infections leading to gastero-intestinal problems is the most likely outcome. Filtration alone, without some form of disinfection, is ineffective. The problem does not end with the microbiota. Rain is seldom buffered by carbonate equilibria and typically falls as dilute carbonic acid at pH 4.5, which is ideal for dissolving toxic metals, such as lead from exhaust fumes or from lead flashing on the roof, and so on. In short, rooftop rainwater may look pure and potable, but is better regarded as good quality "greywater." Such water is usable for showers, baths, evaporative cooling and car cleaning, and is eminently suitable for toilet flushing and kitchen garden irrigation. Provided there is no turbidity, iron, manganese or copper content, rooftop rainwater is also well suited for laundry use.

Kitchen

Drinking water is such a small fraction of domestic water usage (about 2% of the total) that it seems parsimonious to even consider cutting down on it. Perhaps the only comment that need be made is to use it only as needed: It should be part of the general ethos to avoid leaving taps running unnecessarily. The average per capita water usage for drinking and cooking is about 7 liters per day, or perhaps slightly more under very hot conditions, although there is relatively little intercultural variation in this amount.

Ecologically, dishwashers are an abysmal idea, and should never have been invented! However, affluent families seem wedded to the notion of putting dishwashers in the "must have" category, so they are probably here to stay. It must be said that they are capital intensive, polluting, and very wasteful of water. Suggestions to minimize the impact are: Wash only full loads and experiment to find the minimum amount of detergent required to do the job. Preclean dishes with paper napkins and so on. There is a wide range of water efficiency in the models available that should be considered when buying a new one. It is not necessary to own a dishwasher and a large kitchen sink. A small sink is actually better for preparing vegetables and other foods (unless one is slaughtering an entire goat!).

Garbage grinders are also wasteful of water, and are another ecologically dubious household gadget. There is a global trend towards recycling and composting of waste food. If a compost heap is climatically feasible, then it makes good sense to have one in the garden. Nearly all foods can be composted, the main exception being foods with a high fat content, but very fatty foods should not be disposed of in garbage grinders anyway. An excellent way to compost food scraps is to employ compost worms in shaded boxes, a technique that is rapidly growing in popularity, at least in semi-arid areas.

There are few areas where leaky taps[1] are not a significant water loss. It is suggested that all taps, including those seldom used, be checked at least every six months, or more frequently in the case of chemically aggressive, very hard, or turbid waters. It is normally cost effective to replace worn washers, O-rings, packings and faulty fixtures *before* they begin to leak badly. For most purposes an 8- to 10-liter per minute low-usage tap is sufficient.

Bathroom

One has to be careful of the popular generalization that showers consume less water than a bath because, obviously, it depends upon

[1]"Faucets" in Americanese.

how long the shower lasts. About five minutes is all that is required for washing clean; anything more must be regarded as recreational, and indeed, some people shower for 15 minutes or more, in which case the difference in usage between a bath and a shower becomes minor. Normally a full bath uses about 150 to 200 liters of water, as compared to an old-style shower rose of between 10 and 12 liters per minute. A low-flow shower head uses around 7 liters per minute.

Aerating shower heads and taps deliver water at up to 50% reduced flow, as compared to their conventional equivalents, which can be achieved without impairing the effectiveness of washing. Such taps are manufactured for both mains pressure and gravity-feed conditions, but are more costly than conventional models and, having finer apertures, are more prone to clogging by silt or carbonate scale. Also, the rinsing time for washing hair is lengthened. The clogging problem therefore requires sound design choice, to offer taps that are quickly and easily dismantled, cleaned and reassembled. The rinsing problem is solved by the simple expedient of washing the hair when first stepping into the shower, to allow plenty of time to rinse during the rest of the shower.

Deep and narrow washbasins use less water than conventional washbasins, but there is no denying that they look very unattractive compared to their more fashionable counterparts. Moreover, only one encounter with the ridiculously narrow "space-saving" washbasins found in some cramped public facilities (and inexplicably, also in some badly designed homes) is more than enough to bias users for ever against their use! There is still great need for someone to design a washbasin that is both economical in water use, attractive and comfortable to use. In the meantime, insistence upon using the basin's plug, and not leaving the tap running, even when brushing teeth, would make a significant water saving. Overall, the average family generates about 300 liters of bathroom greywater per day, most of which (after reed bed treatment) is eminently suitable for flushing toilets.

Flush toilets are one of the major uses of water in a home, and most households could probably save a significant fraction of this water. The "rules" are:

1. Flush only feces, urine and toilet paper; not diapers, paper towels, tissues and other trash.
2. Fit and use a "half-flush" facility where appropriate.
3. Use a low total volume cistern, of not more than about 7 liters.
4. Repair leaks. A leaking toilet cistern typically wastes between 5 and 15 m^3.year, so even scarcely detectable leaks are worth correcting. In

order to test for such leakage, add a dash of food coloring to the cistern, followed by a few hours without flushing.

A common suggestion is to reduce the volume of a each flush by placing a brick, or a plastic bottle filled with water and weighted, in the cistern. Here one has to be careful, for it is possible to err on the side of too little, as well as too much. There is a possibility that if too little water is used, narrow sewers can become blocked, but the main problem is at the sewage treatment plant (STP). If the sewage is too concentrated, then the plant can become overloaded, in which case insufficient aeration and oxygenation is achieved, leaving the plant odourous and inefficient. In extreme cases, the overloaded plant may become anoxic, and a health hazard. Therefore, before urging the public to reduce flush volumes, some liaison with STP management is required to agree upon a manageable dilution strategy for the raw sewage.

Laundry

Washing machines are large consumers of water, varying between about 28 and 55 liters per person day. The higher figure tends to correspond to the proportion of young children in the family. Avoid small wash loads, use a suds-saver model, and when buying a new washing machine, utilize consumer magazine reviews to select a water-efficient model. Front-loading models typically use about 25% less water than top-loading designs.

WATER CONSERVATION IN THE DOMESTIC GARDEN

Enemy No. 1: Lawns

A lawn surrounding a house is one of the legacies of European colonialism that has been exported to virtually all parts of the world. Unfortunately, lawns are water intensive, typically using between 70 and 90% of all garden water. Moreover, a study in California showed that, despite their high water consumption, most lawns were overirrigated with an additional 90% of the grass water requirement.

Lawns are completely inappropriate for arid and semi-arid environments, but some fraction of the population seems to insist upon large lawns, whether it be in the Arizona desert, the rainless Arabian Gulf, the scorching Indian plains, or the red heart of Australia. Such incongruity is truly worldwide, and it seems like an uphill struggle to convince people that there are very attractive alternatives. If a lawn is considered indispensable, then at least minimize the impact: Water slowly, as infrequently as possible (using timers, and micro-irrigation, such as multiple low-level sprays), and at night. Keep the lawn well aerated, and use wind breaks and shade trees to minimize lawn transpiration. Let the grass grow higher in dry weather (at least 2 centimeters high), and use a coarse drought-resistant variety with a temperature tolerance appropriate to the location. Keep the grassed area to a minimum—careful variation of the edge geometry and placement of trees, shrubs and ground cover can give an impression of a larger grassed area than actually exists, especially if contoured into the landscape with ground cover plants spilling over the lawn perimeter. For greater irrigation efficiency, consider a stepped rather than a sloping lawn. Watering lawns is one of the best ways of recycling cleaning water, roof water runoff and pool water.

If lawns are to be discouraged, then what is the garden to look like? A few cities that have historically been hard pressed to meet their water demands, such as Tucson and Phoenix in Arizona, have pioneered the concept of "xeriscape," which is the art (and science?) of gardening in the desert in an ecologically appropriate and sustainable way. Proponents of xeriscape have met with great success, and are keen to point out that xeriscape is "not just rocks and cactus" (although there is nothing wrong with the striking and varied effects that *can* be achieved with just rock and cactus). There are several basic principles involved:

- reduction or replacement of lawn areas
- use of appropriate plants
- water-conserving zonation
- appropriate irrigation methods
- recycling and/or water harvesting
- soil improvement and use of mulches
- appropriate maintenance

Appropriate Plants

There is a popular notion that arid and semi-arid environments are of low botanical diversity. While this may be true locally, overall, the desert flora covers an enormous range of morphology, including an astonishing variety of water-conservation mechanisms. Many plants in the arid environment have evolved through the unusual crassulanic acid metabolic pathway of indirect photosynthesis, which allows them to photosynthesize at night, when temperatures are lower and the cost in water usage is less. Other very common plant mechanisms of water conservation include thick waxy foliage, hairy surfaces, and spiky leaves, all of which minimize transpiration.

True xerophytes have an extensive lateral root system that may interfere with nearby plants. To avoid root disturbance, allow them plenty of space and use water-conserving mulch around them.

Some care should be taken in species selection, as not all desert flora is water conservative. For example, oleanders are very thirsty, even though they live in desert regions. (They are found in the wadi beds where they can intercept water flowing at or near the surface.) Although common in dry areas, most species of Acacias and some Eucalypts[2] are opportunistic, and will behave as phreatophytes (i.e., transpiring large amounts of groundwater or irrigation water) given half a chance. Obviously, with such species, miserly irrigation can and should be applied.

Each continent, and in some cases each region, has its own native plant species that have adapted to local conditions, and that are naturally water efficient. A study in the Loxton Research Center, South Australia, showed that replacement of exotic plant species with water-conserving natives reduced plant water consumption by about 70%. Currently, only a small fraction of the water-conserving species found in nature are readily available for domestic cultivation, but this is still more than sufficient to provide an enormous choice for the xeriscapist. The most common problem is not how to find water-efficient plants, but how to cut down the choice to a manageable number without the xeriscape becoming overcrowded. In order to give a flavor of the species range available, a few of the many possibilities are listed in the Appendix A.

Water-Conserving Zonation

Popular wisdom recognizes three markedly contrasting xeriscape zones around a house, with the plant water demand in each zone being categorized as: medium, low and very low[3]. The inner zone is well shaded, both with small trees and by the house itself. At the same time

[2]There are about seven hundred species of Acacia and six hundred species of Eucalypts, spanning a huge range of drought and temperature tolerance.

[3]Anyone trying to flood irrigate rice or sugar cane outside their desert home is living on the wrong planet. Don't laugh: Australia, the driest continent, has examples of exactly that!

roof runoff, if not directed straight into tanks, is utilized in depression storage in which downpipes from the guttering direct water into cobble trenches or coarsely mulched areas. This zone provides most or all of the house shade. It is the smallest zone, the easiest area to irrigate with greywater, and the one in which exotics, or other less water-efficient plants (if any) should be planted. The inner zone also includes shade-cloth areas, patios and small kitchen gardens. The middle, or transition zone, is for plants that may require some supplemental water but that are, to a large extent, drought tolerant. The outer zone is generally more sparsely planted with for example, Acacias, and Saguaros, which require no irrigation after their establishment phase. The accent here is on screening with a sparse but attractive desert-scape.

Thus the whole zonation is a transition from a small water-generous core area with maximum shade, irrigated by some form of waste or recycled water, to a more extensive desert perimeter that is more in keeping with an arid ecosystem. Ideally, the total extent of irrigation should be matched by the amount of recycled plus runoff water, although it will probably require several years of learning followed by trial and error in order to achieve such an ideal balance. However, even if irrigation exceeds the amount of saved water, the overall reduction in water use, using xeriscape principles, will probably halve the water usage in the garden. Any water authority seriously contemplating water conservation is advised to develop a range of such zoned xeriscapes, open and free to public scrutiny, and illustrative of the variety and attractiveness attainable by means of xeriscape principles.

Appropriate Irrigation Methods

Domestic irrigation differs from farm irrigation not just in scale, but in priority. Although cost is always relevant, the main consideration for the gardener is minimum effort. The average gardener is not a specialist and has neither the time nor the inclination or the know-how to analyze plant water requirements. Herein lies the biggest single challenge in suburban water conservation: How to stop over-irrigation in the garden quickly, cheaply and easily.

The method of irrigation is very important. With fine-spray irrigation on a hot day with low humidity, over 50% of the water droplet volume can evaporate before it even reaches the ground. After that, further evaporation from exposed leaf surfaces can lose an additional fraction of the applied water. Of course, evaporative cooling from the wetted leaf area may still benefit the plant to some extent, but daytime irrigation is inevitably of low efficiency, and may also cause leaf scorching. Evening or early morning irrigation using timer taps is most appropriate. If a plant cannot survive the daytime heat before evening watering, then it is either too exposed or an inappropriate species. The hot deserts are in low latitudes where prolonged windy conditions are unusual, so gardeners can normally manage to avoid wasteful irrigation during windy conditions. Average gardeners waste about 5% of their water simply by forgetting to turn off the irrigation tap, so timer taps, commonly available from one- to six-hour settings, soon prove their worth in cost and convenience provided that the user (a) remembers to adjust the timer setting according to season, and (b) remembers to switch off the system altogether after rain. Good planning requires that sprinklers water the garden, not the adjacent street, path or other paved areas!

In the arid zone, micro-irrigation is the most appropriate irrigation method of all. The drip distribution system can be covered by the mulch to maintain a relatively cool water temperature. Drip delivery is much slower than in any spray system, so the duration of watering must be correspondingly longer, which also aids in effective infiltration into the root zone. Again, there is a wide range of settings available, with 2 to 8 liters per hour per aperture being the most useful range.

In the case of trees, an important consideration is to prune hard, wherever possible, to reduce the leaf area, and hence the tree's overall transpiration rate. This practice is also worthwhile to increase fruit yields (although fruiting trees utilize more water than nonfruiting trees), to train a more compact shape with better shade qualities, and to improve the tree's appearance. Of course, the first two years of tree establishment requires shade, more frequent watering and fertilizing, followed by a gradual decrease in the frequency of irrigation.

How much to irrigate is the inevitable, tricky question, for there is never a simple answer. Of the many relevant factors, one may consider soil type, plant type (depth of rooting), watering frequency and seasonality. Sandy soils allow rapid infiltration, in which case it is easy to overirrigate, with water seeping deeper than the depth of roots, in which case the deeper water seeps to waste. An exception is the occasional necessity to deliberately overirrigate in order to dissolve the build-up of sodium salts around the root zone and to leach these harmful salts down to the water table.

As a rough guide to watering, it may be assumed that the majority of plants in a given environment use a similar amount of water per unit area of foliar exposure. Translating this into practical terms, the maximum depth to which moisture should penetrate after irrigation is about a third of a meter for shallow-rooted flowers and vegetables, about two-thirds of a meter for larger shrubs and ground cover, and about a meter for trees. However, this rule of thumb could be excessive in some environments (especially in heavy, well-mulched soils, in the case of shallow caliche layers, or in the case of water-miserly specialized plants). A better guide is to experiment to find the minimum that will sustain plants through each seasonal variation, allowing for nonirrigation after wet antecedent conditions, and occasional leaching. The watch word should be: *Water just enough, and no more!* There is an obvious need for local guidance on irrigation in terms of amount and frequency for different planting conditions. It should not be beyond the capability of local water resources to address the difficult problem of "how much to water" in some form of standardized response in the local press, radio or TV broadcasts. If this is not achieved, then the biggest potential water saving by consumers is being missed.

Recycling and/or Water Harvesting

Allowing for wetting losses, about 80% of water that falls on a roof is generally collectable. Sloping roofs assist in collection, while sloping land tends to reduce the excavation necessary for tank water storage. Otherwise, roof runoff can be directed into circular, oval, channel depressions and so on, where cobbles or other decorative porous media can be designed to enhance both the appearance and the water storage potential of the garden. If gutter downpipes are considered ugly, they can be replaced by chains (with or without climbing plants), at between 5- and 10-meter spacing, and other more imaginative water guides, which can become a positive feature of the garden. Some appropriate ground contouring with channels, depressions and mulch covers can make the reuse of recycled water very easy, but such measures require careful consideration at the planning stage, as they are considerably more difficult to incorporate into a mature garden.

Soil Improvement and the Use of Mulches

Most desert soils are almost devoid of organic matter, and hence the second priority, after water, is to supply sufficient organic humus and associated nutrients. Unfortunately, temperatures in desert soils are so high that microbial activity breaks down organic material exceedingly rapidly, with a half life of about a year. Therefore, to establish a rich flora requires continuous replenishment with whatever organic material comes to hand. Practically anything of organic origin can be composted, although the fat content of a compost heap should be kept low. Mature compost can be either dug into the topmost soil layer, ideally to about half the expected root depth of whatever is planted, or spread as a surface mulch (or both, if you have sufficient compost!).

The purpose of a mulch is to cover the surface soil in such a way as to (a) reduce the soil temperature, (b) enhance infiltration into the soil, (c) suppress evaporative moisture loss from the soil, and (d) to inhibit weed growth. Although compost is the best and cheapest mulch, other organic alternatives include straw, wood chips, bark chips, and pine needles. All of these have been used with some success in the desert and semi-desert environment, although care needs to be taken with pine needles, as they rot to create an exceptionally high cation-exchange capacity, which can lead to acidification, and hence to nutrient starvation.

Mulch need not be organic. In addition to woven plastic weed control mat, rock chippings, water-rounded pebbles, sandstone boulders, crushed quartz and scoria are very effective at preventing soil evaporation when spread in a layer of at least 5, and preferably 10, centimeters depth. These nonorganic mulches give scope for variation in ground surface color, allowing pleasing patterns to be made. All mulches, whether organic or mineral, reduce the irrigation water requirement.

Another factor requiring attention is the sodicity of both the irrigation water and the mulch, because high transpiration rates in the desert environment result in high sodium concentrations in the residual soil moisture. Hence good drainage is essential if xeriscape is to succeed.

Appropriate Maintenance

No irrigation system is maintenance free. As with periodic checks for leaks and drips indoors, so it is necessary to minimize water losses by checking the drip and spray irrigation systems for leaks and blockages. As plants grow above ground, so their root system extends below ground, and this requires periodic adjustment of the depth of wetting during irrigation cycles. The xeriscape is an evolving microclimate with great potential for water saving provided the gardener is aware of how little water to use in any given season or situation.

As with household water use, many variables influence water consumption in the garden, so it is hard to generalize. However, as a general guide, converting the garden from a traditional "lawn and borders" format to a more appropriate xeriscape using recycled house water typically results in the order of a 40% reduction in garden water demand. Experience in Arizona has shown that the conversion from traditional to xeriscape-style gardens need not involve a loss in the market value of a property. On the contrary, a well-designed xeriscape may actually enhance the garden value, both aesthetically and financially.

Other Aspects of Water Conservation in the Garden

Car washing with a hose is both very wasteful of water and a source of oil pollution in the street drainage system. Use a car wash that recycles water. (Note to the water authority: Make sure that all car washes *do* recycle their water!) If cars are to be washed at home, do so near trees, hedges, or shrubs that can benefit from the "free drink." Alternatively, divert the runoff into a greywater collection system for subsequent reuse. Wash from a bucket rather than with a hose.

Swimming Pools

In a hot low-humidity environment, the evaporative loss is likely to exceed 2, and possibly 3, meters per year, so in affluent districts, the net evaporative loss can accumulate to many thousands of cubic meters per year. It is also a matter of safety and cleanliness to cover the pool when not in use. If draining the pool for any reason, reuse as much water as possible.

A SUMMARY OF DOMESTIC WATER CONSUMPTION

The details of domestic water usage vary enormously, according to climate, culture, affluence, water availability, and local pricing structures, so it is difficult to summarize the per capita consumption. However, in an attempt to put the above measures into perspective, Table 1 shows an average of published water use data for the drier regions of South Australia, South Africa and the Southwestern U.S.

Water Harvesting

Much research and publicity has been devoted in recent years to the specialized techniques of mist, fog and dew harvesting in coastal areas

Table 1. A summary of average water use in a typical western-style household.

Use	Daily per Capita Consumption, in Litres	% of Total
Garden	175	49.4
Shower and bath	49	13.8
Toilet flushing	35	10.0
Laundry	32	9.0
Leaks and wastage	28	7.9
Dishwashing	18	5.1
Cooking and drinking	7	2.0
Evaporative cooling	5	1.4
Car washing	3	0.9
Swimming pool	2	0.5
Total	354	100.0

Notes:
Showers and baths typically vary between 40 and 52 liters per person per day, and therefore consume about 30% of the indoor water use.
Toilet flush rates in Arizona were standardized to a low 6 litres per flush. In western-type toilets the per capita daily flush rates vary widely between about 23 and 105 liters. The latter figure is grossly extravagant, and capable of a huge reduction. Asian-style toilets tend to use less water and promote a healthier posture, but are losing in popularity to western-style toilets.
According to some studies on laundry usage, the above average value may be an underestimate. Washing machines typically use between 28 and 55 liters per person per day, with the median value tending towards the higher end of the range. There seem to be no reliable data for water consumption by hand washing.
Evaporative cooling is effective in inland, low-humidity environments, but is ineffective in humid areas like the Arabian Gulf (summer conditions).
The pool consumption estimate assumes a pool ownership of 1 in 50 households. The per capita consumptive use is likely to be significantly higher in affluent areas.
Car washing data assume one car per family, with a preponderance of bucket wash as opposed to hose washes or commercial car washes. In some societies, these figures could be much higher.
Miscellaneous appliances, such as a garbage disposal unit, may consume about 4 liters per person day, but very seldom make a significant impact upon the overall water usage.

of south-western South America, monsoon Arabia, and south and south-west Africa. Unfortunately these are exceptional, atypical environments, and the techniques would not work in the majority of hyper-arid areas. On the other hand, large-scale rainwater harvesting using unsealed catchments has been practiced since prehistoric times in such areas as the Negev desert and the dry Alteplano of South America. In more recent times, more efficient sealed catchments have become fashionable. Although very expensive to set up, and only feasible in mediterranean or semi-arid conditions, these are becoming increasingly essential as the primary water resource in isolated drought-prone communities.

TEN STEPS TO NATIONAL WATER CONSERVATION

Reduce Losses

Any water authority that does not put its own house in order cannot possibly gain credibility in fostering water conservation as a way of life. As discussed above, high storage, transmission and delivery loss rate, of up to about 55%, is the biggest single factor in urban water conservation. No system can ever be perfect, but with adequate investment and modern technology there is no excuse for losses in excess of 20%.

Multimedia Resources

Apart from a few hyper-arid areas where water consciousness has always been at the forefront of important issues, water conservation requires a radical shift in public attitudes. In addition to the attitude problem, there is generally great ignorance about water-conserving technology and how it may be applied in *every* home. Water extravagance is generally perceived as being "other people's foibles," and is seldom owned as a common problem that we all have to tackle together. These problems require a sustained high-profile program of interaction with all sectors of society. This cannot be achieved lightly.

Too often we see a national or regional water authority in which the public relations department (or equivalent body with responsibility for conservation) is regarded and resourced almost as an after-thought, an optional extra to be chopped in the first round of financial stringency. What false economy! If water conservation is to be taken seriously, it requires long-term commitment, top-quality professional staff, and modern multimedia resources. Successful conservation requires at least a decade of commitment, which is longer than the term of political office in most countries. This is a serious problem, as it is rare to find a politician with the foresight and honesty to commit funds beyond the next election, or indeed, to find a political system with a mechanism for doing so.

The staff required to promote conservation need to be a combination of technical and media experts, combining both arts and sciences skills. Technical skill, "the hard edge of scientific precision," is required to quantify and analyze the complexities of water technology and to disseminate this knowledge accurately. On the other hand, "softer skills" are required to present multimedia propaganda in an effective way. Experience shows that any attempt to dispense with the technical skills by leaving this department entirely to journalists results in an inaccurate and ineffective message to the public.

Multimedia resources should include: at least access to a VT editing suite, computer graphics capability and good quality sound and vision recording, together with high quality art and publishing resources. These are not cheap, but anything less is a half-hearted commitment to the cause and is a recipe for probable failure.

Political Education

Example should filter from the top down. If high-profile politicians adopt a conservation ethos, it becomes much easier to spread the message to the rest of the community. Just as important is the need to sustain a consistent water conservation message before the public for long enough to change old and deep-rooted attitudes. In order to secure the political will for success, one of the primary targets of water authorities should be to maintain cross-party (if applicable) awareness and support for a long-term program of conservation. Since it is to political advantage, there should be no difficulty in publicizing the advantages and achievements of water conservation!

Public Education

The general public needs to be both informed and reminded frequently of the possibilities, aims and benefits of water conservation. The objectives must be perceived as being realistic and achievable within the means available to all, and at no significant detriment to existing life-styles. Even when both the water authority and the public are well informed of how to go about water conservation, reality does not always follow theory. There is still the final, and in some ways the most difficult, hurdle: inertia.

In order to change people's habits of a lifetime in a fundamental way, the conserving authority must convey its message with enthusiasm, clarity, repetition, encouragement, incentive, assistance and vision. Again, this is no small task. For the objectives to be realized requires a major long-term effort by a fully resourced team of cross-discipline professionals. It must be fully realized at the outset that rewards do not come cheaply, they demand "hard graft" over a long period.

The means of public education range from newspaper, magazine and TV articles, to public lectures (with hard copy for later reference), leaflets, public demonstrations in shopping malls, exhibitions, posters, educational videotapes, advertisements, sponsorship of sports and social meetings, and endorsement of water-conserving policies by celebrities and social groups. Such campaigns are necessarily ambitious, but will be seen by future generations as one of the best investments that contemporary arid-zone communities could have made!

Specialist Groups

The third target for education is the various specialist groups without whose specific information and expertise, the public cannot implement many of the conservation measures discussed above. These groups include architects, builders, plumbers, garden centers, farming groups and extension services. Such groups need to be made acutely aware of their pivotal role in conserving vital water resources, and to be encouraged to take pride in this responsibility.

Don't Tell Me—Show Me

There is a world of difference between being *urged* to adopt conservation measures and being *shown* how to do it, preferably with hands-on experience. Probably the best example in the world is the water conservation program in Tucson, Arizona. In addition to all the lectures and abundance of propaganda literature, there are show houses where all the water-saving measures discussed above are implemented, including greywater recycling and mature xeriscaped gardens. Beyond being a vital research facility, these houses serve as inspection centers where people can see for themselves what can be achieved. Such demonstrations may be promoted as a point of local interest for weekend family visits, although care must be taken to avoid too grand a presentation. The aim is to illustrate achievability, even for low-paid families.

There is a certain fear of dabbling with unfamiliar technology that needs to be overcome before citizens are happy to commit themselves to it. This is cross-cultural and, to be overcome, requires more than just literature. It is a relief, and sometimes a pleasure, for people to be shown that conservation measures are not so difficult after all!

Irrigation

In the rural environment the biggest water wastage is due to poor irrigation practices. At least the farmer, whose stock in trade is agriculture, probably has a wealth of experience with which to judge the optimum irrigation amount. This can be supplemented with modern tensiometers, and moisture probes, in some cases there may be some form of agricultural extension service to assist with estimating the crop water demand. No such assistance is available for the urban/suburban gardener who generally does not have the slightest idea of how much to water his/her garden, and who therefore overirrigates to be on the safe side. Hence there is a need for some index to be published in the daily or weekly press as a guide to domestic irrigation times. If an ethos of conservation is to be instilled into the national psyche, why not announce the current optimum water requirement in the popular media (for perhaps the three xeriscape zones), along with the daily news, weather, UV reading and bush-fire risk?

It is probably safe to assume that large-scale irrigation is still capable of further improvements in water efficiency. This is true of such "advanced areas" as southern California, so is likely to be true of nearly all semi-arid agricultural areas. Overwatering may persist because of laziness, ignorance or unavailability of appropriate technology, but whatever the reason, there is always great need for ongoing encouragement to conserve the water resource.

Availability of Technology

Assuming that conservation publicity is successful, few things could be more frustrating for the public than to find that the equipment needed to conserve water is not readily available. The needs could be high tech or low tech, varying from as simple an item as new tap washers, to highly complex pumping and greywater recycling systems. It is stating the obvious, but before educating the public, attention must be given to the availability of detailed how-to information, availability of off-the-shelf equipment and spares, ease of installation, compatibility (with the local climate, water chemistry and existing reticulation systems), and the permanence of equipment supply. The planning phase of a water-saving campaign may be lengthy, but is an unavoidable necessity for success.

Research

It is helpful if the conserving authority has strong links to a research base in order to determine the local best-practice policies. Soils, geology, water chemistry and climate all influence such factors as irrigation schedules, groundwater recharge options, optimum equipment, and the most appropriate crops and ornamentals. Therefore, for best results, local research needs to supplement international research on more general issues such as greywater reuse and equipment technology. This is beyond the capability of most smaller countries, and hence there is need for greater international cooperation and research to adapt existing technology to regional conditions.

Taxation and/or Water Rates

Economists have established sophisticated methods of pricing and cost analysis, which have earned a well-respected place in conventional management theory. Forget them! The accusation that "economists know the price of everything but the value of nothing" is trite and hackneyed, but may nevertheless be accurate. Seldom, if ever, do pricing structures in arid regions reflect the true value of water, nor should they, since only the richer fraction of society could afford to pay the true price of modern water usage in a water-deficit environment. Whatever equitable pricing system is employed, the differential price above base level should be high enough for even the most affluent sections of society to think twice about watering lawns, and other wasteful practices. The objective should be not to soak the rich but to deter extravagant water usage, for if the highest profile sector of society is seen to be profligate in their use of a dwindling common resource, then the general public's attitude becomes one of cynicism rather than responsibility.

It is also helpful to introduce low-tax incentives on water-efficient equipment and appliances, as opposed to financial disincentives on large sinks, high-flow taps, and so on. Such taxation is generally too complex to be recovered at the retail stage, but may be feasible in the form of wholesale importation levies, subsidies or tax relief for domestic manufacturers.

As yet there are very few examples of effective water-conservation programs around the world. Of the few that have been successfully implemented, none has regretted the effort. Even some of the water-rich cities and regions are being forced to take water-conservation programs very seriously, while, for arid and semi-arid zones the non-availability of water is causing conservation the hard way. Water-conservation programs, for all their difficulties and demands, both material and social, must be given greater prominence. There is no alternative. Profligate use of water now, especially of dwindling groundwater reserves, and its impact upon salinization and other pollution, is creating ever-worse problems for future generations. The time for serious water conservation has come.

REFERENCES

Akad, A.A. 1989. Water conservation in arid and semi-arid regions. *Desalination* 72:185–205.

Akad, A.A. 1990. Conservation in the arid zone countries. *J. Amer. W.W. Assoc.* May:40–50.

Brengle, K.G. 1982. *Principles and Practices of Dryland Farming.* Colorado Associated University Press, CO.

Charles, D. 1991. Squeezing the deserts dry. *New Scientist* 14 Sept. (1786):22–26.

Dandy, G.C. 1987. A study of the factors which affect residential water consumption in Adelaide. University of Adelaide Final Report, Adelaide, Australia.

DeCook, J.K., K.E. Foster, and M.M. Karpiscak. 1988. The W-Index for residential water conservation. *Water Resources Bulletin* 24 (6):1295–1301.

Karpiscak, M.M., K.E. Foster, and N. Schmidt. 1990. Residential water conservation: Casa Del Agua. *Water Resources Bull.* 26 (6):939–948.

Karpisak, M.M., R.G. Brittain, C.P. Gerba, and K.E. Foster. 1991. Demonstrating residential water re-use in the Sonoran Desert: Casa Del Agua and Desert House. *Wat. Sci. Technol.* 24 (9):323–330.

Lynden, E. 1990. The last, precious drops. *Time* (Australia) 12 Nov. (46):48–58.

Syme, G.J., J.F. Thomas, and S.N. Salerian. 1983. Can household attitudes predict water consumption? *Hydrol. and Water Resources Sympos.* Hobart, Australia, p. 52–61.

Tucson Water. 1992. *Beat the Peak: 1992 Conservation Lecture Series.* Tucson Water and the Arizona Daily Star.

UNESCO. 1979. Map of the World Distribution of Arid Regions (with Explanatory Notes). Man and the Biosphere Technical Notes, 7. UNESCO, Paris.

Walters, M.O. 1990. Transmission losses in arid regions. *J. Hydraul. Eng.* 116:129–138.

APPENDIX A. Examples of arid-zone low-water vegetation appropriate for garden use (mostly natives).

Region 1: Australia

TREES AND TALL SHRUBS

Callistemon, Bottlebrush. Numerous hybrids available.

Eremophila, Emu Bush, available in many colors, tubular flowers, extremely drought-tolerant once established, also cold-hardy.

Chamelaucium, Geraldton Wax. For best results, requires pruning after flowering.

Grevillea (Spider Flower) and *Melaleuca,* a very wide range and size across varying soil types.

Banksia, available in shrub or tree form.

Hakea, similar flowers to Grevillea.

Kunzia, suitable for a wide range of soils, including acidic.

Acacia sp., slow growing and prickly, but some very attractive varieties.

SMALL TO MEDIUM SHRUBS

Acacia sp., some low-growing varieties available.

Cassia Nemophila, Desert Cassia, evergreen with bright yellow flowers.

Agonis, Dwarf Weeping Myrtle, red foliage.

Melaleuca, numerous varieties in all sizes, including dwarf varieties.

Templetonia, Cockie's Tongue. Orange red flowers, alkaline soil.

Correa, many species of low winter-flowering plants.

Westringea, Native Rosemary. Attractive hardy plants.

Calothamnus. Red flowering shrub.

Prostanthera, Mint Bush. White and mauve tubular flowers.

GROUND COVERS

Hardenbergia, Native Wisteria. Pink, blue, purple ground cover or climber.

Myoporium, white-flowered, numerous habitats.

Hemiandra, Snake Bush, mauve and white.

Kennedia, Coral Pea. While most are red, the near-black and yellow variety, *K. Negricens,* is dramatically different. Ground cover or climber.

Rhagodia, wide range of salt and pH tolerance.

Region 2: Arabian Peninsula

TREES AND TALL SHRUBS

Maerua Crassifolia (Sarh), grows in rocky habitats. Gnarled appearance. Green/white flowers have no petals.

Acridocarpus orientalis (Qafas), grows at higher altitudes. Showy yellow flowers.

Tecomella undulata (Farfar), large, showy yellow or orange flowers. Described as a tree with "leaves like those of the almond and flowers like roses" by the ninth century A.D. botanist-lexicographer Abu Hanifah ad Dinawari.

Ziziphus spina-christi (Sidr), has cherry-sized edible fruits (nabaq). The cultivated form lacks thorns and has larger fruits.

Nerium mascatense (Haban), pink-flowered oleander that grows in dense clumps in wadi beds. Its need for water may lead the roots to invade drains. All parts are poisonous (even the smoke if they are burnt!), so grow with caution.

SMALL TO MEDIUM SHRUBS AND PLANTS

Cleome species (Muqaybil ash shams). Flowers have curved petals and prominent stamens.

Cassia italica (ishriq), yellow flowers. It has long been an important folk remedy as a purgative.

Carthamus tinctorius (Shawran), safflower. An annual with large, very bright orange to red flowers once widely cultivated as a source of yellow and red dyes; still used as a cosmetic by village women.

Helianthemum lippii (Biqan), rock-rose, favors rocky habitats. Bright yellow flowers.

SMALL PLANTS

Viola cinerea, tiny violet, grows in rocky, dry, limestone soils.

Anagallis arvensis ("uwaynah" in Saudi Arabia), blue pimpernel. Ground cover.

Region 3: Southwestern U.S.

TREES AND TALL SHRUBS

Washingtonia robusta, Mexican Fan Palm.

Mesqite, Prosopis sp.

Parkinsonia aculeata, Mexican Palo Verde.

Cercidium floridum, Blue Palo Verde.

Carnegiea gigantea, Saguaro (the classic western cactus).

SMALL TO MEDIUM SHRUBS

Anisacanthas thurberi., Desert Honeysuckle.

Penstemon sp. (*fendleri, pseudospctabilis, superbus,* etc.), typically red, pink or mauve flower spikes—small to enormous. Cold-hardy, arid alpine.

Leucophyllum sp. (*frutescens,* etc.), numerous species of ornamental sage, some > 2 meters high. Huge temperature range.

Salvia sp., small blue to violet blue flowers. Heat and frost tolerant.

Dalea frutescens, Black Dalea. Autumn burst of rose-purple flowers.

Buddleia marrubeiifolia, Wooly Butterfly Bush. Clusters of orange flowers.

Yucca sp., excellent accent plants, as are *Agave sp.* and *Aloe sp.*

(Prickly pear can be a pest, and is *not* recommended, unless kept under strict control)

GROUND COVERS

Baileya multitiradiata, Desert Marigold.
Carpobrotus edulis, Ice plant.
Cephalophyllum, Red spike ice plant.
Melapodium leucanthum, Blackfoot daisy.
Verbena tenuisecta.

Note: The above are merely typical representatives of the huge selections available. Similar compilations could be made for natives of many other regions. For example, Western China, South and Central America and South Africa all have a particular profusion of drought-tolerant species.

2

The Conservation of Water Through Wastewater Reclamation and Urban Nonpotable Reuse

DANIEL A. OKUN
Kenan Professor of Environmental Engineering, Emeritus
University of North Carolina
Chapel Hill, NC 27599-7400

INTRODUCTION

THE PRINCIPLE GUIDING the modern application of water reclamation and reuse internationally was set up by the United Nations Economic and Social Council in 1958 (UN ECOSOC, 1958): "No higher quality water, unless there is a surplus of it, should be used for a purpose that can tolerate a lower grade."

This principle recognized a growing demand for water by a rapidly increasing world population faced with a relatively fixed annual renewal of freshwater resources. More recently the European Commission in a wastewater directive stated "Treated wastewater shall be reused whenever appropriate" (European Commission, 1991).

This chapter presents the role of water reclamation and reuse in urban water conservation and in water pollution control. While wastewater has long been used for agricultural irrigation, the use of reclaimed water in cities is relatively new. This chapter also describes dual distribution systems, one for potable water and the other for reclaimed water, including examples of their applications in the U.S. and around the world.

HISTORY OF REUSE

The birth of wastewater reclamation and reuse took place in the mid-nineteenth century with the introduction of sewerage systems for conducting household wastes away from communities into the nearest water courses, a practice that exploded with the rapid adoption of water closets. For example, the considerable pollution of the Thames as it passed through London not only caused nauseating conditions in the city but was responsible for repeated epidemics of cholera among those being served with water taken from the Thames. The solution was the construction of a vast interceptor along the Thames, which carried the wastewater downstream for spreading on “sewage farms.” Such “land treatment” became widely adopted by the larger cities in Europe.

The growth of urban sewerage led to increased discharges of untreated wastewaters to water courses, which downstream often served as sources of water supply for other communities. This so-called “indirect potable reuse” resulted in high mortality rates from cholera, typhoid, and other diarrheal diseases. At about the turn of the century, the development of filtration for water supplies and then chlorination sharply reduced enteric disease in the industrialized countries. However, in the cities of Asia, Africa, and Latin America, waterborne disease rates continue to be high and the threats of cholera epidemics are ever present. In addition, the use of these waters for irrigation of market crops continues to be a major hazard to the residents of these cities, and to visitors as well, as generally no treatment is afforded to the wastewaters or to the waters abstracted from the rivers for irrigation. The problems in the developing countries have been well documented (Okun, 1991).

It is understandable that the initial applications of reclaimed water, both in industrialized and developing countries, would be for agricultural and landscape irrigation. The reclamation of water and its use for irrigation offers an economical means of wastewater disposal while providing water for irrigation at low cost. In addition, the irrigation water contains valuable nutrients, which reduce investments in the purchase and application of commercial fertilizers. The only concern was ensuring adequate treatment to protect the health of workers responsible for the irrigation, the people who might be exposed to the reclaimed water, and those who consume the crops that are irrigated with the reclaimed water. A controversial issue is the nature of the regulations that are required for reclaimed water to be used for agricultural production. On the other hand, requirements for reclaimed water for nonpotable purposes in urban areas are well established. Reclaimed water for potable use is governed by drinking water regulations, which are becoming more comprehensive and restrictive for reasons unrelated to the reclamation of wastewaters. In any event, potable reuse does not have public acceptance, whereas nonpotable reuse enjoys wide-ranging, even enthusiastic approval.

The purpose of this chapter is to emphasize the role of reclamation of wastewater and nonpotable reuse to provide additional water supplies for urban areas facing water source limitations.

ROLE OF WATER REUSE

Water reclamation and reuse has evolved as an important option in water resources planning in urban areas for two important reasons. It provides an additional source of water with little health risk and it reduces the cost of wastewater disposal.

Many cities have followed public health principles in the provision of their water supplies; namely, that water for drinking should be obtained from the purest sources available, that the sources should be protected, that adequate treatment should be provided, and that a properly designed and operated distribution system should be available to ensure the safety of the water supply. Conventional water treatment, including filtration and disinfection, have been depended upon to meet modern drinking water quality requirements. Quality guidelines have been adopted internationally (WHO, 1993) while individual countries such as the United States have promulgated their own drinking water standards (EPA, 1976).

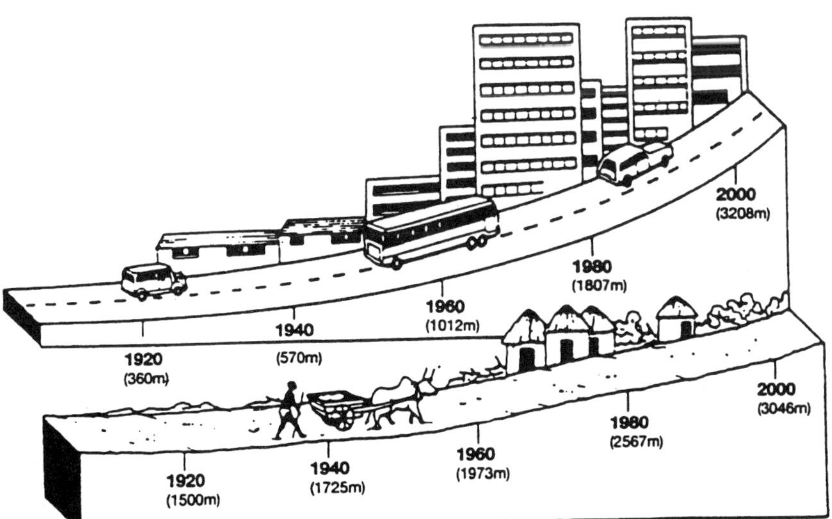

Figure 1. Urban and total world population growth.

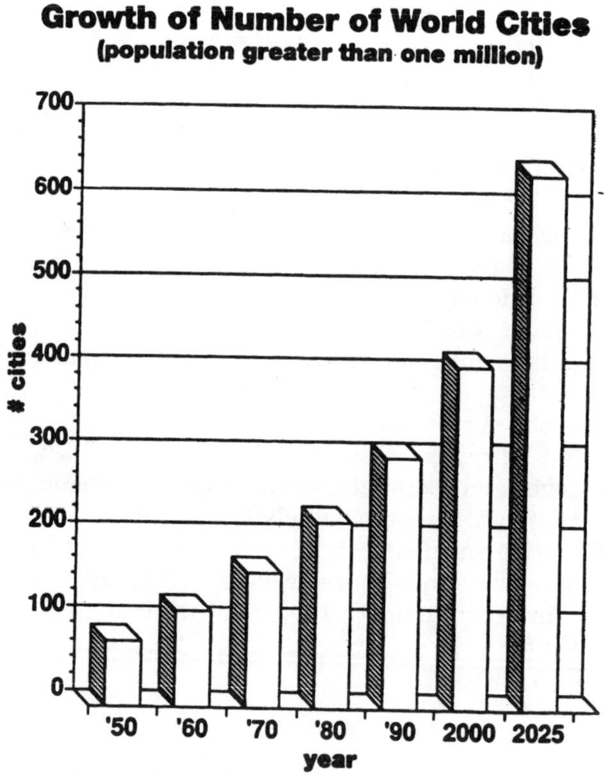

Figure 2. *Growth of cities of greater than one million population.*

The rapid rate of population growth throughout the world and the increasing proportion of that population growth which is in urban settings are illustrated in Figures 1, 2, and 3. This growth has resulted in inadequate sources of water for many urban areas, initially in arid and semi-arid regions but, in the last several decades, in very humid regions as well.

It has begun to be appreciated that polluted sources contain trace concentrations of anthropogenic chemicals, particularly synthetic organic chemicals (SOCs) that are not removed in conventional treatment and that pose health risks when ingested over long periods of time. Also, it was found that waters rich in organic matter, when disinfected with chlorine, produce disinfection by-products that are a health threat when ingested over long periods. More recently, outbreaks of giardiasis

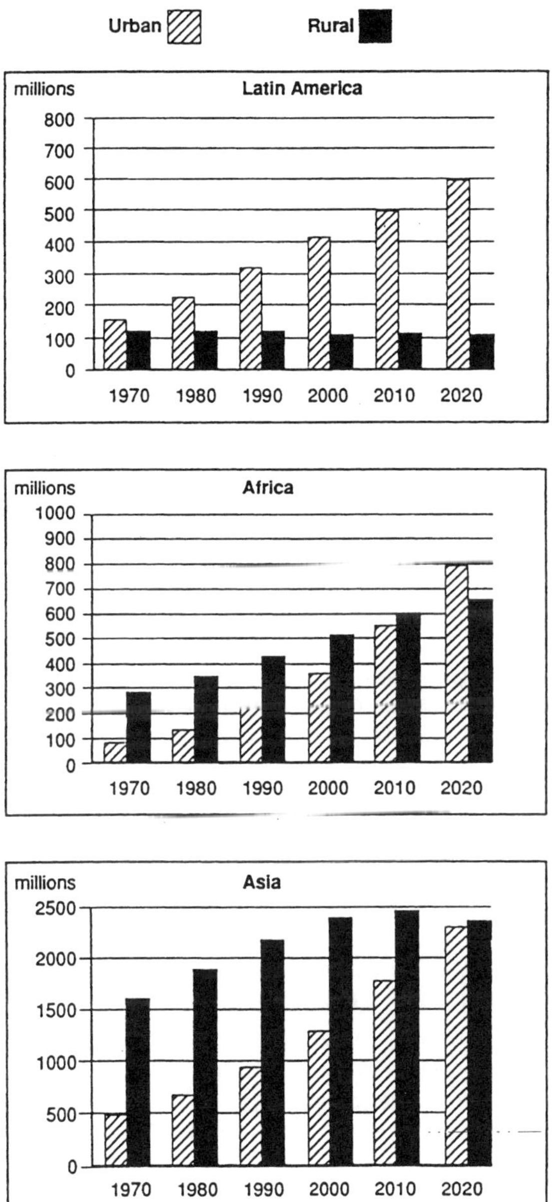

Figure 3. Growth of urban and rural populations in Latin America, Africa, and Asia.

and cryptosporidiosis, caused by protozoa *Giardia lamblia* and *Cryptosporidium parvum,* have provided evidence that conventional water treatment is not always adequate for preventing the spread of these infectious diseases and others that have not yet been well identified. Cities using compromised sources or wastewater need to invest in higher degrees of treatment. Such treatment and the necessary monitoring is very costly, particularly when most of the water used within a city does not need to be of potable quality.

With rapid growth, urban water sources are often no longer adequate in yield. Additional sources are more distant and water rights are not always available. In any case, the cost of developing additional high quality sources is becoming very great. When it is recognized that only a small fraction of the water used in urban areas is required to be of potable water quality, the UN ECOSOC policy mentioned earlier makes sense. Reclaimed water used for the myriad nonpotable purposes in a city conserves the limited high quality waters for potable purposes. While this option has been slow to be adopted, it is now beginning to grow at a rapid rate through construction of dual distribution systems. One of these systems is from a high quality source and is intended for potable purposes. The second system delivers reclaimed water that is also of high quality through treatment but not suitable for long-term ingestion. The water from this system is intended for nonpotable purposes in households, commerce, industry, and public facilities, including amenities such as landscape irrigation, public fountains, lakes, and environmental improvements.

A LOWER-COST APPROACH TO POLLUTION CONTROL

While the need for additional public water supply has emerged as the major reason for investing in water reclamation and reuse in urban areas, the early impetus in cities in the United States came from reducing the cost of wastewater treatment and disposal. With the increasing rigorous requirements for disposal of wastewaters to streams, lakes, and the ocean, particularly for the removal of nutrients, reclamation and nonpotable reuse has often emerged as less costly than treatment for discharge. The initial dual systems were affordable even though they were retrofitted in existing urban areas. Retrofitting existing systems is more costly than installing dual-distribution system lines during construction of new city streets and the buildings they serve. Given that additional water supply is necessary to serve new areas, planning for dual systems is best done before new developments are initiated.

In application, both the need for additional water sources and lower-cost disposal of wastewaters have entered into the adoption of dual systems. The current status of reuse practice is covered in the Environmental Protection Agency's (EPA) second edition of its guidelines for water reuse (EPA, 1992). This publication includes examples of reclamation and reuse; guidelines for uses; U.S. state regulations, standards, and regulations where they exist; and a chapter on international practices.

NONPOTABLE URBAN USES

Nonpotable and industrial uses of reclaimed water justify special attention, particularly compared with uses for agricultural irrigation, in that the value of water in urban use is considerably greater than in agricultural production. The urban and industrial applications have the promise of full cost recovery, which is appropriate for urban and industrial water supply systems. On the other hand, agricultural irrigation, other than for highly specialized market crops, is not likely to recover costs. Urban applications are summarized in Table 1. They include:

1. Irrigation of parks, athletic fields, school yards, highway medians and shoulders, and areas around public buildings and facilities; irrigation of landscaped areas around single- and multi-family dwellings; and irrigation of private landscaped areas around commercial and industrial buildings, golf courses, nurseries, and the like. Irrigation is an attractive urban use because reclaimed water

Table 1. Unrestricted nonpotable urban reuse applications.

Landscape irrigation: Public and private areas
Recreational Irrigation: Golf, tennis, playgrounds, ball fields
Toilet flushing: Commercial/residential buildings
Cleansing: Vehicles, streets, buildings
Commercial air-conditioning
Construction: Concrete, dust control
Industry: Cooling, processing
Fire protection
Environmental enhancement: Stream flow augmentation, parks, ponds, fountains, wetlands

is rich in nutrients, generally significantly reducing the costs for application of commercial fertilizers while reducing runoff of nutrients in stormwater.

2. Toilet and urinal flushing in commercial, industrial, and residential buildings, with its greatest application in multistory facilities. Toilet flushing is a valuable use because it is not a consumptive use; that is, the wastewater can be reclaimed again. Some cities in the U.S. and Japan have mandated that high-rise buildings be served with reclaimed water for toilet flushing and other nonpotable uses.
3. Commercial uses such as vehicle and window washing, water for the application of biocides, concrete production in rapid-mix and onsite concrete construction, and dust control during construction. Air-conditioning in large buildings is a major commercial use.
4. Fire protection since it would allow the potable water lines to be sized for delivery of only the potable water and not for the much greater fire flows. This would help avoid a major problem that arises in almost all urban distribution systems where water pipelines designed to provide fire flows carry little water except during fires, resulting in degradation of water quality because of the long residence time of water in the pipelines. This leads to disappearance of chlorine residuals and permits the growth of troublesome biofilms. To prevent degradation of the water and corrosion of the pipelines, the treated water should be used within twenty-four hours of its production (Geldreich, 1996). With a conventional system, the large size of all pipes in the system, together with the considerable service storage required for fire protection, extends the retention time into weeks. Removing the responsibility for fire protection can reduce the residence time to hours, or a day or two at most. It would permit most of the potable water pipes to be much smaller and made of better materials without joints, reducing the potential for water quality degradation, pipe corrosion, and leakage.
5. Industrial uses, such as when a single industry or group of industries in an industrial area can use a substantial amount of reclaimed water, allows the use of a dedicated line, with the reclaimed water of a quality suited to the needs of the customers. More commonly, industries are scattered throughout a community and the reclaimed water is distributed to all residential, commercial, and industrial customers. Those industries that have special high quality water needs, such as required for special papers, instruments, and so on, would be obliged to provide the additional treatment themselves, much as is done even where they draw from potable water systems.

Industrial uses include makeup water for evaporative cooling towers, boiler-feed water, process water, and irrigation of plant grounds. Evaporative cooling water is the most widely used application in industry. Industries that have found reclaimed water entirely suitable for their purposes include pulp and paper, chemicals, steel manufacture, textiles, petroleum and coal products, and others where the product poses no health risk to consumers.

6. Environmental enhancement includes fountains, reflecting pools, and augmenting the flow in urban streams that go dry as a result of urban development, and preserving or creating wetlands.

Urban irrigation and air-conditioning are consumptive uses and tend to be highly seasonal, requiring storage facilities of considerable size to conserve reclaimed water produced during wet periods to meet requirements during dry periods. Other urban uses, such as toilet flushing and industrial use, have little seasonal variations and are generally not consumptive, decrease requirements for seasonal storage, and increase the amount of water available for reclamation and reuse.

In the United States, particularly in the arid and semi-arid areas in the southwest and in humid Florida, landscape irrigation has been by far the major urban use. Industrial uses within a city have been negotiated on a case-by-case basis and are increasing in number. Vehicle washing and the use of reclaimed water for construction have become popular. Toilet and urinal flushing with reclaimed water is just beginning in the U.S. but is widely practiced in Japan. As is to be expected, reuse applications are site-specific.

NONPOTABLE REUSE IN THE U.S.

The first dual distribution system in the United States was built in 1926 to provide water supply for a rapidly growing Grand Canyon Village in Arizona on the south rim of the canyon where rainfall is sparse and little freshwater is available. Water for the village was originally brought in by tankers over roads and rails. A spring, about one km (0.6 miles) deep in the canyon, was then developed from which water was pumped to the village. The wastewater produced was too valuable to be discarded. Under the auspices of the U.S. Public Health Service, the National Park Service, and the railway company that built the El Tovar Hotel (the first large structure in the village), the wastewater began to be treated and reclaimed for landscape irrigation and toilet flushing in the hotel in 1926 (Garthe and Gilbert, 1968). As the village grew, the system was renovated and enlarged, the last time in 1989 (Fleming, 1990). In addition to the original uses for landscape irrigation and toilet flushing in the visitor area, uses have now extended to serving the permanent village for staff employees. This includes schools and service buildings as well as vehicle washing, cooling water makeup, construction, and other nonpotable uses.

Since this initial project, many reclamation and nonpotable use projects in urban settings have been developed for single or groups of large users not necessarily involving comprehensive dual reticulation systems. One of the most noteworthy was the construction in 1942 of a 2.4-m (96-inch) pipeline 7.2 km (4.5 miles) from the Baltimore Back River-activated sludge plant to deliver about 4.5 m^3/s (100 mgd) to the Sparrows Point plant of the Bethlehem Steel Company. The nonpotable water was for process use, cooling water being provided by once-through passage of water obtained from a nearby water course. This reclamation program significantly reduced the demand on Baltimore's limited upland supply and simultaneously reduced wastewater discharge from the city to local receiving waters (Okun, 1973). In the years following, many power plants began to use reclaimed water for cooling tower makeup. A related power plant use was that for the TECO power plant, near Tampa, Florida. This plant uses seawater for once-through cooling and almost the entire production of a small wastewater reclamation plant for stack gas scrubbing to reduce air pollution.

Water Quality Requirements

Requirements for reclaimed water to be used for nonpotable purposes were first addressed in California where the so-called "Title 22" regulations were adopted in 1968 and last revised in 1978 (CSDHS, 1978). Their successful application without untoward incidents led to their adoption by other states. To date, there are no federal regulations for nonpotable reuse. However, EPA's guidelines (EPA, 1992) recommend treatment and quality for a wide range of uses, as shown in Table 2.

The most stringent nonpotable guidelines are for unrestricted urban use where large populations are likely to be exposed and for irrigation of so-called "market crops." In brief, the treatment includes the filtration of secondary effluents followed by disinfection to produce water with average turbidity equal to or less than 2 NTU, and a maximum not to exceed 5 NTU, with no detectable fecal coliform in 100 mL, and a minimum chlorine residual of 1 mg/L after a minimum contact time

Table 2. Quality guidelines for nonpotable reuse applications (adapted from EPA, 1992).

Type of Use	Treatment	Reclaimed Water Quality
• Unrestricted urban uses • Food crops eaten raw • Recreational impoundments	• Secondary • Filtration • Disinfection	• pH = 6–9 • ≤ 10 mg/L BOD • ≤ 2 NTU[a] • No detectable fecal coli/100 mL[b] • ≥ 1 mg/L Cl_2 residual[c]
• Restricted access area irrigation • Processed food crops • Nonfood crops • Aesthetic impoundments • Construction uses • Industrial cooling[d] • Environmental reuse	• Secondary • Disinfection	• pH = 6–9 • ≤ 30 mg/L BOD • ≤ 30 mg/L SS • ≤ 200 fecal coli/100 mL[e] • ≥ 1 mg/L Cl_2 residual[c]
• Groundwater recharge into nonpotable aquifers	• Site-specific and use-dependent Spreading: primary (minimum) Injection: secondary (minimum)	• Site-specific and use-dependent

[a]Should be met prior to disinfection. Based on a 24-hour period. Should not exceed 5 NTU at any time
[a]Based on 7-day median value. Should not exceed 14 fecal coli/100 mL in any sample.
[c]After a minimum contact time of 30 minutes.
[d]Recirculation cooling towers.
[e]Should not exceed 800 fecal coli/100 mL in any sample.

of 30 minutes. With regard to microbiological quality, this is the equivalent of drinking water standards, so that inadvertent ingestion would pose little health risk. There are no requirements for organic chemicals but the water would not be ingested for long periods of time so that trace contaminants, including disinfection by-products, are not an issue.

Treatment for Water Reclamation

Research by the Los Angeles County Sanitation Districts demonstrated that a high quality secondary effluent treated with small doses of either a coagulant (such as alum or ferrous salts), a polymer, or both followed by conventional sand filtration and chlorine disinfection can readily and continuously provide a reclaimed water that will meet

Table 3. Suggested guidelines for treatment of unrestricted nonpotable urban reuse.

Treatment:	Secondary (biological) treatment (effluent BOD and SS ≤ 30 mg/L) Filtration Disinfection
Quality:	pH: 6 to 9 BOD_5: ≤ 10 mg/L Turbidity: ≤ 2NTU (24-hr. average) ≤ 5 NTU (at any time) Fecal coliform: Nondetectable Chlorine residual: ≤ 1.0 mg/L after 30 minutes ≥ 0.5 mg/L in distribution system
Monitoring:	pH: weekly BOD_5: weekly Turbidity: continuous Fecal coliform: daily Chlorine residual: Continuous

requirements (LACSD, 1977). A practice that is widely adopted is the maintenance of a chlorine residual throughout the reclaimed water distribution system to serve as a monitor and to prevent the formulation of biofilms in the pipe lines, which can reduce the flow and quality of the reclaimed water. Some of the practices adopted in California are listed in Table 3.

Reclamation plants are often located near the water markets rather than near the points of disposal, as is conventional practice for treatment plants to avoid costly pumping of the reclaimed water, particularly where the urban areas to be served are at high elevations. The sludges produced are commonly discharged to trunk sewers leading to treatment plants for final disposal. A typical flow profile is shown in Figure 4. This practice permits reclamation plants to be located more centrally in urban areas because they do not produce the odors or the noise that accompany sludge handling.

Storage

As in any water distribution system, hourly demands for reclaimed water do not match the rate of output of the treatment facilities. To meet the variable demand throughout the day, storage tanks, known as service or operating reservoirs, are required. Also, as in potable water distribution systems, the tanks need to be covered to protect

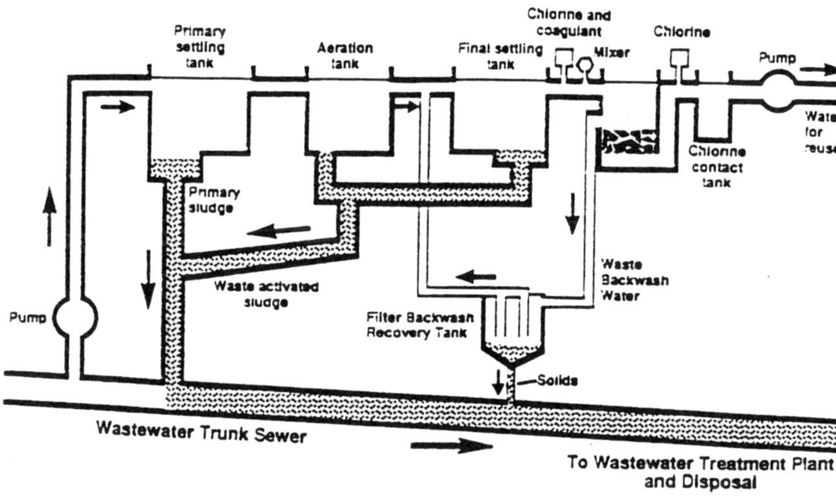

Figure 4. Typical flow profile of reclamation plants in Southern California.

against air-borne contamination and the growth of algae. To maintain water pressure in the reclaimed water distribution system, the tanks must be elevated or located on high ground. Service storage requirements generally need to be 15 to 25% of the average daily demand. Additional storage is required if water for fire protection is to be provided from the reclaimed water system. When the reclaimed water system is to be used for fire protection, the tanks must be higher than is required for potable water. To reduce the cost of elevated storage in a dual system, it is feasible to arrange for pumping from the potable water service reservoir to the higher reclaimed water service reservoirs during periods of high demand, as during fires, when the capacity of the reclaimed water elevated tank may not be adequate. The reason for having the reclaimed water provide fire protection is not to save water, as very little water is used for fire protection, but to reduce the size of pipes in the potable water distribution system to maintain a higher quality of that water.

To meet a significant demand for irrigation, several months of seasonal storage may be required, the amount depending on the length and severity of the dry seasons. For potable water systems, where the source of water is a river with variable flow, storage is provided in aquifers or in impounding reservoirs. In such instances, the storage requirements are high because the demand is greatest during the dry

Figure 5. *Surface storage for reclaimed water for Irvine Ranch Water District.*

period when the river flows are at a minimum. With reclaimed water, less storage is required because the quantity from the source is relatively constant while only the seasonal demand varies.

Seasonal storage of reclaimed water in aquifers underground is highly advantageous and may be done by injection or through spreading. Where such underground storage is not feasible, surface storage impoundments may be employed. Because it is not feasible to cover such reservoirs, this option suffers from the disadvantages of losses through evaporation and degradation of water quality, particularly when the nutrients have not been removed. This is generally the case when a major use for the reclaimed water is irrigation and the nutrients are economically valuable. Figure 5 illustrates surface storage for the Irvine Ranch Water District. Water drawn from the reservoir is refiltered and rechlorinated. Reclaimed water storage to meet variations in diurnal demand is in covered service reservoirs and retreatment is not required.

Dual distribution systems are continually finding wider use in the United States, with many states being obliged to develop regulations for urban reuse. Two such states are Washington and North Carolina, both among the wettest states in the U.S. Dual systems have become economically attractive to individual cities in humid regions where new sources of water are costly and because of the political and environmental difficulties involved in building impoundments and abstracting waters that originate on land under the domain of other local governments.

EXAMPLES OF DUAL SYSTEMS

Several examples of dual systems are described below, followed by examples from other countries.

St. Petersburg, Florida

Planning of the first major comprehensive urban dual system in the United States to serve all customers was initiated in St. Petersburg, Florida, in 1969 by the late Lloyd Dove, then director of Public Works, after extensive discussions with the author (Okun, 1969a). Papers in the same vein were presented by the author at about the same time (Okun, 1968a, 1968b, 1969b).

The initiative for the St. Petersburg system came principally from the economies that might be realized as a result of new Florida regulations requiring nutrient removal prior to discharge of effluents to Tampa Bay and the Gulf of Mexico. The lower cost of tertiary treatment, filtration and chlorination, and federal grant funds made reclamation and reuse more economical. Because the initial objective was wastewater disposal and landscape irrigation, users were not metered. Instead a monthly flat rate of $15 per hectare ($6 per acre) was designed to encourage maximum use of the reclaimed water. However, demand for the reclaimed water in the summer exceeded its availability, largely because of waste, so meters would need to be installed.

Filtration and chlorine disinfection facilities were added at the four activated sludge treatment plants serving different quadrants of the city. The reclaimed water reticulation system was sized to include provision for fire protection, although the potable water system already included fire protection. Covered service reservoirs for the reclaimed water at the treatment plants were introduced after difficulties had been experienced using open water storage facilities. Covered reservoirs were needed due to the proliferation of algae, a not surprising outcome when nutrient-rich waters are exposed to sunlight. During periods of low demand, the excess effluent was injected into deep wells penetrating saline aquifers that underlay the city. The original intention was to create a Ghyben-Herzberg freshwater bubble in the aquifer from which the reclaimed water might later be withdrawn for use during high-demand periods. However, the design of the injection wells proved to be unsuitable as the reclaimed water spread out widely under an impermeable stratum, and was not readily recoverable. Studies are being undertaken to reexamine the design of the wells with the intention of permitting recovery of the reclaimed water.

Many urban nonpotable uses have developed: air-conditioning; make-up water for cooling towers at a resource recovery power plant; vehicle washing; garden nurseries; landscape irrigation; and watering of golf courses, tennis courts, and baseball fields. In addition, the reclaimed water system is used for fire protection with additional hydrants, colored differently from the potable water hydrants, throughout the city, adding to its security.

As the situation in St. Petersburg evolved, the major benefit to the city became the reduction in the demand on its high quality water source. Water for St. Petersburg and other water utilities in the vicinity comes from a limited groundwater aquifer with well fields some 50 km (30 miles) north of the city in Pasco County. Excessive community withdrawals from these well fields have contributed to the drying up of lakes in the so-called "Land of Lakes" region. St. Petersburg is the only city in the region whose demand on the well field has decreased despite its growth in population, as shown in Figure 6 (Johnson, 1995). In Florida alone, more than twenty cities have established dual systems by retrofitting reclaimed water lines throughout their service areas.

Irvine, California

At about the same time that St. Petersburg inaugurated its system, the relatively new city of Irvine in southern California, which had begun with a conventional water system, initiated installation of reclaimed water lines for all new construction. This was driven in part by the need for control of wastewater disposal to the sea. The major factor, however, was the high cost of water purchased by the Irvine Ranch Water District (IRWD) from the Metropolitan Water District of Southern California (MWD) and the expectation, very soon realized, that the cost of MWD water would increase significantly.

While Irvine began by installing reclaimed water lines during new construction, it has since found it economical to return to the older parts of the city and retrofit reclaimed water pipelines. Incidentally, even though federal funds for construction of community wastewater

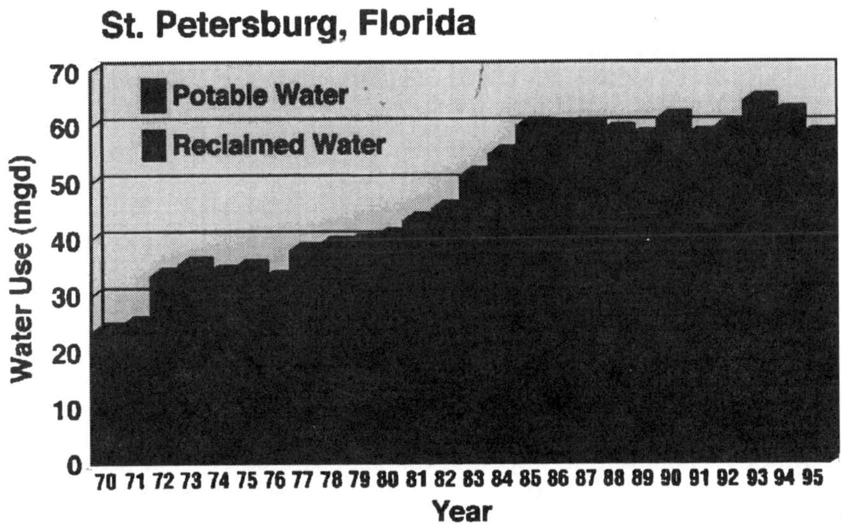

Figure 6. Impact of nonpotable reclaimed water use on potable water demand in St. Petersburg, Florida.

treatment plants were available, IRWD did not use them. All investments in wastewater treatment and dual systems were undertaken because they were economical. Irvine was the first city in the U.S. to use its reclaimed water for toilet flushing on a large scale, a requirement for all its new high-rise office buildings (Young et al., 1994). Irvine's reclamation plant is typical of many in southern California in that it is located near the water markets. The reclamation plant is not obliged to handle the solids produced, which are conveyed to a terminal plant as shown in Figure 4. Such satellite plants are acceptable in residential and commercial areas because they do not involve the odors, noise, and truck traffic that accompany sludge handling.

The IRWD meets its seasonal storage requirements by using surface storage which, because reclaimed water is rich in nutrients, stimulates algae growth and requires refiltration and disinfection for all water drawn from these reservoirs. Diurnal variations in demand are met from covered service reservoirs making refiltration unnecessary.

San Jose, California

The city of San Jose in northern California has built a dual system in a regional effort with six other utilities. The project, South Bay

Water Recycling, involves in its first stage 100 km (60 miles) of reclaimed water distribution system, up to 1.5 m (60 inches) in diameter, designed to deliver about 80,000 m^3/d (21 mgd) to serve San Jose, Santa Clara, and Milpitas. Reuse is for landscape and agricultural irrigation as well as commercial and industrial cooling and processing. It is intended, by the year 2000, to add about 30 km (20 miles) of pipeline, providing about 40,000 m^3/d (10 mgd) (Rosenblum, 1997). The impetus for the project was to restrict wastewater discharges into South San Francisco Bay but, as it developed, the benefits in making customers "drought-proof" has become the key selling point for bringing customers on-line.

Japan

Water reclamation in Japan developed differently from the United States because less than half of its population was sewered and because of the great density of population and limited water resources. The principal objective of the reclamation program in Japan was to reduce water demands through provision of reclaimed water for toilet flushing (primarily in large buildings such as multifamily residential units), commercial buildings, factories, and school buildings (Murakami, 1989). Tables 4, 5, and 6 show the various types of reclaimed water uses in Japan.

High-rise buildings being retrofitted for flush toilets and new buildings offer excellent opportunities for introducing dual plumbing. The program began by recycling wastewater produced in a building or group of buildings with a reclamation plant in the basement of one of the buildings. It was soon recognized that large-scale reclamation works and an area-wide reclaimed water reticulation system as part of a dual system would be more effective and economical than numerous small reclamation facilities (Okuno, 1994). Table 7 shows a comparison at a later time when the total reclaimed water usage in Japan had increased

Table 4. Uses of reclaimed water in Japan (source: Murakami, 1989).

Use	%	1,000 m^3/d	mgd
Nonpotable in dual systems	40	110	29
Industrial	29	77	20
Agricultural	15	40	11
Stream flow augmentation	12	32	8
Snow removal	4	11	3
	100	270	71

Table 5. Uses of reclaimed water in dual systems in Japan (source: Murakami, 1989).

Use	Percent
Toilet flushing	37
Cooling water	9
Landscape irrigation	15
Car washing	7
Washing and cleansing	16
Flow augmentation	6
Other	10

Table 6. Types of buildings using reclaimed water in Japan (source: Murakami, 1989).

Buildings	Percent
Schools	18
Office buildings	17
Public halls	9
Factories	8
Hotels	4
Others (residences, shopping centers, etc.)	44

Table 7. Comparison of types of water reuse in Japan and California (source: Asano et al., 1996).

Type	Japan	California
Agricultural irrigation	13%	60%
Industrial	41	2
Environmental	32	23*
Snow removal	6	0
Urban	8	1
Groundwater recharge	0	14
Total reuse m^3/day	100 million	432 million

*Includes urban landscape irrigation.

almost four-fold in less than ten years. It shows a much heavier investment in reuse for industrial and environmental purposes than for agricultural irrigation, which dominates reuse in California (Asano et al., 1996).

The Bureau of Sewerage of the Tokyo Metropolitan Government has been one of the leaders of water reclamation in Japan. Effluent from the Ochiai Wastewater Treatment Plant is filtered and chlorinated for distribution to sixteen high-rise buildings for use in flush toilets. Tokyo has also pioneered work in utilization of wastewater in heat pumps, taking advantage of the differences between the temperatures of wastewater and the atmosphere in winter and summer.

A new so-called "Twenty-First Century" Ariake Wastewater Treatment Plant is to be located on the waterfront of Tokyo Bay and to have multiple facilities, including a wastewater reclamation plant, a heat recovery and supply station, a recreational center, and international championship tennis courts. The treatment is to include an anoxic-oxic process, a fixed-bed biofilm process, an ozonation process, and low-pressure reverse osmosis (TMG, 1993).

In Osaka, a city with twelve wastewater treatment plants, four have been provided with advanced treatment for reclamation with a major objective of enhancing the water environment. A particularly attractive use for the reclaimed water is to provide water for moats around the Osaka Castle Park in downtown Osaka. Other objectives include the restoration of urban stream flow and enhancement of the urban environment, including a beautifully landscaped Sewerage Science Museum (Sazaki, 1996).

Australia

An interesting development of a dual distribution system is that for a new community, Rouse Hill, to accommodate about 250,000 people near Sydney, Australia. All households are being fitted with both potable and reclaimed water services, including toilet flushing. Water for fire protection is provided from the reclaimed water system. This was not done as a water conservation measure. It was intended to reduce the size of the potable water pipelines to preserve drinking water quality in the potable water distribution system; reclaimed water lines are larger in diameter than the potable water lines (Ius, 1994; Long, 1994). Diurnal storage is provided in covered elevated tanks. The reclaimed water storage tank is higher than the potable water tank, but of smaller capacity. Provision is made to permit the potable water to be pumped into the reclaimed water tank. One of the features of the system is that wetlands are used to improve the quality of excess effluent from the reclamation plant before discharge into nearby streams.

Singapore

Singapore is an island city/state with a dense, growing population of some three million people. It receives heavy rainfall averaging 2500 mm (100 inches) per year but has limited water resources because of its small size of only about 540 sq. kilometers (210 sq. miles). For these reasons, water reclamation became absolutely essential. Several secondary activated sludge treatment plants discharge their effluents to the sea. At one location, a portion of the municipal effluent was withdrawn from the outfall and given additional (tertiary) treatment to serve the industrial needs of the Jurong Industrial Estate. The treatment involves conventional sand filtration and chlorination. The reclaimed water is pumped for service storage to a covered tank on a hilltop. A major housing development was later built for a population of about 25,000 people in fifteen-story buildings, all the toilets being served with reclaimed water. Originally, the operation of the facility was the responsibility of the estate, the Singapore Public Utilities Board. However, a more professional agency took over this responsibility.

Other Countries

Sao Paulo, Brazil, soon to be the world's second largest city, is severely short of water and has been discharging untreated or partially treated wastewater into nearby rivers. Also, about half its population is not yet served with sewers. Completion of a unit of a major activated sludge treatment plant revealed that the effluent was of higher quality than the river into which it was discharged. SABESP, the state authority responsible for water supply and sewerage, believed the effluent to be of too high quality to be wasted and therefore undertook pilot studies for more advanced treatment for reclamation primarily for serving some 30,000 industries in the region (Okun and Crook, 1989).

India is a country where every large city is short of water. Water service is seldom available twenty four hours per day. For example, Chenai (Madras), a city of 4.5 million people, takes its water from rain-fed reservoirs and groundwater aquifers but currently suffers a shortfall of about 50% of its supply as shown in Figure 7 (Nair and Selvam, 1996). The Madras Metropolitan Water Supply and Sewerage Board is taking steps to meet the demand by managing its groundwater

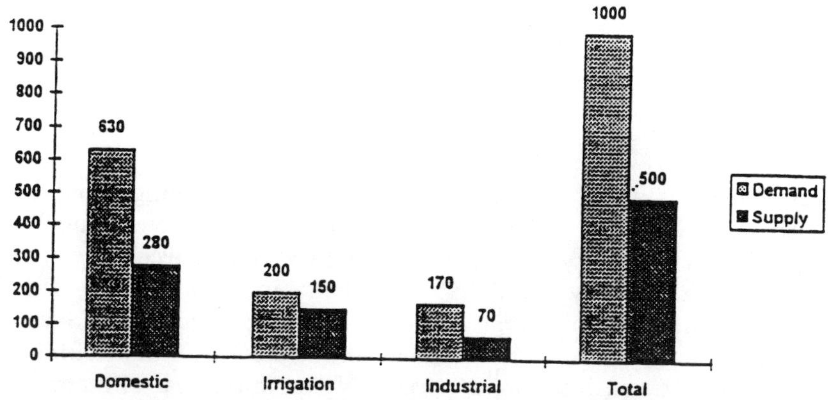

Figure 7. Fresh water supply and demand in Madras, India (in 100 m³/day, 1996).

supply and by using reclaimed water for large industries, agro-forestry and agricultural, and landscape irrigation. This is typical of most water supply organizations. Although most of the shortfall is in its domestic supply, no mention is made of urban reuse other than landscape irrigation. For the long-term, planning for reuse seems to require more attention to urban nonpotable reuse.

Israel has extensive reclamation programs almost entirely dedicated to agricultural use. A feature of the reclamation program is a major water carrier from Tel Aviv to the Negev Desert for irrigation. Reclaimed water quality practices are similar to those in the U.S. except that they have not yet begun to consider the potential for reclamation for urban use.

The situation with regard to reclamation and nonpotable practice is substantially different in the oil-rich countries of the Middle East where water may be more valuable than oil. With little restriction on energy and the availability of funds for high-tech capital investments, their reclamation practices focus on extensive use of more costly treatment such as thermal distillation and membrane filtration. Such treatments are widely used elsewhere in the world but generally only for high-value uses, not for irrigation or toilet flushing.

The EPA guidelines for water reuse (EPA, 1992) describe briefly the reuse situation in seventeen countries throughout the world. The factors that help determine approaches to both urban and agricultural reuse in developing countries, as well as industrialized countries, are also presented.

REUSE FOR AGRICULTURAL IRRIGATION

Reclaimed water produced for urban nonpotable reuse may have a market in nearby areas for agricultural irrigation, particularly for market crops grown to serve the urban population and for export. As indicated in Table 2, quality guidelines for such irrigation should be the same as for unrestricted urban uses. The use of untreated or inadequately treated wastewater for irrigation of market crops has been responsible for high endemic rates of diarrheal disease in the population. Also, it is responsible for disease outbreaks among travelers who consume fruits and vegetables irrigated with wastewater and, occasionally, unwary customers in industrial countries who buy such products from their own supermarkets.

In 1989, the World Health Organization (WHO) adopted health guidelines (Table 8) for the use of wastewater in agriculture and aquaculture that are not nearly as rigorous as those shown in Table 2. For the irrigation of crops likely to be eaten uncooked, the WHO recommendations call for a fecal coliform limit of less than 1000 fecal coliform per 100 mL and no more than one intestinal nematode per liter (WHO, 1989). The attractiveness of this lower quality standard is that it can be attained by the use of relatively low-cost stabilization ponds, an appropriate practice for developing countries. However, in the larger cities and urban areas of both industrialized and developing countries, stabilization ponds are seldom feasible because of the limited and costly land required and because the effluents from such treatment facilities are not easily disinfected. Such reclaimed water would not be suitable for distribution for other nonpotable purposes within a city. In large cities, conventional wastewater treatment is more appropriate even for disposal of wastewater into rivers and lakes. The additional tertiary filtration and disinfection for unrestricted nonpotable use would not be significantly more costly.

For large cities not yet sewered and without any treatment facilities, stabilization ponds might be an appropriate first stage in meeting wastewater treatment and reuse requirements, with the understanding that they would soon need to be replaced by conventional treatment facilities. Meanwhile, the nonpotable application and exposure of the public would need to be limited to irrigation of lands restricted to the public and for nonfood crop irrigation due to the low quality of effluent from stabilization ponds. In any case, the WHO guidelines were not intended for urban reuse. Finally, reclamation for urban use promises cost recovery, which is not likely for agricultural applications.

Table 8. WHO water quality guidelines for agricultural reuse[a] (source: WHO, 1989).

Category	Reuse conditions	Exposed group	Intestinal nematodes[b] (arithmetic mean no. of eggs per liter[c])	Fecal coliforms (geometric mean no. per 100 ml[c])	Wastewater treatment expected to achieve the required microbiological quality
A	Irrigation of crops likely to be eaten uncooked, sports fields, and public parks[d]	Workers, consumers, public	≤ 1	≤ 1000[d]	A series of stabilization ponds designed to achieve the microbiological quality indicated, or equivalent treatment
B	Irrigation of cereal crops, industrial crops, fodder crops, pasture, and trees[e]	Workers	≤ 1	No standard recommended	Retention in stabilization ponds for 8–10 days or equivalent helminth and fecal coliform removal
C	Localized irrigation of crops in category B if exposure of workers and the public does not occur	None	Not applicable	Not applicable	Pretreatment as required by the irrigation technology, but not less than primary sedimentation

[a]In specific cases, local epidemiological, sociocultural and environmental factors should be taken into account, and the guidelines modified accordingly.
[b]*Ascaris* and *Trichuris* species and hookworms.
[c]During the irrigation period.
[d]A more stringent guideline ($\leq$ 200 fecal coliforms per 100 mL) is appropriate for public lawns, such as hotel lawns, with which the public may come into direct contact.
[e]In the case of fruit trees, irrigation should cease two weeks before fruit is picked, and no fruit should be picked off the ground. Sprinkler irrigation should not be used.

RECLAMATION FOR POTABLE REUSE

Reclamation of wastewater for indirect potable reuse is being proposed in the U.S. for augmenting water supply. This would be done by recharging aquifers that are drawn upon for drinking water or by augmenting surface supplies by discharging reclaimed water into reservoirs used for potable water supplies or by discharging reclaimed water into rivers drawn upon downstream for water supply. One justification for purposeful potable reuse is that it will provide a better quality water than that which is unintentionally reclaimed when water is drawn from a river into which untreated or poorly treated wastewaters are discharged upstream. The practice of drawing water supplies from polluted streams was accepted many years ago when filtration and chlorination were thought to be sufficient treatment to render the most polluted waters adequately suitable for public drinking water supplies. However, with the explosive growth of the production of synthetic organic chemicals (SOCs), many of which have been found to be mutagenic, carcinogenic, and/or teratogenic, drinking water standards have become considerably more stringent. In addition, the chlorine used for disinfection reacts with natural organics to form several classes of disinfection by-products, which have also been found to be carcinogenic after ingestion over long periods of time. From the 1960s to the 1990s, the number of contaminants being regulated in drinking water in the United States has grown five-fold and the same pattern is occurring in Europe and has been adopted by WHO in its drinking water guidelines (WHO, 1993).

The purposeful introduction of wastewaters into a high quality water supply is no longer acceptable. Where wastewaters are discharged to water supply sources, efforts are being made to increase the level of wastewater treatment prior to discharge into rivers. However, providing more extensive treatment and monitoring of the water that is withdrawn for potable purposes becomes an obligation.

The problem with making reclaimed water safe for continuous ingestion is that only about 10 to 15% by weight of the organic compounds in drinking water have been identified and the health effects of only a few of have been characterized (National Research Council, 1980). Moreover, new SOCs are being introduced into the marketplace continually and their monitoring in drinking water and their removal is both technically difficult and financially quite costly. Furthermore, while the more well-known microbial contaminants may be under control in the industrialized countries, the scourge of cholera, typhoid, and other enteric diseases are still rampant in the developing countries. In the industrialized countries, other pathogens such as the protozoan parasites *Giardia* and *Cryptosporidium,* as well as viruses such as infectious hepatitis have begun to emerge. While technology does exist for rendering wastewaters potable, to use thermal distillation and high quality molecular membranes for treating water for irrigation, toilet flushing, vehicle washing, and other urban nonpotable purposes hardly seems to be economically warranted.

Most purposeful augmentation of potable water supplies with reclaimed water has occurred through groundwater recharge. In such instances, passage through the vadose zone helps reduce the concentration of some of the organics and microorganisms. However, the capacity of soils to remove organics and microorganisms may be exhausted over time, each chemical at a different rate. The effluent from such soil treatment may eventually become worse than the wastewater charged to that aquifer.

Nevertheless, some cities have embarked upon studies for potable reuse projects without evaluating the nonpotable option in a realistic manner. The vision of a reticulation system with pipe lengths twice those of a conventional system seems to lead to the rejection of the nonpotable option. Given that dual systems are being built without subsidy is evidence that they are not inherently too costly. Even if they were somewhat higher in cost, customers have endorsed reclaimed water for nonpotable purposes while being strongly opposed to its use for drinking (EPA, 1992). Experience has also demonstrated that where the microbiological quality of reclaimed water is much the same as that required for potable water, there is little health risk even if the water were inadvertently ingested for short periods of time. Acceptance of dual systems by the profession is indicated by the publication of Manual 24 on dual systems in 1983 by the American Water Works Association with a second edition published in 1994 (AWWA, 1994).

INSTITUTIONAL ISSUES

The production and distribution of reclaimed water may be the responsibility of either a community's water supply or sewerage agency. A new agency might also be established that would work closely with the established water supply and sewerage agencies. If the purpose of developing a reclaimed water program is to obtain additional water supply, the water supply agency would be the likely lead. If, on the other hand, control of water pollution is primary, then the sewerage agency would take the lead.

The preferable approach by far is to combine the three services in a single agency. This has the following advantages:

1. The staff necessary to manage the treatment facilities and reclaimed water distribution systems would be drawn from water supply and sewerage experienced personnel in the preexisting agencies.
2. Integrating the water supply and sewerage services would yield both economies and efficiencies of scale because of the provision of common services such as administrative buildings, computers, personnel management, vehicle and equipment maintenance, and laboratory services.
3. The assessment of the costs and benefits would be internalized. The added cost of treatment for reclamation beyond what might be needed for disposal (if there is additional cost, since the treatment for reclamation would be less costly than treatment for disposal in industrial countries) along with the cost of the second distribution system would be offset by the savings in the potable water service.
4. Planning for the future requires integration of water and sewerage services. Economies can be achieved by regionalization within watersheds so that the wastewater of one community (conventionally discharged to a river that provides the drinking water supply of the downstream community) can become the reclaimed water that serves both communities while preserving the high quality water for the potable needs of both.
5. The financing of a new reclaimed water service is much more feasible if it builds upon the existing water supply and sewerage infrastructure and income. In developing countries where only minimal sewerage exists, the potential income from drinking and reclaimed water services provides opportunities for cost recovery not available if only the sewerage and its treatment for disposal is contemplated.

A reclaimed water service is not likely to be feasible in a city if its water supply service is not already sustainable with charges for the service adequate to meet continuing costs of operation, maintenance, and reasonable expansion. In the developing countries, few sewerage agencies are now viable and they are not likely to be in a position to undertake reclamation in the absence of a sound water supply organization.

CONCLUSION

It is not difficult to envisage a future when a potable water distribution system will deliver water drawn from high quality sources and/or from plants with membrane or other sophisticated treatment, with high quality pipe materials being used for the potable service. This is especially true since the amount of water needed to be of drinking water quality will be relatively small. Such expensive treatment would be affordable if provided only for the water intended for drinking. However, it might be difficult to justify high quality water for conventional systems where most of the water is used for nonpotable purposes, such as landscape irrigation, and toilet flushing. The reclaimed water would serve most of the myriad nonpotable needs.

Wastewater reclamation and urban nonpotable reuse, through the use of dual distribution systems, permits adding to limited high quality sources of water while imposing little public health risk. It is an option that municipal and governmental officials and their consulting engineers have an obligation to consider among the several alternatives available for providing additional water supplies for rapidly growing urban areas throughout the world.

REFERENCES

Asano, T., M. Maeda, and M. Takaki. 1996. Wastewater reclamation and reuse in Japan: Overview and implementation examples. *Water Science Tech.* 34:11, 219.

AWWA. 1994. *Dual Water Systems (M24),* Second Edition. American Water Works Association, Denver, CO.

CSDHS. 1978. Wastewater Reclamation Criteria. California Administrative Code, Title 22, Div. 4, California State Department of Health Services.

EPA. 1992. *Guidelines for Water Reuse.* U.S. Environmental Protection Agency, EPA/625/R-92/004.

European Commission. 1991. Council Directive Concerning Urban Wastewater Treatment, 91/271/EEC, 21 May 1991, OJ No. L 135/40, 30 May 1991.

Fleming, P.A. 1990. Water supply, reclamation and reuse at Grand Canyon: A case study. *Proceedings of Conserv 90,* Phoenix, AZ, p. 129.

Garthe, E.C. and W.C. Gilbert. 1968. Wastewater reuse at the Grand Canyon. *J. WPCF* 40:1582.

Geldreich, E.E. 1996. *Microbial Quality of Water Supply in Distribution Systems.* Lewis Publishers—CRC, New York, London, Tokyo.

Ius, R. 1994. Developing appropriate design parameters for a domestic dual water supply system. *Proceedings of the Recycled Water Seminar,* Australian Water and Wastewater Association, Artarmon, NSW, Australia.

Johnson, W. 1995. Director of Public Works, City of Petersburg (personal communication).

Long, G.C. 1994. Implementation of a dual water supply system at Rouse Hill. *Proceedings of the Recycled Water Seminar,* Australian Water and Wastewater Association, Artarmon, NSW, Australia.

LACSD. 1977. Pomona virus study. Los Angeles County Sanitation Districts. California Water Resources Control Board, Sacramento, CA.

Murakami, K. 1989. Wastewater reclamation and reuse in Japan. *Proceedings of the 26th Japan Sewage Works Assn. Annual Technical Conference—International Session,* Tokyo.

Nair, S.S. and L.P. Selvam. 1996. India-Madras: Renovated sewage for agriculture, industry and groundwater recharge. World Bank, Washington, DC, 23–24 September.

National Research Council. 1980. *Drinking Water and Health,* Vol. 2. National Academy Press, Washington, DC, pp. 252–253.

Okun, D.A. 1968a. The hierarchy of water quality. *Environmental Science and Technology* 2:672.

Okun, D.A. 1968b. The future of water quality management. *J. Institute of Water Pollution Control* (Britain) 67(2):127.

Okun, D.A. 1969a. Alternatives in water supply. AWWA Annual Conference, Denver, CO, 61:215.

Okun, D.A. 1969b. New developments in urban water supply. *J. Japan Water Works Assn.* 414:54.

Okun, D.A. 1973. Planning for water reuse. *J. AWWA* 65:617.

Okun, D.A. 1991. A water and sanitation for strategy for the developing world. *Environment* 33(8):16.

Okun, D.A. and J. Crook. 1989. Water reclamation and reuse in Sao Paulo, Brazil. Report for the Pan American Health Organization, Washington, DC, July.

Okuno, N. 1994. Full-scale implementation of wastewater reuse in Tokyo. *Proceedings of the International Conference on Water and Wastewater,* Beijing, July, p. 1071.

Rosenblum, E. 1997. South Bay Water Recycling, City of San Jose, CA, (personal communication).

Sazaki, T. 1996. Creation of a new water environment. *Osaka and Its Technology* 29:45.

TMG. 1993. How Can Sewerage Make the World More Beautiful? Bureau of Sewerage, Tokyo Metropolitan Government.

UN ECOSOC. 1958. *Water for Industrial Use.* UN Report No. E/3058ST/ECA/50. United Nations Economic and Social Council, UN, New York.

WHO. 1989. Health guidelines for the use of wastewater in agriculture and aquaculture. Report of a WHO Scientific Group, Technical Report Series 778, World Health Organization, Geneva, Switzerland.

WHO. 1993. 1993 Guidelines for drinking water quality. Vol. 1, Recommendations, Geneva, Switzerland.

Young, R.E., T.R. Holliman, and J. Parsons. 1994. Using reclaimed water for bathroom flushing—A case history. *Proceedings of the Water Reuse Symposium,* AWWA/WEF, p. 721.

3

Rainwater Harvesting and Recycling

K. D. SHARMA
Senior Hydrologist
Central Arid Zone Research Institute
Jodhpur 342 003, India

INTRODUCTION

RAINWATER HARVESTING CAN be applied in many locations in the world but is of particular importance in countries with arid and semi-arid (ASA) lands, where it may often provide the only feasible solution for an improved water supply. Its potential depends on climate, hydrology, landscape, and vegetatation (Lee and Visscher, 1992).

Climate

The harsh climate in the ASA regions is characterized by (Jones, 1981):

- low total quantity of annual rainfall
- strong variations in rainfall through space and time
- high temperature and evaporation levels

Average rainfall in the Indian ASA regions (Figure 1) range from 50–750 mm. It varies from less than 100 mm in the northwest to 450 mm at the eastern boundary of the arid zone in Rajasthan, from less than 300 mm to 500 mm in the arid zone of Gujrat, and from 200 to 450 mm in the Haryana-Punjab region (Sharma and Vangani, 1992). In the semi-arid zone, it varies between 500–750 mm. Much of it (78 to 96%) is received during the southwest monsoon season in June to September. The mean number of rainy days varies from 8–17 in the west to 25 in the east.

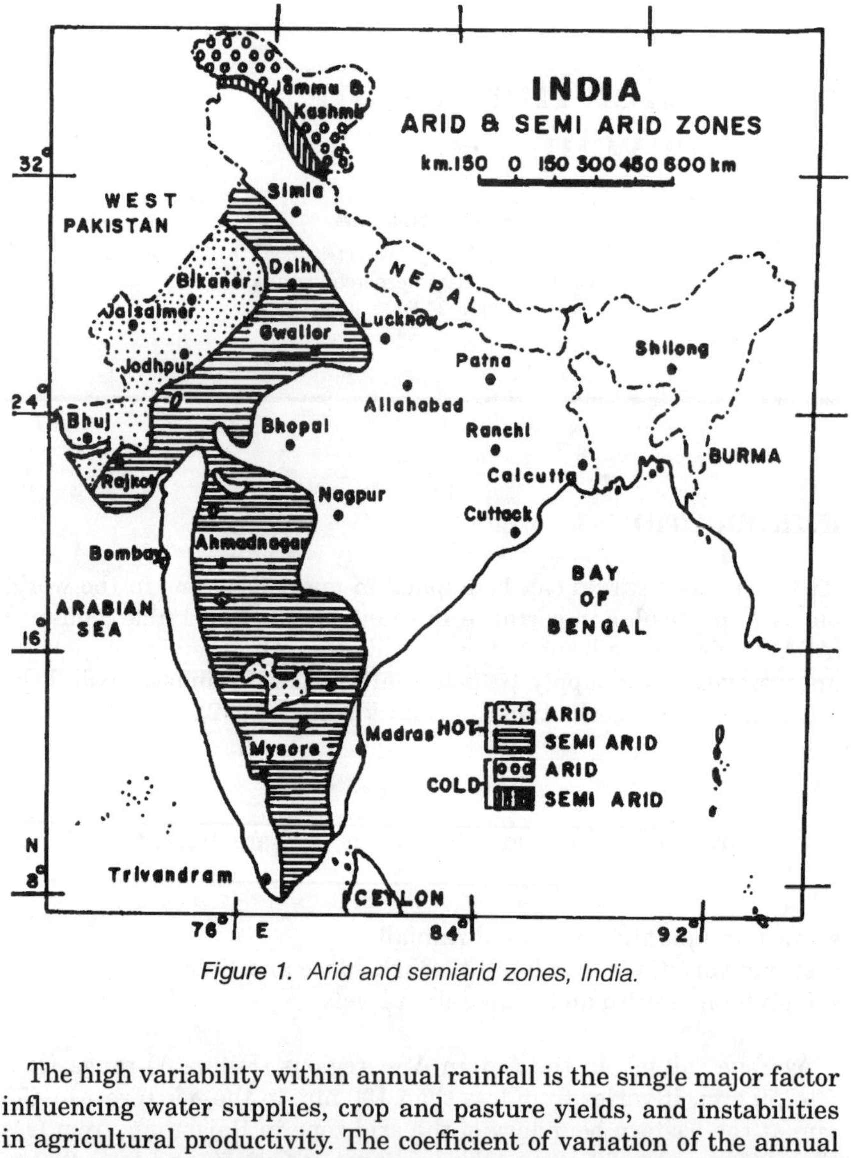

Figure 1. Arid and semiarid zones, India.

The high variability within annual rainfall is the single major factor influencing water supplies, crop and pasture yields, and instabilities in agricultural productivity. The coefficient of variation of the annual rainful ranges from 37% in the eastern part to more than 59% in the western part in Rajasthan, 50% in Punjab, while in the Kachchh region of Gujrat it varies from 60 to 70% (Krishnan, 1968).

The prevalence of clear sky and high solar radiation results in high temperatures during the day time in the summer, which may average from 40 to 43°C in May, with the peak value often reaching 44 to 49°C. The temperature falls below 0°C in many locations during winter, and frost occurs for a few days. The potential evapotranspiration often exceeds 1500 to 2300 mm per year as a result of high insolation, strong wind regime, and high temperature.

Hydrology

In the ASA environment, runoff rates rise and fall rapidly due to high rainfall intensities, steep slope gradients, and low vegetation cover on the slope surface. However, runoff is not generated uniformly due to the often short, sharp, intense bursts of rainfall, and variations in slope gradients and surface type along flow lines (Sharma, 1992). The local population has good insight and knowledge of the hydrological characteristics and, therefore, is a very important source of information to assess local conditions. River flow is ephemeral, and almost all surface flow is stormwater that rushes off the land during and immediately following rainstorms. However, there is a slow, subsurface flow of water in the river bed itself which depends on the porosity, the continuity of the pores, and the hydraulic gradient. Flow rates are usually in the order of magnitude of 10^{-2} to 10^{-3} m s^{-1} for coarse sandy river beds and down to 10^{-9} m s^{-1} for compacted clay (Jones, 1981).

In the ASA regions the availability of groundwater is characterized by deep water levels, high salinity, negligible recharge, and uneconomic cost of exploitation. About 65% of the area in the Indian arid zone either has more than 15 m depth of water table, contain more than 3 g L^{-1} salts, or has discharge potential of less than 1 LS^{-1} (Das, 1994). Besides, the utilizable groundwater is being exploited at a faster rate than the recharge and thus the groundwater reserves are being depleted at an alarming rate.

Landscape and Vegetation

Generally, the ASA landscapes are comprised of rocky top slopes and isolated outcrops (slopes > 45°), medium slopes (10–20°), low slopes (1–10°), and river channels. Not all of these landscape types may be present within a region. Below any rock outcrops, the soils are generally coarse and stony. They are usually sandy loams on the steeper slopes, changing to loams and silt loams as gradients decrease. On the low slopes, alternating wet and dry soil conditions can result in mineral concentration a few meters down in the soil column in the form of iron pan, laterite, or murram. These conditions may impede downward drainage and result in more frequent and rapid runoff.

Vegetation cover ranges from high-density evergreen and deciduous tree and plant associations on the aprons surrounding rocky outcrops and along the drainage lines, to thorny shrubs on the more gentle low slopes.

Under conditions of low and erratic rainfall, the soils of ASA regions are often in a delicate balance with nature. As pressures on soils increase from farming, clearing of vegetation for firewood collection, charcoal production, and livestock grazing, the structure and nutrient content can rapidly degrade. The consequence can be intense gullying and soil loss on even the lowest gradient slopes. Surface reservoirs may rapidly fill with soils washed off a hillside or dams might be breached by torrential flows.

RAINWATER HARVESTING SYSTEMS

Since time immemmorial, rainwater harvesting systems in the form of surface and subsurface storage have been widely used in ASA regions. However, with the advant of pipe water supply, the traditional rainwater harvesting sytems have been neglected and abandoned. For example, all surface water sources dried up, the water table dropped, salinity increased, and yield from wells decreased considerably in most parts of the ASA regions due to consecutive droughts (1986–1991). This has once again drawn the attention to the role of rainwater harvesting structures in drinking water supply systems (Anonymous, 1988). Besides, the greatest opportunity for increasing water supplies is to improve existing water systems and, thus, make more water available without a complete new installation. A few more prevalent rainwater harvesting systems are discussed here.

Rooftop Rainwater Harvesting

Rooftop rainwater harvesting comprised of the rooftop as the catchment area, connected by gutters and pipes to a storage container. The most suitable rooftop surfaces are corrugated iron sheets, tiles, and asbestos sheets. Thatched roofs pose problems as the runoff is less and generally of a low quality (Yaziz et al., 1989).

PREVALENCE

Simple rooftop rainwater harvesting (Figure 2) is used traditionally in most ASA communities. This system provides an on-site source of

Figure 2. Rooftop rainwater harvesting system.

water supply next to homes or public buildings. In the Indian arid zone, all the old buildings, forts, and palaces have rooftop rainwater harvesting systems.

DESIGN CONSIDERATIONS

The actual size of the system depends on system cost, amount of rain to be collected, expectations and need of owners, and level of external support (Gould, 1991). The volume of rain and its distribution over the year, the size of the catchment area, and the projected supply ultimately determine what size tank can be filled and how often (Rathore, 1988). Usually, the worst scenario will present itself during the longest dry season. The size of catchment area and tank should provide sufficient supply for the users during this period. The volume of tank can be calculated by

$$V = t \cdot n \cdot cpc + E$$

Where V is the volume of tank, t is the number of days, n is the number of people, cpc is the consumption level per capita per day, and E is the evaporation loss. The necessary catchment area can be determined by dividing the tank volume with the accumulated average rainfall value per m^2 over the preceding wet months, multiplied with the runoff coefficient that can be set at 0.8 for galvanized iron or tile roofs (Michaelides and Young, 1985; Michaelides et al., 1986).

The roof should be smooth, made of nontoxic substances, and sufficiently large to fill the tank with the available rainfall conditions. Existing roofs of houses and public buildings can be used for a rooftop catchment system. In some cases, enlarged or additional roofed structures may be built.

The function of guttering is to protect the building by collecting the water running off the roof and direct it via a downpipe to the storage tank. The gutter should have a uniform slope of 0.5% and be large enough to collect the heavy runoff from high intensity rains. With all roof catchment tanks, the first 20 to 30 L of rainwater running off the roof, which contain large quantities of leaves and bird droppings, should be discarded to keep the water potable (O'Brien, 1990).

There is a wide variety of tank shapes and sizes ranging from small jars, cylindrical or square standing tanks, to hemispherical excavated tanks. A range of material is used for water tanks including polythene, galvanized iron, brick, stone, masonry, reinforced concrete, and ferrocement (Lee and Visscher, 1992). Most practitioners now agree that 2 to 100 m^3 ferrocement tanks provide the easiest and cheapest solution in most situations (Edwards et al., 1984). Rooftop rainwater harvesting systems, like all water supply systems, require good periodic management and maintenance to ensure that the quality and reliability of the water supply is high. Proper management and maintenance of the system as well as water use management will ensure a troublefree longer service of the rooftop rainwater harvesting system.

Tanka

Tankas, the most prevailing rainwater harvesting system in the Indian arid zone, is a local name for a covered underground cistern. These have been constructed for centuries in individual households, near religious centers, schools, and public places.

PREVALENCE

The first known construction of Tanka in the Indian ASA regions was during the year 1607 near Jodhpur (Vangani et al., 1988). Thereafter, almost every house in the rural area within the Indian desert got a Tanka constructed to meet the day-to-day drinking water demand. In the recent past, development of these structures has also been reported from other developing countries such as Botswana, Ghana, Kenya, Yemen, Sri Lanka, Thailand, and Indonesia (Hofkes, 1986; Lee and Visscher, 1990; Pacey and Cullis, 1986).

DESIGN CONSIDERATIONS

The traditional Tankas are made by digging a hole of 3.00 to 4.25 m diameter in the ground and plastering the base and sides with lime mortar of 6 mm thickness or cement mortar of 3 mm thickness (Vangani et al., 1988). The top is covered with locally available brush wood, thorns, and dried grass mats (Figure 3). The catchments are made in a variety of shapes and sizes using locally available sealing materials such as pond silt, murram, coalash, and gravel.

Tankas constructed by traditional methods are temporary. The useful life of such structures is about three years. They are subjected to seepage through the bottom and sides. Moreover, the catchment areas are not commensurate with the amount of rainfall received and runoff generated. The brush wood/thorn/grass lid does not prevent the pollution of water and evaporative losses. In general, these are small and the stored water is inadequate for drinking throughout the year.

Vangani et al. (1988) designed an improved Tanka of 21 m^3 capacity, which gets filled up with annual dependable rainfall of 125 mm (Figure 4). The harvested water is adequate for a family of six people throughout the year for drinking and the structure has a useful life of at least twenty-five years. Table 1 shows the design parameters for Tanka in the Barmer district, an arid region in India. About 10,000 such structures are successfully functioning in the region.

The cost of harvested rainwater in the Tanka varies from Rs. 0.05 to Rs. 0.07 per liter (42 Rs. = 1 US $). With a growing population the

Figure 3. A traditional Tanka covered with thorns.

Tanka alone is not capable of meeting the increased demand for water. However, it can be viewed as a complementary source of an integrated drinking water supply scheme involving the other water sources. The Tanka is a vital source for meeting the drinking water demand of the rural population, especially for the people living in scattered hamlets.

Nadi

Nadis are small excavated or embanked village ponds that harness the meager precipitation, to mitigate the scarcity of drinking water in ASA regions (Figure 5). Water from Nadis is consumed by humans and livestock and is available for two months to a full year after rain, depending upon the catchment characteristics and the amount, intensity, and distribution of water received. This is an ancient practice and the Nadis are the most important resource in the Indian ASA regions (Sharma and Joshi, 1981). The first recorded masonry Nadi was constructed in 1520 near Jodhpur.

Figure 4. An improved Tanka.

PREVALENCE

Since Nadis are the vital water resource in the Indian ASA regions, each village has one or more depending on the demand of water and the availability of suitable sites. As a case study, the number of Nadis in Kachchh, Nagaur, Barmer, Jodhpur, and Jaisalmer districts were found to be 2245, 1436, 592, 1314, and 1822, respectively (Vangani and Singh, 1994). In the Nagaur district, the dependency on Nadis amounts to 13.2% throughout the year and 32.1% as a group percentage (Table 2).

A Nadi consists of two components: catchment and water impoundment areas. Their characteristics vary as per the amount of rain-

Table 1. Design parameters for Tanka (Vangani et al., 1988).

Parameters	Range of Values
1. Mean annual rainfall (mm)	187–321
2. Dependable rainfall (mm)	134–253
3. Catchment area (m^2 m^{-3} of runoff)	20–37
4. Catchment area for 21 m^3 capacity	420–777
5. Diameter of catchment area (m)	23–32

Figure 5. A village Nadi.

fall and environment. The Nadis range from 1.5 to 12 m in depth, 400 to 700,000 m^3 in capacity, and have drainage basins of various shapes and sizes (8 to 2000 ha).

Figure 6 depicts the relationship between the catchment area and the storage capacity of Nadis under various landforms in the Indian arid zone. The highest storage capacity per unit of catchment area was observed on dune complex and sandy plain areas having 1–2% slope (Table 3). In younger alluvial plains and rocky/gravel pediments, the Nadi volume per unit of catchment area increases with the increasing slope. The storage capacity of Nadis under various rainfall zones is given in Table 4. In sandy areas, the highest storage capacity of the Nadis was found in the 300–350 mm zone, because of stabilization of sand with grasses. Sharma et al. (1989) showed that the buried pediments generated the highest runoff (55 mm), followed by the older alluvial plains (29 mm), rocky/gravel pediments (16 mm), and interdunal plains (13 mm). Thus, the first two landforms are the potential catchments for rainwater harvesting in the region.

Among factors limiting proper utilization of Nadis, sedimentation, water pollution, seepage, and evaporation are the most important.

Table 2. Dependency of drinking water sources (Chatterji et al., 1985).

Nadi water Well water	No well water	Up to 6 months	Up to 12 months	Group (%)
No Nadi water	7.0	—	16.9	24.0
Up to 6 months	6.9	4.4	0.7	28.4
Up to 12 months	13.2	0.1	5.0	20.5
Group (%)	32.1	13.2	25.5	100.0

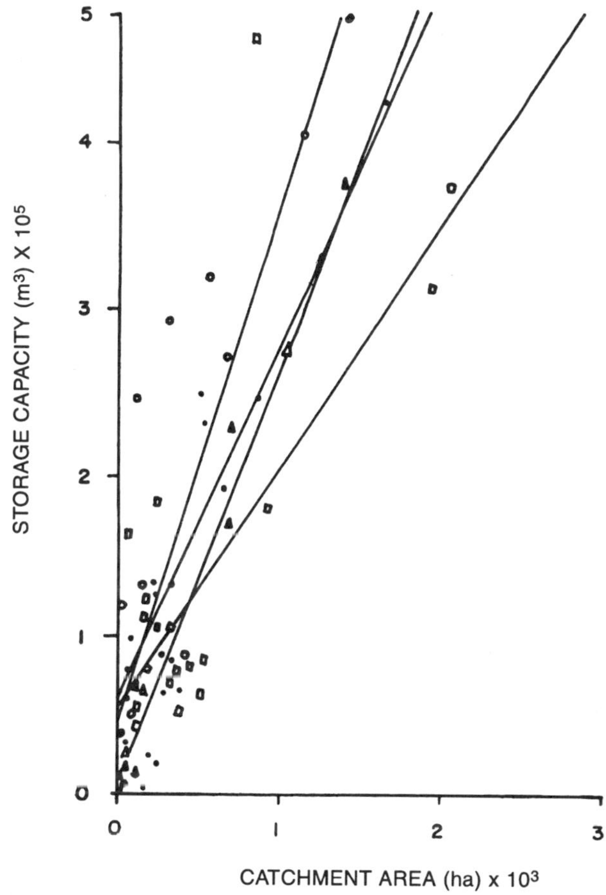

Figure 6. Storage capacity of Nadi in relation to catchment area in sandy plain (○), dune complex (▲), younger alluvial plain (□) and rocky/gravel pediment (.).

Sharma and Joshi (1987) found wider variations in the annual sediment deposition under various environmental settings (Table 5). The older alluvial plain has the highest sedimentation rate, while the younger alluvial plain has the least. Sediment yield from rocky/gravel pediment, dune complex, and sandy plain lies between these two extremes. Sediment input to the Nadis can be reduced appreciably either by promoting vegetation in the catchment or by constructing silt traps at the inlet. The sediment deposition decreased by 65, 94, and 77% from the blown sand, younger alluvium, and older alluvium areas,

Table 3. Effect of catchment slope on storage capacity of Nadis (Sharma and Joshi, 1981).

	Nadi volume (m^3) per ha of catchment area				
	Slope group (%)				
Physiography	< 1	1–2	2–3	3–4	> 4
Dune complex	54.3	108.2	45.7	29.0	13.8
Sandy plain	110.6	154.2	121.1	—	—
Younger alluvial plain	466.4	578.5	731.8	—	—
Rocky/gravel pediment	51.4	421.5	724.7	945.9	1236.5

respectively, by promoting vegetation in the catchments (Sharma and Chatterji, 1982). However, in catchments having biotic interferences, the silt traps at the inlets reduced the sediment deposition by 71% under similar rainfall and catchment characteristics.

Due to poor maintenance and improper utilization, the Nadi water is highly polluted and present health hazards in the form of guinea worms, water hyacinth, mosses, and algae. Diseases of stomach and skin, infectious diseases, and guinea worms are associated with the consumption of Nadi water. To minimize the pollution of water, the entry of animals and human beings into impoundments must be restricted. Hand pumps and wind mills may be used for the withdrawal of water. Also, the catchments should not be used for defecation by the villagers and movement of livestock must be checked.

Nadis usually have large surface areas compared to the volume of water stored. Therefore, heavy losses occur since evaporation is a function of surface area. On the other hand, the seepage increases with the

Table 4. Effect of rainfall on Nadi volume/catchment area ratio under different physiographic settings (Sharma and Joshi, 1981).

	Nadi volume (m^3) per ha of catchment area as a functional rainfall zone (mm)				
Physiography	250–300	300–350	350–400	400–450	450–500
Dune complex	55.3	110.5	53.0	37.1	—
Sandy plain	120.4	128.0	131.1	137.3	—
Younger alluvial plain	—	—	—	—	—
Rocky/gravel pediment	491.7	518.9	785.5	1644.7	2264.7

Table 5. *Rate of sediment accumulation in Nadis in different environments (Sharma and Chatterji, 1982).*

Environment	Nadi volume (m^3) × 10^3		Sediment yield (cm^3 m^3 runoff)		Annual sediment production rate (m^3/ha)		Annual reduction in Nadi volume (%)		No. of observations
	Range	Mean	Range	Mean	Range	Mean	Range	Mean	
Sandy plain	(2.6–603.8)	125.1	(337.0–27,030.0)	4790.0	(0.4–8.4)	3.1	(0.5–4.9)	1.9	36
Dune complex	(0.4–62.8)	23.5	(36.0–90,910.0)	9200.0	(0.1–15.7)	4.8	(0.7–6.1)	2.2	18
Younger alluvial plain	(12.6–256.6)	130.1	(360.0–10,200.0)	2790.0	(0.3–12.1)	2.7	(1.8–13.2)	3.4	30
Older alluvial plain	(9.9–335.1)	119.1	(8403.0–32,260.0)	21,870.0	(5.2–26.4)	18.4	(1.2–9.0)	5.3	5
Rocky/gravelly pediments	(9.9–628.3)	131.8	(910.0–62,500.0)	12,260.0	(1.5–28.3)	14.3	(1.6–28.0)	7.8	11

depth of stored water. The highest evaporative losses were observed in the dry season during March/June when the demand for water was also high. High evaporation also occurs in October and November after the rainy season. Seepage losses were greatest during the rainy season in July/September when the Nadis were full. Evaporative losses varied from 55 to 80% of the total losses under various environments (Sharma and Joshi, 1983). The highest losses due to evaporation (2473 mm y) and the least to seepage (66 mm y) occur in rocky/gravel pediment due to large surface area and impervious bed (Figure 7). The highest seepage occur in the older alluvium (1577 mm y) followed by the sandy plain. At present, no technique is perfect for controlling the evaporation in Nadis. Different types of treatments and linings are available to reduce the seepage losses and may be adopted depending upon their

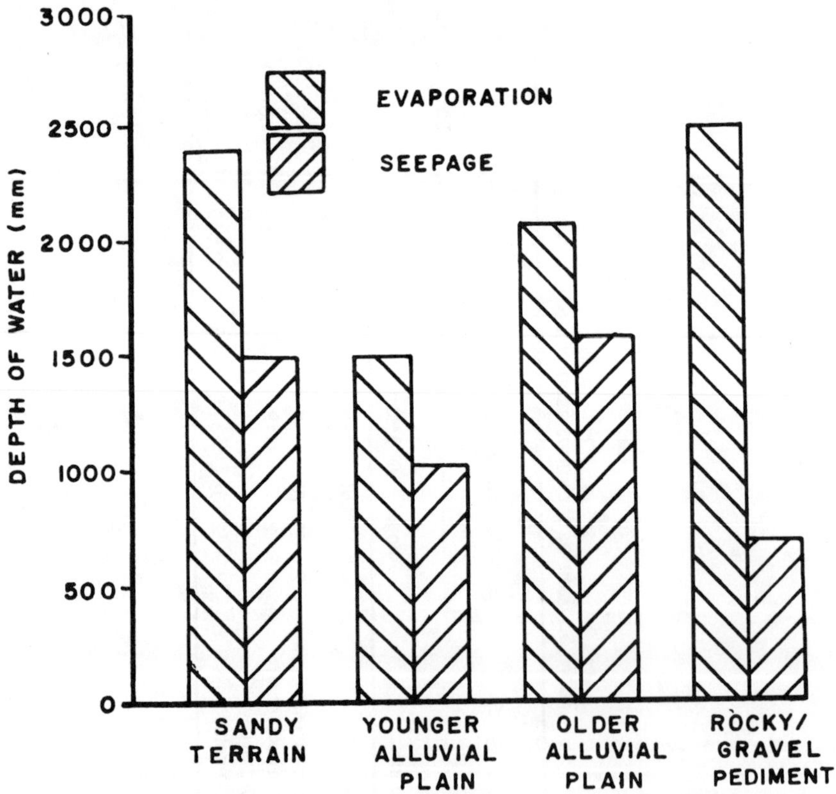

Figure 7. Annual evaporative and seepage losses in Nadi.

Table 6. Optimized Nadi characteristics to minimize evaporative and seepage losses (Sharma and Joshi, 1983).

Physiography	Optimized Depth (m)	Optimized surface area ($m^2 \times 10^3$)	Water availability (month)
Dune complex	2.5	29.1	4.8
Sandy plain	2.0	27.1	8.3
Younger alluvial plain	5.0	161.0	12.0
Older alluvial plain	3.0	96.3	12.0
Rocky/gravelly pediment	6.0	126.5	12.0

cost effectiveness. Sharma and Joshi (1983) minimized these losses by optimizing the depth and surface area of Nadis under various environments (Table 6). These could act as guidelines for the new construction in the ASA regions.

Dams/Reservoirs

Where water flows regularly into a small valley, earth dams can be built to retain water on site. Earth dams are raised banks of compacted earth, approximately 2 to 5 m in height, built at the downstream end of a hollow. They often include a clay core with stone aprons and a spillway to discharge excess runoff. The water storage volume of earth dams generally ranges between 1000 and 50,000 m^3.

PREVALENCE

In the Indian arid zone, the Luni–Jawai forms the major drainage system. Several reservoirs along the rivers have been constructed to cater to the needs of water for human and livestock consumption and also for irrigation. There are 865 minor to medium reservoirs with a total storage capacity of 1096.63×10^6 m^3 in the Luni basin (Anonymous, 1990).

DESIGN CONSIDERATIONS

The runoff in the catchment area can be estimated by

$$q = \mathrm{C\,i\,A}$$

Table 7. Runoff coefficients in ASA regions (Sharma, 1990).

Rainfall (mm)	Terrain	Runoff coefficient
< 200	Rocky/gravel eroded surface	0.2–0.3
	Interdunal plain	0.1–0.2
200–400	Rocky surface with gravel mantle	0.2–0.3
	Rocky surface without gravel	0.3–0.4
	Sandy loam to loamy sand	0.15–0.2
	Above with compaction	0.3–0.4
400–600	Hills and uplands	0.4–0.6
	Sandy/alluvial plain	0.2–0.4
> 600	Compacted and smoothed soil	0.3–0.5
	Soil on < 10% slope	0.0–0.3
	Rocky natural surface	0.2–0.5

where q is the runoff rate, C is thc runoff coefficient, i is the rainfall intensity, and A is the catchment area. The runoff coefficient is the proportion of rainfall that will result in runoff. For different types of catchment areas, different coefficients apply (Table 7). The required catchment, A, for a given demand can be identified as follows:

$$A = \frac{D + E}{Q}$$

where D is the demand, E is the evaporation loss, and Q is the runoff amount.

The materials to be used are mainly clay for the dam wall and large stones for protection against erosion (USDI, 1977). Standard guides are available for design and construction of various components of the small earth dams in the ASA regions and must be adhered to (Fowler, 1977; Singh, 1983).

REDUCING EVAPORATION FROM WATER SURFACE

Evaporation from a free water surface is a major constraint in developing local water supplies. Mono-molecular films or layers, extensively investigated, have not been found effective in ASA regions due to the disruption of film by wind. Instead, shading of water surfaces with polyethylene sheets successfully reduced evaporation by 91% (Khan, 1992). Evaporation reduction with floating material ranged from 37% to 82%. Foam rubber sheet, polyethylene sheet, and bamboo reduced evaporation by 74, 66, and 53%, respectively. The cost of saved water under different treatments varied between Rs. 2.54 and Rs. 4.44 m^{-3} (Khan, 1992).

Subsurface Water Harvesting

Subsurface water harvesting systems exploit water already infiltrated and concentrated through natural hydrological processes into the sand rivers that fill valleys in ASA regions. River beds are traditional watering points. People use the surface flow during rainy seasons and dig shallow unprotected wells into the bed each dry season once surface flow has dried up.

PREVALENCE

Subsurface and sand dams are common in the ASA regions in India, Kenya, and Tanzania (Nilsson, 1988). The material in the river bed should be mostly sand, the coarser the better. A dam of 3 m depth in a river bed of 30 m width stores around 5000 m^3 of free draining water if the soil is coarse sand. This provides a minimum supply of 110 Lpd for 300 people over a 150-day period.

DESIGN CONSIDERATIONS

Each of the subsurface dams has a different design and construction method. The amount of extra water available to well users at the dam location depends on the depth, width, length, and hydraulic gradient of the sand river (Hanson and Nilsson, 1986). The subsurface dam is an impermeable barrier, made from either clay or masonry, across the river bed from the surface down to an impermeable layer below. Another type, the raised sand dam, may be made from masonry. This dam raises the level of the river bed, catching sand from upstream and creating a larger groundwater reservoir. The dam is raised in successive stages of 50 cm each to achieve this effect (Gezahegne, 1986).

The clay dam is best built where a layer of clay underlies the sandy river bed. A masonry dam can be built where a bed rock layer occurs between 2 and 4 m below the surface of the river bed. The raised sand dam is built where bed rock is exposed or near the river bed surface, and raised in 50 cm increments after each rainy season.

Subsurface water harvesting systems provide a large, concentrated source of water for communal use, thereby extending the life or improving the yield of traditional wells used by rural people. The stored water quality is generally good and evaporative losses are low in comparision to open reservoirs. Traditionally, the communities in ASA regions have used this type of storage, and therefore the improvements in sand river supplies are most acceptable. However, it is the most site-specific of the water harvesting technologies.

Anicut, Checkdam, and Percolation Tank

These are constructed across an ephemeral stream with fill section to intercept the runoff from local catchments and store it for optimum utilization (Khan, 1992). The stored water behind the structure is used for drinking, irrigation, and recharging of the downstream wells (Figure 8). In most instances, cultivation takes place in the submergence area in the dry season. In the Gujrat State of India, the checkdams not only recharge the sweet water into the surrounding area but also check the saline ingress towards the fertile and valuable croplands on the Saurashtra coast.

PREVALENCE

These structures are suitable in hilly and uneven topography where ephemeral streams are available and good runoff production catchments are present. Also, the subsurface formations must be receptive

Figure 8. A checkdam recharging the groundwater.

to the enhanced groundwater recharge. The anicut, checkdam, and percolation tanks are widely adopted in the hard rock and basaltic terrain of southeast Rajasthan, Gujrat, Maharashtra, Madhya Pradesh, and in the Deccan plateau regions in India (Anonymous, 1988).

DESIGN CONSIDERATIONS

In general, these structures are designed on the basis of fifty years' recurrence interval of the rainstorm. Basically, there are two components of the system: an earthen embankment and a masonry spillway. Construction takes into account the catchment area, expected runoff, designed storage, and the area to be benefited. A thorough and detailed knowledge of the geological, hydrological, and morphological features of the area is necessary for selecting the sites. Local guidelines and design parameters have been developed for their construction by the Soil Conservation/Irrigation Departaments. These guidelines are region-specific and must be followed according to site conditions. The above structures enhance the groundwater recharge. In a case study, the groundwater recharge was enhanced up to 35% over a period of three years in sandstone terrain by the construction of an anicut on an ephemeral stream in western Rajasthan (Sharma and Kalla, 1980). In general, the average recharge is 50% of the storage in any normal year. In basaltic formations, a percolation tank normally influcnces about 1.5 to 2.0 km^2 area and recharges about 0.15×10^6 m^3 of runoff during the normal rainfall years. The enchanced recharge is variable ranging from 0.032×10^6 to 0.182×10^6 m^3, representing about 50% of the storage capacity (Anonymous, 1988).

RUNOFF RECYCLING

Better and efficient utilization of stored rainwater is essential for proper management of soil and crop. Lack of sufficient available water in the root zone during significant periods of crop growth is common in dry farming areas. During prolonged dry spells, the stored runoff water should be applied to the crop by an efficient method of irrigation that gives maximum return per unit of water (Mann et al., 1981). The traditional methods of irrigation such as flooding, check basin, and border irrigation are of little consequence, as losses are high and efficiency of utilization is low. The principal of application, therefore, is to cut down conveyance and evaporative losses and to force water deep into the soil profile, where roots are extensive and can utilize the moisture better.

A supplemental irrigation of 84 mm depth with the harvested runoff to direct seeded pearlmillet on flat land resulted in 337% higher yield over unirrigated crop at Jodhpur (Singh, 1994). Similarly, in an experiment on different levels of supplemental irrigation (no irrigation, 50 mm, 100 mm, and irrigation at 60% depletion) at Jodhpur, Singh (1994) found that one supplemental irrigation of 50 and 100 mm increased the grain yield of pearlmillet by 27 and 57%, respectively, over unirrigated control. A supplemental irrigation of 75 mm given 60 days after sowing resulted in 81% higher grain yield than control. It was also found that a yield increase of 100% of rainfed pearlmillet could be obtained by giving one supplemental irrigation of 75 mm at 60% moisture deficit.

Singh (1990) reported that on semi-arid deep black soils receiving an annual rainfall of 500 mm, 10% of annual rainfall can be harvested into storages. Using this water for one supplemental irrigation resulted in a 64% increase in the grain yield of wheat.

Every 50 mm supplemental irrigation increased the grain yield of wheat by 400 kg ha^{-1} (Verma et al., 1986). The yield of black gram also increased from 35 to 124% with one to three supplemental irrigations of 50 mm each, respectively.

Studies revealed that about 20 to 25% of annual rainfall can be harvested into farm ponds in the red soil region receiving 550–850 mm annual rainfall. One supplemental irrigation of 50 mm significantly increased the grain yield of groundnut, cowpea, chilly, and ragi (Hedge et al., 1981). Two supplemental irrigations at critical stages increased the yield by about 200%. In the semi-arid region of Andhra Pradesh-India, the sorghum yield increased by 79% with one supplemental irrigation. In pigeonpea, one supplemental irrigation of 20 mm increased the grain yield by 93% (Venkateshwarlu, 1981). The effect of supplemental irrigation from harvested rainwater in the ponds on the yield of various crops is given in Table 8. As shown, the yield of various crops increased significantly over control with one supplemental irrigation of 50 mm at most critical stages of the crop growth.

Chittaranjan et al. (1981) observed a benefit to cost ratio as high as 3.4 with one supplemental irrigation of 50 mm in sorghum. However, the safflower did not show any response to the supplemental irrigation. In the red soil region of Telangana, various crops responded to the supplemental irrigation to the extent of 100% (Kameswar Rao et al., 1981).

Khadin

Khadin is a system of growing crops on harvested and stored rainwater in the farm soils by constructing an earthen bund across the gentle

Table 8. Effect of supplemental irrigation on crop yield (Venkatesnwarlu, 1981).

		Grain yield (t · ha^{-1})	
Location	Crop	Without Irrigation	With Irrigation
Varanasi	Barley	2.6	3.4
Ludhiana	Wheat	1.9	4.1
Agra	Wheat	2.2	2.7
Bijapur	Sorghum	1.7	2.4
Bellary	Sorghum	0.4	1.4
Sholapur	Sorghum	1.0	1.8
Rewa Upland	Rice	1.6	2.8
	Wheat	0.6	1.9
Anand	Tobacco	1.2	1.8

slope of the farm in the valley bottom (Kolarkar et al., 1980). This practice dates back to the fifteenth century and is being followed in the 100 mm rainfall zone in the Indian desert.

In the extreme western desert tract in India, many large rocky areas, rugged hillocks, and plateaus form the runoff-producing catchments for the adjoining flat valley bottoms, which are main farming lands in the region. A large earthen bund/dam is constructed across the gentle slope of the valley bottom to facilitate the spreading of harvested runoff. Water accumulates to a depth of 60 to 90 cm, gradually seeps into the ground, and recharges the adjoining wells. If the accumulated persists, it is drained out through the sluce before the crops are sown, as per the volume of runoff received in the Khadin. The soils of Khadin are very fertile due to frequent deposition of fine sediments. Depending upon the type of catchment, the ratio of farm to area varies from 1:12 to 1:15 (Kolarkar et al., 1983).

At present, local varieties of crops in most of the Khadins are grown and fertilizer use is negligible. Thus there is a considerable need to improve upon the system to increase productivity through agroforestry and agrohorticulture systems.

COST-BENEFIT ANALYSIS

Financial and economic issues are critical to the introduction of water harvesting systems. The earlier literature indicated that the price levels of water harvesting systems are high compared to other water supply schemes. However, recent studies indicate that a more balanced picture arises when taking into account the service level provided and

Table 9. Construction cost of rooftop rainwater harvesting (Vangani et al., 1988; Lee and Visscher, 1992; Khan, 1994).

System	Volume (m^3)	Cost per m^3 of tank volume in Rs.
Cement jar	1	800
Granary tank	8–10	544–992
Standing tank	5.5–13.5	800–2400
	20–30	800–1408
Masonry standing tank	50	2240
Ferrocement tank	70–80	352–832
Masonry underground tank	21	381
	200	275
	300	624

the true pricing mechanism. Apart from systems cost and benefit-to-cost analysis, other aspects need to be considered including affordability, willingness to pay, provision of subsidies, and financing mechanism (Lee and Visscher, 1990).

Cost estimates show a wide range due to different pricing in different countries. Therefore, cost figures from the literature should only be used as an indication. Up-to-date breakdown of costs should be made locally for each technology. Costs of rooftop rainwater harvesting given in project documents are summarized in Table 9. Where roofs are not suitable for rainwater harvesting, cost of construction of new roofs and gutters must be added to the cost of the tank. This cost ranges from Rs. 122 m^{-2} for a 36-m^2 roof to Rs. 374 m^{-2} for a 80-m^2 roof (Lee and Visscher, 1992). Costs of gutters range from Rs. 58 to Rs. 128 m^{-2} (1988 prices). Cost for surface and subsurface water harvesting systems (Table 10) also shows a considerable variation.

Table 10. Cost of surface and subsurface water harvesting systems (Lee and Visscher, 1992).

Description	Volume (m^3)	Storage Cost per m^3 of volume in Rs.
Ground tank	17	611
Subsurface dam	3500	80–125
Rock dam	13,000	51
Small earth dam	30,000	61
Medium earth dam	60,000	67

Pacey and Cullis (1986) observed that household rainwater tanks were on average 13 to 20 times more expensive than hand-dug wells and 6 to 10 times more expensive than pipe water supplies. Communal rainwater tanks were 3 to 10 times more expensive than hand-dug wells and 2 to 5 times more expensive than pipe supplies. Omwenga (1984) reported that per capita construction cost of rooftop rainwater harvesting was 3.8 to 11.0 times higher than protected springs and 1.3 to 6.4 times more expensive than dug and drilled wells. However, pipe supplies were 0.7 to 3.9 times more expensive than rooftop rainwater harvesting. O'Brien (1990) calculated the cost of a cistern project as Rs. 1696 per capita, compared to Rs. 1504 per capita for a deep well with handpump.

Assessing Benefits

In comparing technologies, not only the costs but also the benefits need to be taken into account. The benefits of water harvesting systems are both direct and indirect. Therefore, many systems cannot be valued in conventional cost-benefit terms. Benefits may include:

- resource conservation through flood control and increased groundwater recharge
- employment and/or building up of local skills through water harvesting system construction
- increased local self-reliance on water supply
- increase in local organization capacity and community cooperation in developing and managing their own resources
- time saved in collecting water, which can be invested elsewhere
- lower management and maintenance needs and costs

These benefits need to be taken into account when a system is being selected to provide an acceptable service level on a sustainable basis.

CASE STUDIES

Case studies are success stories of rainwater harvesting systems. Two case studies, one related to rooftop rainwater harvesting (Rathore, 1988) and the other to subsurface water harvesting (Raju, 1994), are given in the subsequent sections.

Rooftop Rainwater Harvesting

The village Jaislan in the Nagaur district of arid India, is located in the saline groundwater tract. The village covers a geographical area of 868 ha and has 879 inhabitants living in 133 households. Drinking water has always been a problem in the region from the settlement of the village about 250 years ago. The villagers transported the drinking water from another village at a distanct of 3 km in earthen pitchers on their head, or on camel back and bullock cart. There would be a long wait during the summer and it took about 10 to 12 h for the turn of an individual to fetch water. The water, considered very precious, was kept under lock and key and was judiciously used.

The villagers were not aware of rainwater harvesting technology about 150 years ago. It was only by chance that a villager dug a small pit in his field to store water during the rainy season. The harvested rainwater did not last long due to seepage in sandy soils. Then he tried a paste of clay and ash as sealant and was successful in retaining the water. This paved the way for other people to adopt this technique.

In 1906 another villager tried rooftop rainwater harvesting. An underground water storage tank was linked through drain pipes to his rooftop with the help of villagers. He paid only for the mason, whereas the labor and material were contributed by the villagers. This was a successful experiment and became an important non-saline potable water source in the village. Inspired by this, two more rooftop rainwater harvesting systems were constructed between 1923 and 1926. The water was supplied to the entire village during the summer when acute scarcity of water was felt. The number of rooftop rainwater harvesting systems gradually increased. Presently, there are about 100 such systems working successfully in the village. As a result, the entire village has become self-sufficient in meeting the drinking water needs on a sustainable basis.

Subsurface Water Harvesting

Due to continuing drought conditions in the Kachchh district, Gujrat, India since 1982, the withdrawal of groundwater is more than the recharge. As a result of continuous overdraft, the water levels have declined, many open wells have become dry, and seawater has intruded the coastal aquifers. The total area affected by sea water intrusion is 15,550 ha in 244 villages.

Shree Vivekanand Research and Training Institute has constructed rainwater harvesting systems since 1987 with the assistance from voluntary organizations and state governments. In all, 58 checkdams, 48 percolation tanks, 2 subsurface dams, 39 recharge wells, and 42 storage tanks had been constructed until July 1993. The total capacity of the recharge structures is 1.44×10^6 m^3. These can take two or three floods in the normal rainfall year, thus increasing the recharge capacity to about 3×10^6 m^3 in a year of good rainfall. The area likely to be benefited is of the order of 3301 ha in twenty villages.

The rise in water level over a period of six years is in the order of 6 m with a maximum decline in salts in the order of 920 ppm. This water harvesting project has given encouraging results and proved the effectiveness of recharge wells and subsurface dams as recharge structures. The cost of water works out to be Rs. 7.33 m^{-3}.

CONCLUSION

Rainwater harvesting systems, both traditional and improved, have met the drinking water demand in ASA regions since the ancient times. However, they require proper design, construction, and maintenance in order to serve for a longer period of time the growing population in the region. These systems not only create an additional water supply source but also act as supplements to the existing resources. In the developing world, they are the only source of drinking water at many interior and scattered areas and, therefore, must receive proper attention.

REFERENCES

Anonymous. 1988. *Seminar on Water Harvesting Systems and their Management.* National Drinking Water Mission, Bangalore, India.

Anonymous. 1990. *Inventory of Irrigation Tanks in Western Rajasthan.* Department of Irrigation, Jodhpur, India.

Chatterji, P. C., N. S. Vangani, and M. L. Sharma. 1985. Unmanaged drnking water sources in arid areas of Rajasthan: A case study of Nagaur district. *Annals of Arid Zone* 24:47–55.

Chittaranjan, S., M. S. Rama Mohan Rao, and S. Selvarajan. 1981. Runoff harvesting and recycling on black soil for increasing crop production. *Extension Bulletin 3,* CSWCRTI Research Center, Bellary, India.

Das, S. K. 1994. Groundwater of the Indian arid zone. In: *Sustainable Development of the Indian Arid Zone.* R. P. Singh and S. Singh (Eds.), Scientific Publishers, Jodhpur, India, pp. 43–47.

Edwards, D. B., K. Keller, and D. Yohalem. 1984. A workshop design for rainwater roof catchment system. WASH Technical Report No. 27, Water and Sanitation for Health Project, Arlington, U.S.A.

Fowler, J. P. 1977. The design and construction of small earth dams. *Appropriate Technology* 3:4–6.

Gezahegne, W. 1986. Subsurface flow dams for rural water supply in arid and semiarid regions of developing countries. Publication Series B, No. 18, Tampere University of Technolgy, Tampere, Kenya.

Gould, J. E. 1991. Rainwater catchment systems for household water supply. *Environmental Sanitation Review No. 32,* Asian Institute of Technology, Bangkok, Thailand.

Hanson, G. and A. Nilsson. 1986. Groundwater dams for rural water supplies in developing countries. *Groundwater* 24:496–506.

Hedge, B. R., T. C. Channappa, and B. K. Ananda Ram. 1981. Water harvesting and recycling in the red soils in Karnataka. *Indian Journal of Soil Conservation* 9:107–115.

Hofkes, E. 1986. *Small Community Water Supply in Developing Countries.* IRC International Water and Sanitation Centres, The Hague, The Netherlands.

Jones, K. R. 1981. *Arid Zone Hydrology.* FAO, Rome, Italy.

Kameswar Rao, V., K. P. R. Vithal, and K. Vijayalakshmi. 1981. Rainwater management in red soils of Telangana. *Panel Discussion on Soil and Water Conservation in Red and Black Soils,* Banagalore, India, pp. 58–66.

Khan, M. A. 1992. Development of surface water resources. In: *Rehabilitation of Degraded Arid Ecosystem.* A. S. Kolarkar, D. C. Joshi and K. D. Sharma (Eds.), Scientific Publishers, Jodhpur, India, pp. 136–143.

Khan, M. A. 1994. Central Arid Zone Research Institute, Jodhpur, India (personal communication).

Kolarkar, A. S., K. N. K. Murthy, and N. Singh. 1980. Water harvesting and runoff farming in arid Rajasthan. *Indian Journal of Soil Conservation* 8:113–119.

Kolarkar, A. S., K. N. K. Murthy, N. Singh. 1983. Khadin—a method of harvesting water for agriculture in the Thar desert. *Journal of Arid Environments* 6:59–66.

Krishnan, A. 1968. Distribution of arid area in India. In: *Proceedings of Symposium on Arid Zone.* S. P. Chatterji, B. B. Roy, and S. P. Das Gupta (Eds.), Sree Saraswati Press, Calcutta, India, pp. 1–15.

Lee, M. D. and J. T. Visscher. 1990. *Water Harvesting in Five African Countries.* IRC International Water Supply and Sanitation Center, The Hague, The Netherlands.

Lee, M. D. and J. T. Visscher. 1992. *Water Harvesting—A Guide for Planners and Project Managers.* IRC International Water Supply and Sanitation Center, The Hague, The Netherlands.

Mann, H. S., B. V. Ramana Rao, and G. G. Rao. 1981. Quantification of rainfall and its utilization efficiency in Rajasthan. *Journal of Arid Environments* 4:107–113.

Michaelides, G. and R. J. Young. 1985. Provision in design and maintenance to protect water quality from roof catchments. *International Journal of Environmental Studies* 25:1–11.

Michaelides, G., M. Allybokus, and R. J. Young. 1986. Optimized design and water quality studies of rooftop rainwater catchment project in Mauritius. *Water Supply* 4:117–122.

Nilsson, A. 1988. *Groundwater Dams for Small Scale Water Supply.* Intermediate Technology Publications, London, U.K.

O'Brien, L. 1990. Report on RWH project in Togo. *Raindrop* 3:5–6.

Omwenga, J. M. 1984. Rainwater harvesting for domestic water supply in Kisii, Kenya. Water and Sanitation Publication No. 17, Tampere University of Technology, Tampere, Kenya.

Pacey, A. and A. Cullis. 1986. *Rainwater Harvesting: The Collection of Runoff in Rural Areas.* Intermediate Technology Publications, London, U.K.

Raju, K. C. B. 1994. Shree Vivekanand Research and Training Institute, Mandvi, India (personal communication).

Rathore, M. S. 1988. Traditional drinking water systems in Rajasthan—a study of underground tanks. In: *Development Ecology and Rural Drinking Water Supply in the Rajasthan Desert: Aspects and Alternatives.* P. S. Shekhawat (Ed.), Institute of Development Studies, Jaipur, India, pp. 287–311.

Sharma, K. D. 1990. Runoff coefficient. In: *Watershed Management,* Central Arid Zone Research Institute, Jodhpur, India, pp. 49–52.

Sharma, K. D. 1992. Runoff and sediment transport in an arid zone drainage basin. Ph.D. Thesis, Indian Institute of Technology, Bombay, India.

Sharma, K. D. and P. C. Chatterji. 1982. Sedimentation in Nadis in the Indian arid zone. *Hydrological Sciences Journal* 27:345–352.

Sharma, K. D. and D. C. Joshi. 1981. Nadis—the vital water resources of the Indian arid zone. *Journal of Arid Environments* 4:247–251.

Sharma, K. D. and D. C. Joshi. 1983. Optimization of Nadi characteristics to minimize evaporative and seepage losses. *Journal of Arid Environments* 6:277–281.

Sharma, K. D. and A. K. Kalla. 1980. Effect of Nala bunding on Vindhyan sandstone aquifer in western Rajasthan. *Indian Journal of Soil Conservation* 8:44–49.

Sharma, K. D. and N. S. Vangani. 1992. Surface water resources in the Thar Desert and adjoining arid regions of India. In: *Perspectives on the Thar and the Karakum.* A. Kar. R. K. Abichandani, K. Anantharam, and D. C. Joshi (Eds.), Department of Science and Technology, New Delhi, India, pp. 130–140.

Sharma, K. D., S. Singh, N. Singh, and A. K. Kalla. 1989. Role of satellite remote sensing for monitoring of surface water resources in an arid environment. *Hydrological Sciences Journal* 34:531–537.

Singh, G. 1990. Rainwater harvesting and recycling for sustainable agriculture production. In: *International Symposium on Water Erosion, Sedimentation and Resource Conservation.* CSWCRTI, Dehra Dun, India, pp. 157–168.

Singh, R. P. 1983. Farm ponds. Project Bulletin No. 6, AICRDA, Hyderabad, India.

Singh, R. P. 1994. Management and recycling of runoff water in the Indian arid zone. In: *Sustainable Development of the Indian Arid Zone.* R. P. Singh and S. Singh (Eds.), Scientific Publishers, Jodhpur, pp. 233–239.

USDI. 1970. *Design of Small Dams.* United States Department of Interior, Washington, DC, U.S.A.

Vangani, N. S., K. D. Sharma, and P. C. Chatterji. 1988. Tanka—A reliable system of rainwater harvesting in the Indian desert. Technical Bulletin No. 33, Central Arid Zone Research Institute, Jodhpur, India.

Vangani, N. S. and S. Singh. 1994. Surface water of the Indian arid zone. In: *Sustainable Development of the Indian Arid Zone.* R. P. Singh and S. Singh (Eds.), Scientific Publishers, Jodhpur, India, pp. 33–42.

Venkateshwarlu, J. 1981. Maximization of crop production in dryland. *Indian Journal of Soil Conservation* 9:124–140.

Verma, B., S. Chinnamani, S. N. Bhola, D. H. Rao, S. N. Prasad, and C. Prakash. 1986. Twenty-five years of research on soil and water conservation in ravine lands of Rajasthan. Technical Report, CSWCRTI Research Center, Kota, India.

Yaziz, M. I., H. Guting, N. Sapari, and A. W. Ghazak. 1989. Variations in rainwater quality from roof catchments. *Water Research* 23:761–765.

4

Natural and Recycled Water Quality in the Arid Zone: Real and Perceived Dangers

GORDON STANGER
32 Dutchman Drive
Hallet Cove SA 5158, Australia

THE MILLENNIUM PICTURE

"CLEAN WATER FOR all by 1990." That was the slogan for the "International Drinking Water Supply and Sanitation Decade." Such a bold objective looks unattainable from the perspective of the late 1990s. The world has not only failed to provide clean water for all, it has even failed to maintain the status quo. Although some countries have made striking advances, overall, the rising global population has outstripped programs for water quality improvement, with about a third of humanity still having to cope with unsafe water supplies. In many cases the poorer countries are slipping backwards, polluted water becoming an ever more pressing issue of public health. Sadly, the trillion-dollar third world debt has paralyzed many water quality improvement programs, with the poorest of the poor being the victims, and billions of infections arising from waterborne vectors each year.

Is this global picture mirrored in the arid zones? Arid regions encompass both the richest (e.g., U.S. and the oil-rich Arabian Gulf countries), and the poorest (e.g., sub-Saharan Africa) of societies, so there is great diversity in the technical capacity of the arid zone to deal with problems of water pollution. We may also question the problems themselves: Are there water quality issues that are specific to arid zones? Again, there are no simple answers, as the problems are as much a function of specific environmental conditions as of the pollutants themselves. However, we can say that the arid zone is more sensitive to long-term pollution than many wetter ecosystems. Where water is at a premium, with few or no alternative resource options to those already at risk, the impact of pollution becomes more serious. As a rough approximation, most groundwater pollution in the arid zone is effectively permanent.

Pure pristine water is virtually a thing of the past. Water in active hydrologic circulation contains ubiquitous traces of atmospheric pollution in the form of micro-industrial pollutants and dust, and hence dissolved major inorganics. In addition, the essential requirement for fertilizer in the case of humus-deficient arid soils gives rise to high nutrient levels in the groundwater. At the same time, the almost universal rise in living standards over the past half century has given rise to consumerism, the proliferation of landfill sites, and the genesis of harmful leachates, some of which have now seeped into actively utilized aquifers. Even if best-practice environmental management is globally and immediately implemented, leachate migration is a long-term insidious problem that will inevitably deteriorate water quality for many decades to come.

A major problem faced by most countries is an almost total lack of discernment, on the part of legislators, to distinguish which pollutants are of a genuinely serious nature. Very few politicians have the necessary technical background to know which parameters constitute a health hazard as opposed to being merely aesthetically objectionable. The result is that most countries blindly conform to World Health Organization (WHO) water quality criteria, while maintaining a woefully unbalanced perspective on their relative importance. For example, substantial financial resources are devoted to avoiding or eliminating nitrate from drinking water supplies, while the much more serious problem of biological hygiene, at well heads and in surface water, is commonly underfunded or even completely ignored.

Throughout the arid zone, the foremost problem is not technical but political, in that legislators either cannot, or will not, deal with water pollution issues in accordance with their true health risks. For example, in one Arabian Gulf country, the equivalent of many millions of dollars is spent each year by the government's water authority to analyze huge numbers of water samples for unimportant inorganic parameters, while employees of that authority are actively discouraged from analyzing, or even raising questions about, biological contaminants. Is the public well served by such unwise bias? That same country has few, if any, cases of illness from chemical water pollutants, but has at least half a million cases a year of gastro-intestinal illness, much of which is waterborne, and occasional instances of which are fatal, with babies and infants being most at risk.

VULNERABILITY OF THE ARID ZONE

Although it is difficult to generalize, a few characteristics of arid regions render them particularly vulnerable to water quality problems. Of these, the most obvious is that of intense evaporative concentration of salts. Several compounding processes give rise to unacceptably high salt concentrations, notably of sodium, chloride, and sulphate ions.

Rainwater is generally thought of as pure but it is, in fact, just very dilute sea-water, with, typically 7 or 8 mg L^{-1} of chloride ion (more close to the oceans, less in more continental areas). From the time that a raindrop forms to the point where it has reached the land surface and percolated to a sufficient depth for evaporative loss to be minimal, its evaporative concentration is generally at least an order of magnitude higher in deserts than in moister climates, so even "pristine" groundwater has effectively been degraded before it is ever utilized.

Since water is so scarce in arid zones, its use, for domestic and irrigation purposes, frequently involves recycling. Thus, in arid lands it is not uncommon for ground-water to be subjected to four or five cycles of land application. The proportion of water that is consumptively used is a function of climate, but to give just one example from Arabia, the typical components of irrigation water use are: 40% transpiration, 10% direct evaporation, 10% channel losses, and 40% leachate to groundwater. Hence, residual groundwater salinity approximately doubles with each cycle of irrigation. Although there is some need for modification of these proportions by more efficient water conveyance and the judicious use of halophilic crops, the salinity buildup imposes a severe constraint upon arid land irrigation that is seldom encountered in more favored climates.

The next most significant effect is that of changing water tables. All water tables fluctuate in response to seasonal recharge and recession, but in arid regions over the past few decades, the imperative of developing much-needed water supplies, both for domestic use and agricultural development, has caused an unprecedented and dramatic decline in water tables in such regions as the southwestern U.S., parts of Mexico, sub-Saharan and North Africa, eastern Arabia, the Arabian Gulf, huge areas of India and Pakistan, western China, and many other marginal lands. Apart from the obvious costs of pumping from ever greater depths, and declining quantities of available groundwater, there are subtle, and not-so-subtle, ramifications for water quality.

First, many aquifers are stratified, with the freshest water overlying brackish or even saline layers.[1] Even in aquifers that generally yield good quality water, there are usually pockets of poorer quality water in depressions within the base of the aquifer. Hence, as the fresh water is skimmed off, there tends to be an increase in salinity which may vary according to local conditions.

Second, depression of the water table in coastal areas to below sea level reverses the hydraulic gradient, thereby inducing seawater intrusion into the aquifer. If sufficient recharge occurs into the aquifer, this intrusion can, to some extent, be counteracted. The catch is that arid regions have no excess water available to allocate for this purpose. Saline intrusion, and its devastating effect upon coastal agriculture, has been well known and documented for at least the past forty years. Unfortunately, although hydrogeologists have repeatedly warned about this slow and insidious process, few governments around the world have had the courage and resolve to deal with the problem. Consequently, agricultural production is falling throughout the arid coastal lands at the very time when it needs to be increased.

Third, by depressing the water table, many aquifers that were previously hypoxic become exposed to atmospheric oxygen for the first time. In most cases, this is not a problem but some natural toxins, which are immobile under hypoxic conditions, oxidize to a very mobile form under oxidizing conditions. Hence uranyl, selenate, arsenate and some heavy metals ions all have the potential to dissolve in groundwater, thereby forming a serious health hazard. The most publicized example is the dispersion of selenium in Kesterson National Reserve in California, (Presser and Ohlendorf, 1987) caused by the draining of selenium-rich marine sediments. The resulting genetic mutation and toxicity to wildlife was horrific, but hundreds of similar examples also occur in which human water supplies are either affected already or could be affected in the future. Such toxins are seldom analyzed in the course of routine water monitoring, while the problems are generally slow to develop and difficult to identify until serious damage has occurred.

Fourth, acidification is becoming an increasing problem worldwide. In like manner to the toxins mentioned above, oxidation of sulphides in the ground results in high sulphate concentrations and increased

[1]Obviously, the denser, more saline groundwaters sink to the lowest permeable horizon.

acidity. By itself, acidity is not necessarily a problem. Some people even prefer slightly acidic drinking water, but the problem lies in the propensity of acidic waters to dissolve a wide range of metals, with consequences ranging from unsightliness of stained laundry to "metallic tastes," and subchronic to acute metal poisoning. Sources of sulphides in the ground are very diverse, ranging from metal sulphide mines in many an exposed mountain range to low-lying coastal sediments, as in parts of eastern Australia. Mountainous regions generally define watersheds, in the sense of the uppermost boundaries of catchment areas, whereas coastal lowlands commonly define the lower extremities of a catchment, so acidification has the potential to occur anywhere within an aquifer system.

If falling water tables are detrimental to water quality, is the converse true for rising water tables? Unfortunately not! Wherever intensive irrigation raises the water table to within about a meter of the arid land surface, there arises the dual problem of waterlogging of root systems and evaporation of soil moisture, with concomitant increasing groundwater salinity. The depressing salt-encrusted irrigated land of the Indus valley is one of the most frequently cited examples, but there is scarcely a country within the arid and semiarid regions of the world that has not suffered similar salinization, albeit on a smaller scale. By contrast, Australia seems to be almost unique in suffering a different form of dryland salinity. This geologically ancient landscape—the flattest continent—has been eroded to form very low hydraulic gradients over most of its surface. The combination of slow groundwater drainage and intense evapotranspiration has resulted in near-stagnant ground-waters accumulating very high salinities. Indeed, much of the most fertile and productive agricultural land, from the Murray-Darling basin to Western Australia, is underlain by groundwater of comparable salinity to seawater. Prior to settlement, there was a natural balance between rainfall and its interception and transpiration by natural vegetation, in which recharge to the water table was almost nil. By removing this natural scrubland to convert it to farmland, well-meaning settlers put in train an environmentally disastrous sequence of events, the full effect of which is still to appear. Without perennial vegetation to mop up the rainfall, an increasing fraction of sporadic rainfall has been able to infiltrate below the annual crop's root zone, thereby raising the salt-laden water table to the surface. In this way, millions of hectares of the best agricultural land are under threat, or have already reverted to halophylic scrub.

WATER QUALITY ASSESSMENT

The major aspects of water quality are, in decreasing order of importance to health: biological, chemical, radiological and physical.[2] This order is quite different from the public perception, in which the appearance of the water is more important than biological protection (an important misconception for "Public Awareness" departments to correct). Citizens invariably demand that their water supply be "pure," although they have no technically meaningful concept of what "pure" actually means.

Biological protection from pathogenic bacteria, viruses, protozoa and helminths is of paramount importance, and has been since the earliest discovery of micro-organisms by Koch and Pasteur in the late nineteenth century. Unfortunately, biological analysis of a water is still less precise, less repeatable, more time-consuming and very much more labor-intensive than "complete" chemical analysis. More accurate methods of bacterial and viral analysis, such as labeled immuno-fluorescence, are on the horizon, but are still experimental. Hence it will be a few years before they can be incorporated into routine analysis. In the meantime, pathogen identification is only feasible on a few selected ("suspect") samples, using the index organism *Escherichia coli* as proxy for more difficult isolates. The ideal of listing a suite of identified pathogens, in much the same way as chemical parameters are listed today, will eventually be realized. When that happens, we will see a radical shift in emphasis away from major ion chemistry to a more health-related combination of key ions and pathogens. This change is a long time in coming, but when it is achieved it will be an enormous advance on current practices.

No Ministry of Health, in any country, has the groundwater expertise and resources necessary to prevent and combat biological pollution of groundwater. Therefore it is inappropriate to regard biocontamination of water as "strictly a Ministry of Health problem." Moreover, with the proliferation of sewage effluent and sanitary landfill sites, the extent of biocontamination of groundwater is, and will continue to be, a growing problem that should be high on the agenda of all water-related agencies. Therefore, no water laboratory that deals solely with ground-water chemistry to the exclusion of microbiology can be regarded as adequate for the exacting needs of modern water protection. This should be obvious, yet there are still many countries, both developed and undevel-

[2]i.e., Color, turbidity, smell, taste and temperature.

oped, where the balance of analytical capability is absurdly weighted towards chemistry rather than microbiology.

Water chemistry itself has undergone a revolution over the past decade. There are two interrelated reasons for this. First, the proliferation of toxic and carcinogenic chemicals in both surface and groundwater has at last (and somewhat belatedly) become a matter of public concern. PAHs, PCBs, THMs,[3] herbicides, pesticides, fungicides, and so on, are ubiquitous throughout both surface and groundwaters at very low concentrations, and it is now appreciated that bio-concentration of these traces can occur throughout the food chain, terminating in humans. Research is still active to establish the importance of various concentration pathways, and to assess the physiological effects of this galaxy of chemicals. Second, and concurrent with concern about these chemicals, has been a greatly improved analytical capability, enabling the discrimination and measurement of their concentrations to remarkably high precision. Routine analysis to sub-ppm[4] concentrations is now commonplace, while in many cases, routine measurement to sub-ppb levels is becoming increasingly common. The requisite instrumentation is very expensive, but laboratories that can afford such sophistication are now capable of recognizing and monitoring pollutants that were impossible to deal with even a decade ago.

Is such precise chemical measurement necessary? From the health point of view the answer is a qualified yes. For example, the human tongue can easily detect chlorophenols (and a range of other compounds, such as many other halogenated aromatic hydrocarbons) at a concentration of 1 ppb, so a few liters of such compounds, if spilled, are enough to degrade the taste of groundwater throughout an entire catchment.

It is appropriate that a substantial range of "priority pollutants" has been specified as having acceptable concentrations at about the ppb level, but it should be appreciated that these limits are highly tentative and err greatly on the side of conservatism. A typical acceptable limit is defined as: "That concentration which, if ingested over a lifetime, is estimated to cause a single death (generally through cancer) in a population of $10.^5$" This is a risk of 0.00001. Compare this with the risk of dying from cancer anyway, about 0.3, and we see

[3]Polycyclic aromatic Hydrocarbons, Poly-Chlorinated Biphenyls and Tri-Halo-Methanes.
[4]1 ppm = 1 part per million, 1 ppb = 1 part per billion = a concentration of 10^{-9}. One part per billion of contamination can be envisioned as, for example, a teaspoonfull of chilli sauce dispersed throughout an Olympic-sized swimming pool.

that the additional health risk from most groundwater "pollution" is minimal. Also, it is illogical to worry about the health risk from many trace chemical pollutants in water when the risks from smoking and from many foods and beverages[5] are thousands of times greater. Even more important than the chemical pollutants are the biopollutants. Hence a ranked list of health risks (as opposed to consumer acceptance and agricultural suitability), in approximately decreasing order of importance, is: pathogens (far more important than all the chemicals put together), heavy metals, organic carcinogens, metalloids, nitrate, fluoride, and sulphate.

In order to meet safe (international) water quality standards, modern analytical technology has advanced to facilitate the measurement of an enormous range of compounds, most of which were previously below detection limits. Although toxic and carcinogenic, these are generally maintained in water supplies at remarkably safe concentrations, and hence their mere presence at such concentrations is not cause for concern.

Over the last half century, millions of water samples from around the world have been analyzed for the major ions and for not much else. While such analyses have their place in terms of geochemical provenance studies, and in initial assessment of their suitability for drinking water, they are either of trivial significance to health or completely unrelated to health in any way. The current realization is that many minor and trace elements and compounds are of far greater significance than "the majors," and hence a major shift in laboratory retooling is necessary.

BIOLOGICAL ASPECTS OF IRRIGATING WITH WASTEWATER

We have reached a stage where most arid-zone water resources are in overdraft. Hence the quality of recycled water either already is, or will become, of comparable importance to the quality of "pristine" water. Over the past few decades attitudes towards the recycling of wastewater for crop and municipal irrigation have changed from considering it something of a novelty to an essential and integral part of water resource management in the majority of arid countries. This trend will inevitably become even more important into the next century, so it is

[5]e.g., α-Benzo-pyrene from cigarettes, nitrosamines from cooked meats, aflatoxins from mushrooms and cereals, etc.

surprising that there seems to be so little appreciation of the differing quality criteria between potable and irrigation waters. The microbiological aspects are considered first.

Bacteria

Sewage wastewater treatment has traditionally been directed towards BODs (Biological Oxygen Demands), suspended sediments and the presence of pathogenic organisms. In countries where strict standards apply, the finished wastewater has even been chlorinated to high concentrations (up to 4 mg L^{-1} Cl° in some countries), which is somewhat of an "overkill" with respect to bactericidal requirements. Provided that solid organic residues are filtered out, as is almost universally the case, the risk of bacterial shielding from oxidizing agents in finished waters is very small. Rather, any bacterial risk is from subsequent regrowth of biofilms in the wastewater reticulation system, in which case bacterial shielding and even vigorous regrowth can occur in such genera as *Acinetobacter, Aeromonas,* and *Pseudomonas.* Some concern has also been expressed about the detection of viable, but nonculturable *Salmonella spp.,* and possibly of *Vibrio cholerae,* in chlorinated wastewaters (Watkins and Cameron, 1991).

While such concerns are real and should be investigated, several factors strongly mitigate any bacterial health risk from irrigation reuse of adequately treated wastewaters. First, the high standards of most modern secondary and tertiary treatment works strongly attenuate any residual bacterial population. Second, irrigation onto dry surface soils strongly desiccates any exposed *E. coli,* and hence, presumably, bacteria in general. This feature has been demonstrated even in temperate climates (Hunter and MacDonald, 1991), and is especially true of hot desert conditions. Third, in the case of spray irrigation, any residual bacteria are subject to both desiccation and solar UV disinfection, resulting in short survival times (relative to bacterial survival in the soil). An example for fecal coliforms given by Feachem et al. (1983) is, "normally less than 15 days, but occasionally up to 30 days."

Viruses

Although viruses and bacteria are of prime importance to potable water quality, the significance of viruses to irrigation wastewater quality is much less, and is subordinate to the risks from helminths and some parasitic protozoans. This does not mean that viruses are not present. Indeed, many viruses are effectively immune from low levels of chlorination and, in the case of the smaller viruses, may be able to pass through normal filtration processes with minimal attenuation. As with bacteria, most viruses are readily killed by desiccation, high air temperatures, solar UV exposure, and low humidity or soil moisture. Again, an example of survival time on soil and crop surfaces, given by Feachem et al. (ibid. 1983), for enteroviruses at up to 30°C, was "normally less than 15 days, but occasionally up to 60 days." A few viruses such as Hepatitis A have much higher survival rates in the environment. It is emphasized that no unequivocal viral infection has been reported from agricultural workers using secondary or treated wastewater, even though one might have expected such workers to be at high risk of infection from spray irrigation (aerosol transmission). In the light of such findings, and unless future research identifies a more tangible risk, strict viral standards, such as Arizona's 1 PFU[6] per 40 liters, may be unnecessarily cautious.

Helminths

Helminth survival and transmission are probably the most significant risks in wastewater use. They are also the pathogens that primary treatment works are least capable of removing. Moreover, helminth eggs are largely, if not entirely, immune from chlorination and are capable of survival for many months under warm, moist soil conditions. Conventional treatment can remove helminth eggs, but require very efficient filtration, or multiple stabilization lagoons. A faster but more complex alternative is to use a single-stage lime-accelerated sedimentation system (Gambrill et al., 1992). With any such treatment the risk of intestinal nematode infection, even for agricultural workers, is still low. In the case of irrigation water being recycled for a second or subsequent time, the principal danger from nematodes is to the crop being irrigated, not to the ultimate crop consumer. The Engelberg guideline for recycled water, as recommended by the World Bank, the UNDP,[7] and as adopted by the majority of arid countries, consists of < 1 viable intestinal nematode egg (ova) per liter. Compared to raw sewage water, this is a reduction of five or more orders of magnitude, and hence may be regarded as a highly conservative limit.

[6]"Plaque Forming Unit"; a practical proxy measure for viral concentration.
[7]United Nations Development Program.

Protozoan Parasites

A particular problem with pathogenic protozoans is their substantial resistance to chlorination and their ability to survive for long periods (in some cases up to about a year) in warm, moist environmental niches. This is particularly the case with *Cryptosporidium spp.,* which are killed by desiccation but are otherwise robust, with a thermo tolerance of up to 60°C. Hence any absence of index bacteria, such as *E. coli,* is not necessarily an indication of the absence of protozoans. Of the various cases of waterborne Cryptosporidiosis that have been reported, all were associated with drinking water (Smith, 1992), although, in principle, one would expect spray irrigation to be a potentially infective pathway. The remaining potentially significant pathogens include *Naegleria fowleri,*[8] *Entamoeba histolica,* and *Giardia intestinalis,*[9] none of which has high-risk infective pathways during agricultural use. On the other hand, survival of some of these protozoans within moist soils could constitute environmental reservoirs from which infective pathways could evolve. This is most likely in the case of flood irrigation. Finally, protozoans can harbor and protect bacteria that would otherwise have been oxidized.

Overall, therefore, it is particularly important to remove helminths, and to a lesser protozoans, from wastewater prior to land application.

CHEMICAL ASPECTS OF IRRIGATING WITH WASTEWATER

Although enhanced nutrient concentrations associated with reused wastewater can have some advantages, the chemistry of irrigated wastewater is not simple and there are several potentially deleterious aspects to consider.

Nitrate

Cyanobacteria and termites are both capable of elevating groundwater nitrate to concentrations in excess of the WHO-recommended limit for drinking water (of 10 mg L^{-1} of nitrate-N), but it is unusual for

[8]Fortunately, cases of amoebic meningoencephalitis are very rare, but *N. fowleri* and other amoebae are of sufficient global distribution so that they must be regarded as a potential threat in hot stagnant waters.

[9]*Giardia* is often associated with cool temperate waters, but its prevalence in such seasonally hot deserts as Central Australia is proof of its significance in the arid zone.

natural nitrate in desert regions to be a problem. Most nitrate is introduced to the environment either as fertilizer or as the dissolved nutrient in irrigation wastewater. From the agricultural perspective, wetter environments have the advantage of a humus-rich soil, in which most if any applied nitrate is locked up as a near-surface organic reservoir. This is not so in arid regions, where, under optimum conditions, about half of the applied nitrate is utilized by crops, about a quarter is lost as nitrogen gas following biodenitrification, and the remainder is quickly and easily leached to the water table, where it can end up in potable water supplies.

At low concentrations nitrate is beneficial to plants, whereas at high concentrations it tends to accelerate growth beyond the plant's capacity to synthesize fructose, with the result that fruit becomes large, but has poor flavor or an insipid taste. American studies (Bouwer and Idelovitch, 1987) have shown that application rates in excess of 90 mg L^{-1} are detrimental to nitrate-sensitive crops, while concentrations in the range of 60 to 80 mg L^{-1} NO_3^- are more than sufficient to meet the nitrate demand of most crops. Such a range is below that of raw wastewater and may even be richer in nitrate than many treated wastewaters, (wastewaters typically vary between about 5 and 180 mg L^{-1} NO_3^-). Hence with wastewater treatment plants of low denitrifying capacity, there is a significant risk of subsequent deterioration in crop quality. As a general rule for nonnitrate-sensitive crops, an application limit of about 90 mg L^{-1} NO_3^-, equivalent to about 20 mg L^{-1} of total nitrogen, is a safe limit for irrigation. As an additional consideration it is environmentally dubious to introduce high nitrate into groundwater by way of irrigation leachate. At the other extreme, some municipal wastewaters are so rigorously treated to reduce nitrogen species that nitrate fertilizer again has to be added at the end-user stage. In such cases there is an obvious need for dialogue between the relevant parties in order to maintain a mutually acceptable level of nitrogen delivery.

Boron

Boron is generally associated with high salinity, and hence the more obvious problems of incipient salinization are more likely to be noticed before crops succumb to boron toxicity. However, there are a few rare arid environments where boron is exceptionally concentrated and therefore becomes a limiting factor for agriculture. Perhaps the most novel of these are oil-field formation waters, which vary enormously in salinity but which commonly have relatively high boron concentrations of between 5 and 10 mg L^{-1} B. This constitutes a chemical hazard even for the more boron-tolerant crop, and hence surface recycling of such waters for agriculture is seldom a viable option.

Heavy Metals

The hydrochemistry of limestone-rich environments more or less precludes the possibility of widespread acidity, and hence of heavy metal mobility, but there are huge areas of desert that are not so fortunately buffered; hence where dissolved metals have the potential to reach deleterious concentrations. Physiologically, crops tend to concentrate any metals in two areas. First, studies in China have shown a marked enrichment in the root zone (Yediler et al., 1994); second, heavy metals are concentrated in seeds and other edible parts of many crops (Barman and Lal, 1994). Therefore, wherever the geology identifies sulphide mineralization, care should be taken to monitor the irrigated wastewater for metals.

Organic Chemicals

Phenolic compounds and other soluble oil derivatives (BTEX) have some capacity to transfer to crops, thereby imparting a slight deleterious taste. This is particularly the case where the finished water is chlorinated. However, domestic wastewater is almost always low in such organics, and if the wastewater treatment is adequately oxygenated, then most, if not all, of these distasteful organics should be depressed to below organoleptic detection levels. It is normal and appropriate for industrial wastewaters to be kept separate from domestic wastewaters, and for irrigation wastewaters to be restricted to the latter. Adherence to this principle should minimize any problems from either organic or metallic contaminants.

BOD

Biological Oxygen Demand (BODs) may have values ranging from zero in well-oxygenated systems (and most natural arid-zone groundwaters), to over 1000 mg L^{-1} in poorly treated or grossly overloaded sewage treatment plant outflows. The normal range of BODs does not have any significant effect upon mature crops, although high BODs may cause poor germination of seeds, or inhibit the early growth of seedlings in a variety of crops (Nashikhar, 1994). The BOD is often regarded as the primary index for assessing the quality of treated water, but there is poor correlation between the BOD and some other deleterious parameters. Hence a low or even zero BOD is no guarantee that pathogens and refractory organic pollutants (those that are resistant to biochemical or chemical breakdown under oxidizing conditions) have been reduced to a safe level (Juanico, 1993).

PUBLIC AWARENESS OF WATER QUALITY ISSUES

Globally, the general public is naive and confused about water quality, and in particular about the significance of pollutants. Thus, the public becomes alarmed by unimportant threats to water quality, while real threats go completely unnoticed. Any realistic assessment of groundwater pollution requires a clear perspective of relative risks, which is sometimes lacking even by those involved in water, and which is virtually absent from the general public's thinking. In this respect, the news media, whose stock-in-trade is alarmist reporting, is usually unhelpful and cannot be trusted to provide a mature and balanced view of pollution issues. A few examples of typically poor public perceptions and misunderstandings are listed in Table 1.

It becomes increasingly important that these, and many other related issues, be explained to the public. It is unfair to scare people over unimportant pollutants, but equally, it is unwise to leave them ignorant about the issues that really matter. The criticality of water quality and its protection requires that inhabitants of arid and semi-arid lands, as much as of any other environment on the planet, be well educated in issues of water quality.

A welcome change by some water authorities is recognition of the importance of a public information/awareness service. This is potentially a powerful and vital tool in the fight against water pollution. It is not an extravagant or optional extra (as viewed by many a ministry around the world), nor is it a media channel for boasting about real or imagined achievements (another common misconception). Rather, it is the primary tool for counteracting ignorance, prejudice, and misinformation about water issues, and should be active on all fronts, ranging from the schoolroom, to homes, industry and the "corridors of power." For these reasons, the public awareness service needs strong direction from water experts and should not be left in the hands of media specialists, many of whom have only a superficial understanding of the issues in water technology.

Table 1. Myth versus reality in public awareness of water quality issues.

Myth	Reality
1) Waste can be discarded into water courses. Flash floods will wash it away.	This was the prevailing and justifiable attitude for many centuries, but is no longer tenable in this age of rapidly increasing population and consumerism. One person's refuse becomes another person's pollution.
2) All pollution can be cleaned up, if necessary.	A few cases of groundwater pollution can be solved quickly and easily, but most are persistent, difficult and expensive to clean up. Some, and especially the "DNAPLs"[10] cannot be cleaned up by any known technology, and must be regarded as destroying the aquifer into which they have leaked. The only "solution" is to relocate the downstream water users.
3) Water that *looks* crystal clear is "clean and healthy."	Most substances of concern (pathogens, toxins and carcinogens) are colorless, odorless and tasteless. Conversely, turbidity from clay, and color from iron, manganese, tannins and other organic acids are nearly always harmless. Thus taste and appearance are extremely misleading in terms of health.
4) Rainwater is pure and is preferable to mains water for potable supplies.	All rainwater collection systems are biologically contaminated, while very few are adequately disinfected. Roof collection systems are particularly prone to pathogen infection.
5) All dissolved substances are harmful.	Most natural solutes are harmless, even though some such as sodium chloride may affect taste. Even most pollutants are harmless at trace concentrations. A few toxins, such as the physiological trace elements Zn, Se, Cu and F, are even beneficial at low concentrations.
6) Pollution thresholds should all be "zero."	There is no such thing as a zero threshold. A glass of tap or bottled water probably contains at least a few molecules of almost every chemical known to man. Apparent zero concentration only indicates a relatively insensitive limit of analytical detection. A zero threshold for pollutants is both unrealistic in principle and hopelessly unattainable in practice.

Table 1. (continued)

Myth	Reality
7) Pollution from mining and other extractive industries is far from population centers and hence is unimportant.	Groundwater flows inexorably down-gradient and, if left uncontrolled, many plumes of pollution will eventually flow from mountainous areas to coastal or interior communities. These effects may be very slow, but could have an impact long after the original industry has finished or relocated. Such effects are insidious and need constant monitoring.
8) Water pollution is caused by "other people."	All of us have a direct and indirect adverse impact upon the aquatic environment (surface and subsurface). Sewage, domestic refuse, batteries, sump oil, discarded appliances, cars, etc., all have to be disposed of somewhere. Moreover, agriculture and industries that provide all our consumer items are themselves major polluters. Few, if any, individuals bear the true cost of groundwater pollution from waste.

[10]DNAPL is the industries' acronym for "Dense Non-Aqueous Phase Liquids."

Perhaps the most insidious problem is that the process of population degrading the water quality is so slow, albeit inexorable, that resource managers and/or their political paymasters are tempted to take the easy way out of pretending that there is no problem. Few quality-related issues are amenable to quick fixes. Most cases of water pollution have built up over decades and, even if treatable at all within reasonable cost-benefit constraints, will similarly require decades to be put right. The "quick fix" in time for the next election is a pipe-dream, and politicians more than anyone else need to be made aware of this fact.

THE SUSTAINABILITY OF WATER QUALITY

A key point in the Environmental Summit in Rio de Janeiro (1991) was the need for sustainability of water resources. The main emphasis was given to the quantitative sustainability of our water resources, but relatively little was heard about the long-term maintenance of adequate water quality. Yet in arid regions, it is water quality that is as much under threat as the existence of the resource itself. In the six years since the Rio summit, the degradation of both surface waters and aquifers in arid countries has actually accelerated under the inexorable pressures of population growth. We still seem to have failed to grasp the fundamental reality that ultimate sustainability is achievable *only* if the population is stabilized— a problem that no country in the world has been able to tackle successfully, although China has made some progress in that direction. Conversely, the alternative is for both the per capita water use to decrease and the overall water quality to degrade as the population increases. Since arid-zone surface waters and groundwaters are among the most ecologically sensitive of our planetary water resources, many communities have already suffered a significant loss of water quality, with the great majority of cases involving increasing salinity.

With so many aquifers around the world approaching, or already at, their limit of sustainable utilization, prevention of further aquifer pollution becomes absolutely critical. Sewage effluent, landfill leachate, industrial waste, salinization, oil, coal and metal ore extraction all can, and generally do, have a deleterious impact upon groundwater quality. Communities in arid regions, many of which are entirely dependent upon groundwater, cannot afford to lose a single aquifer to water pollution. The imperative of water quality protection may seem expensive but, compared to the alternative costs of remediation, or worse, of aquifer abandonment, it is money well spent and expertise well utilized.

REFERENCES

Barman S. C. and M. M. Lal. 1994. Accumulation of heavy metals in soil and cultivated vegetables, and weeds grown in industrially polluted fields. *J of Environmental Biology* 15 (2):107–115.

Bouwer H. and E. Idelovitch. 1987. Quality requirements for irrigation with sewage water. *J of Irrigation and Drainage Engineering* 113 (4):516–535.

Feachem R. G., D. J. Bradley, H. Garslick, and D. D. Mara. 1983. *Sanitation and disease: Health aspects of excreta and wastewater management.* World Bank Studies in Water and Sanitation, #3, J. Wiley.

Gambrill M. P., D. D. Mara, and A. A. Silva. 1992. Physicochemical treatment of tropical wastewaters: Production of microbiological safe effluents for unrestricted crop irrigation. *Water Science and Technology* 26 (7–8):1449–1458.

Hunter C. and A. MacDonald. 1991. The occurrence of coliform bacteria in the surface soils of two catchment areas in the Yorkshire Dales. *Journal of Irrigation Water Engineering and Management* 5:534–538.

Juanico M. 1993. Alternative schemes for municipal sewage treatment and disposal: A case study. *Ecological Engineering* 2 (2):101–118.

Nashikhar V. J. 1994. Effect of domestic wastewaters with various BOD levels on the germination and early seedling growth of some crops. *Environmental Ecology* 12 (3):507–509.

Presser T. S. and H. M. Ohlendorf. 1987. Biogeochemical cycling of selenium in the San Joaquin Valley, California, USA. *Environmental Management* 11 (6):805–821.

Smith H. V. 1992. Cryptosporidium and water: A review. *J. Inst. Water Engineering and Management* 6 (Aug.):443–450.

Watkins A. and S. A. Cameron. 1991. Recently recognised concerns in drinking water microbiology. *J. Inst. Water Engineering and Management* 5 (Dec.):624–628.

Yediler A., P. Grill, T. Sun, and A. Kettrup. 1994. Fate of heavy metals in a land treatment system irrigated with municipal waste-water. *Chemosphere* 28 (2):375–381.

5

Sustainable Groundwater Allocation Through Simulation/Optimization Techniques

NARA M. SOMARATNE
Ministry of Water Resources
P.O. Box 2575, Ruwi 112
Sultanate of Oman

J. F. PUNTHAKEY AND R. M. WILLIAMS
Department of Land and Water Conservation
10 Valentine Avenue
Parramatta NSW 2150, Australia

MUSHTAQUE AHMED
Department of Soil and Water Sciences
Sultan Qaboos University
P.O. Box 34, Al-Khod 123
Sultanate of Oman

INTRODUCTION

THE SHORTAGE OF water is a limiting factor to economic and social development in many parts of the world. Arid regions are dependent on groundwater and the amount of groundwater that can be extracted varies from place to place. Although a small amount of potable water is sufficient to meet the physiological needs of man and animals, high production is necessary for extensive irrigation development (Caponera, 1996). The challenge is to use water in such a way as to maintain a productive form of resource use in harmony with the environment. This is the rationale behind the call for a sustainable groundwater allocation.

Countries worldwide are becoming increasingly concerned about achieving sustainable development. Sustainability when related to water resources planning must be considered if positive net benefits derived from the use of water resources is to be maintained into the foreseeable future (UNESCO, 1994). Rapid depletion of groundwater can create acute water shortages with serious social, economic, and environmental implications (Al-Ibrahim, 1990). Often the environmental damages are often beyond the natural hydrological cycle to self-correct. Competition for water increases as the human population increases, a competition that demands a sound allocation policy. Therefore, water use must expand in a manner that does not damage the environment. Such an expansion must protect the two most critical environmental resources, land and water. In order to address this situation, Gorantiwar and Smout (1995) emphasized the need for optimal allocation of land and water resources through different types of irrigation water management.

When groundwater is abundant, users tap the resource directly, without conflict or need for management or allocation policy. However, at times and places such as arid and semi-arid regions where water is scarce, there is no question about the crucial importance of integrated demand management aimed at efficiency, equity, and long-term water security using tools of allocation and planning. This may include not only allocation of the existing resources over existing users, but also the development of techniques to enhance the resources (supply management) or to reduce the uses (demand management) (Conley and Midgley, 1988; De Jong, 1996). Groundwater and nonconventional water resources (desalinated and treated sewage effluent) are the major components of the water supply in arid climates. Therefore, it is necessary to develop an integrated management policy to allocate both as a single resource. Groundwater allocation with respect to area and time, considering meteorological conditions, hydrology, hydrogeology, as well as economic, social and environmental factors was originally solved by trial and error for differing scenarios of water demand (Tsiourtis, 1995). However, simulation modeling through the use of linear programming now provides an optimal solution for water allocation among the differing competing activities over a number of time periods. This chapter examines factors that should be considered to ensure sustainable groundwater allocation. In addition, it addresses how simulation and optimization models help in allocating groundwater. Some of the common sustainability constraints are also discussed.

WATER USERS AND RIGHTS

It is of prime importance to identify existing water rights before any subsequent allocation is made. Historically, and to some extent culturally, a number of differing water rights have evolved. The concept of absolute ownership is widespread in Islamic custom although the nature and extent of the concept varies from one doctrine (*madhab*) to another. The extreme view is that man's land is his own property, he is permitted to dig a well as deep as he wants and to pump unlimited quantities of water without bearing any responsibility, even if his neighbor's well goes dry. This right is generally qualified by the need to follow custom (*urf*). If a person does with his property what is permitted by *urf*, then he bears no responsibility if it inflicts damage upon his neighbor. If his action is not in line with the *urf* of the area, on the other hand, he is responsible for the damage. The harshness of the absolute ownership rule is partly mitigated by the customary right of man and animal to drink from a privately owned well (*haqq al shafah*). The ancient sharia law of the *Hanbali* school (Caponera, 1978) holds the view that a man may not dig a well so deep that it causes his neighbor's to go dry if his neighbor has a preexisting well. This would appear to be analogous to the modified form of prior appropriation applicable to groundwater in the United States, where six categories of legal doctrines govern conflicts in the different states.

The rule of absolute ownership gives each owner an absolute right to pump groundwater under his land for use at any location without any responsibility, even for using water wastefully, to neighboring owners. The doctrine is based on the outdated assumption that the movement of groundwater is unknown and unpredictable.

The responsible use rule is the same as absolute ownership except that waste is prohibited and water must be used on land overlying the aquifer unless it can be exported without affecting other overlying owners. Therefore, a groundwater user in a reasonable use state may divert all the water he can pump as long as it is used on overlying land without waste but otherwise has no obligations to his neighbors. According to the restatement rule, a well owner may withdraw water for his use on his own or nonoverlying land unless other users are unreasonably affected. The new California rule (Hutchins, 1974) was an offshoot of the reasonable use rule in the eastern states but it further developed the reasonable use rule in significant ways. Under the correlative rights doctrine, owners of land overlying the groundwater have equal (correlative) rights for benefit of overlying land in such quantity as is reasonably necessary provided the supply is sufficient. If insufficient, each owner is entitled to a fair and just proportion for on-site uses which has traditionally been a proportion based on the owner's area for the total area of land overlying the aquifer.

Many western states of the U.S. have ostensibly applied the prior appropriation doctrine as formulated for surface water. This is a water allocation system based on the principles that "first in time is first in right" and that "beneficial use shall be the basis measure and limit to the use of water." However, the concept of prior appropriation is somewhat at odds with the nature of groundwater and some states have softened the doctrine's protection of senior users so that later appropriators are able to change the condition of the water occurrence, lower the water table, artesian pressure or water level if the prior appropriator can reasonably exercise his water right under the changed conditions.

In general, most sophisticated groundwater management systems have been developed in the western states of the U.S. and the Murray Basin in Australia where conjunctive use of groundwater and surface water is common and frequently necessary. This approach has been adopted by many countries where water resources are considered as a common good or a national wealth and not a part of the private property of the land owner. More experimentation in groundwater law is expected as groundwater management is a volatile field characterized by different and often conflicting approaches. However, the recent trend is towards developing integrated management models with applications at local and regional scales.

ALLOCATION FOR SUSTAINABLE MANAGEMENT

The purpose of water management is to plan for sustainable water use for human activity and for production while maintaining sufficient flows for water table-dependent environmental purposes. Sustainable groundwater management requires first and foremost that the abstraction from an aquifer cannot over a prolonged period be larger than the recharge to the aquifer. Renewable water resources should not be used at excessive rates that jeopardize future use. Nonrenewable resources should be used in ways that maximize the long-term availability, with full consideration of possible long-term substitutes. Unmanaged water use by individuals or organizations is not permitted. In addition, the interests of one user versus another and of different types of water use must be reconciled. Finally, the water itself (both quality and quantity) may be exhausted or applied in excess and, therefore, must be protected. Similarly, ecosystems and other natural resources associated with water, such as land, soil, and vegetation, must be protected to maintain the quality of water.

Use of water resources should be consistent with the desired general and regional economic development. This is specially true since the primary purpose of water management is to allocate, in a specific hydrologically or geographically defined area, the limited water resources among various competing water users. This requires familiarity with relevant socioeconomic development policies to ensure that the water allocation decisions support the government's development objectives, in harmony with all other policies (De Jong, 1996). Further, it should generate outputs that are socially and economically sound as well as valued by the community. The movement of resources from lower- to higher-valued uses should be facilitated wherever possible. The benefits of water use activities should overweigh the costs. Priorities should be given to projects that maximize the overall benefits.

Resource users should be exposed as much as possible to external market forces. Where feasible, they should be given incentives to use water more efficiently. The relative scarcity or real value of the resource should be made clear in order to encourage sound decisions about developing and using the water.

The notion of fairness in allocating water must also be considered. Fairness is intrinsically linked to the question of economic efficiency. In principle, the "some for all" rather than "all for some" rule (Abrams, 1996) applies here. For example, while economic efficiency might rule out allocating a small volume of water to a large number of properties, equity may rule out allocating high volumes of water to a few properties.

In allocating water, the often competing claims and rights of all existing and potential future users must be considered. Private and public benefits should be recognized and cost apportioned equitably between users and government in proportion to the benefits received. Where possible, conjunctive use of water and drought management (Utton, 1987) must be pursued. The conjunctive use of water represents the combined and integrated management of surface and groundwater for optimal productive allocation efficiency. It has the potential to contribute to improved agricultural performance, sustainability, and equity in a variety of ways. In addition, nonconventional water resources, particularly in arid regions, need to be integrated with conventional water resources and be managed as one resource.

In order to maintain both human activity and the natural environment, the type and intensity of water use should suit the resources involved (land, soils, and water) and the natural ecosystems that depend on them. Diversity of naturally occurring species should be maintained to preserve the quality of these resources. For this purpose, deliberate allocation of water for environmental uses is needed (Pigram, 1991). The direct and indirect impacts of development should be identified and controlled.

Community Involvement

Development and use of water resources should contribute to the quality of life of the whole community. The responsibility for action by communities and government should be defined. In addition, collective management of aquifers must be practiced by farmers. For this purpose, farmers must be informed and educated about aquifer management principles. The impacts of strategies should be identified in advance, and causes and effects should be as specific as possible to provide confidence in the projected results. However, strategies should allow action to be taken as soon as possible, even if further investigations or research are required. Some of these principles are hard to meet and others conflict with one another. However, effective management of all water resources requires that all principles are at least considered. A good management plan seeks to arrive at a reasonable balance among the users of water resources.

Incentives for Better Management Efficiency

Groundwater allocation in Murray geological basin, New South Wales, Australia, was introduced (DWR, 1991) to ensure better management and efficient use of groundwater. The incentives for better management efficiency include announced allocation, carryovers, borrowings, transfers, and supplementary entitlements for conjunctive irrigators.

The concept of *announced allocation* in groundwater management is implemented with an annual announced allocation for all groundwater users with volumetric entitlements. In New South Wales, allocation greater than 100% may be available in some cases. The announced allocations are on groundwater management zone basis and for a specific period of five years. The levels of announced allocation may be reviewed at the end of the period depending on the groundwater in the total entitlement levels for each zone. When entitlements approach the sustainable yield levels, the announced allocation returns to 100%. These initiatives substantially increase the volume of groundwater available to be used in a succession of water years and are most beneficial for those properties where groundwater entitlements are currently fully utilized.

Carryover is based on total property entitlement of a normal year. The entitlement level on which carryover is based will not be increased to reflect the new five-year allocation available under the announced allocation scheme. The carryover threshold could be introduced (e.g., 20%) with an annual carryover rate (10%). The purpose of this initiative is to encourage farmers to use as much of their entitlements as possible. It also provides an ongoing credit for any small component of unused entitlements. These credits accumulate up to a carryover threshold level and are available for use in drier seasons or whenever commodity prices justify increasing planting areas.

Borrowing from future allocations is also allowed to enable farmers to complete their irrigation requirements in any particular water year. Overdraws can be limited to fixed percentage (20%) of the entitlements.

Traditionally, water entitlements in Australia were tied to specific parcels of land (Pigram et al., 1992). Later, *water transfer arrangements* have been introduced in order to attain more efficient water use and productivity. In New South Wales, transfer arrangements can be used within the same groundwater management zone where required. However, temporary transfers are permitted (DWR, 1991), provided: the properties are within the same management zone; transfers are between existing licenses with a history of groundwater use; volume of transfers is limited to the entitlements applying to vendors' properties; and the effect of the transfer will have limited, if any, impact on surrounding groundwater users.

APPLICATION OF SIMULATION/OPTIMIZATION TECHNIQUES

Groundwater flow models are often used for management of groundwater resources. Such models are based on the conservation equations of mass of water combined with empirical laws such as Darcy's law to formulate water flow fluxes. The general conservation equation requires that the sum of all fluxes into a control volume is equal to the amount stored plus the amount delivered or minus the amount abstracted by sources and sinks in that volume. The conservation equation is a second-order partial differential equation of the form:

$$T_x \frac{\partial^2 h}{\partial x^2} + T_y \frac{\partial^2 h}{\partial y^2} = S \frac{\partial h}{\partial t} + R(x,y,t) \qquad (1)$$

where T_x and T_y are the transmissivities (m^2d^{-1}), h is the piezometric head (m), S is the storage coefficient, x and y are the cartesian coordinates (m), and R is the recharge (+) or discharge (−) rate (md^{-1}). The parameters, boundary conditions, and initial conditions for time-dependent problems are required to obtain solution to Equation (1) in a deterministic fashion. Generally, this computation relies on numerical methods, which discretize the aquifer into cells or elements.

In practical applications, however, the parameters and boundary conditions are never completely known. Instead, head measurements are available. In such situations, the model is used in an inverse way. The unknown parameters and boundary conditions are varied and fitted such that the model reproduces the measured heads. This procedure is called the calibration of a model. In general, it is an iterative procedure which may even go so far as to adjust the underlying conceptual model (Kinzelbach and Li, 1997). A calibrated model can then be used for extrapolation to new management options such as the effect of new groundwater allocations on existing users. It may also be used to compute the water balance of the region or subregions under a given pumping scenario.

Major setbacks exist in the use of groundwater flow simulation models while deriving the comparative best alternatives from different management scenarios. This process must be completed through a trial-and-error procedure that cannot guarantee the achievement of the best alternative (Theodossiou and Tolikas, 1995). This is why flow simulation models, although essential tools in aquifer management, cannot be used by themselves for the solution of many problems. That is, in order to determine the efficient allocation plan, flow simulation models need to be coupled to optimization models.

The combined use of groundwater flow simulation and optimization models based on linear programming formulation has been practiced (Gorelick, 1983; Chau, 1989) and is now becoming common in aquifer management (Punthakey et al., 1992, 1994). Simulation-optimization models permit the determination of the hydraulic response of an aquifer in accordance with the set objectives (such as water demands) and constraints (such as restriction on drawdowns or pumping rates).

Consider a simple case in which an existing renewable groundwater supply, W, is to be allocated among U water users (u = 1, 2, ... U). This is simply a resource allocation problem, in which the allocation X_u to each user u is to be determined so that total allocation does not exceed the total renewable resource. Assuming the objective to be satisfied is maximization of net economic benefit (sum of benefits, B_uX_u, minus, pumping costs, C_uX_u), the model can be written as:

maximize

$$\sum_{u=1}^{U} B_u(X_u) - C_u(X_u) \tag{2}$$

subject to

$$\sum_{u=1}^{U} X_u \leq W \tag{3}$$

where B_u and C_u are the benefit and cost per allocated unit, respectively.

Optimal values of each allocation, X_u, can then be determined using Lagrangian multipliers (UNESCO, 1994).

Case History 1

The first case involves groundwater overdraft in the Lower Gwydir Valley in New South Wales, Australia. This valley experienced regional declines in groundwater levels of 2 m and as much as 20 m locally during the 1992–93 summer irrigation season. Although the management policy allows for and anticipates short-term groundwater overdraft during drought periods, the increased public concern over water level declines led to a review of the allocation policy of the valley. The approach was to examine the long-term groundwater withdrawal potential of the aquifer system by developing a groundwater flow model to simulate management scenarios for groundwater allocations during normal and drought years. The three management scenarios simulated were as follows (Somaratne et al., 1994):

1. What would have been the groundwater condition in 1990, if normal allocations according to the existing allocation policy had been pumped since 1980?
2. What would be the groundwater condition in the year 2000, if normal allocations were pumped from 1990?
3. What would be the groundwater condition in the year 2000, if drought year allocations were pumped from 1990?

Groundwater extraction for the last and second scenarios was assumed to be the full normal year allocations. In the case of Scenario

3, groundwater extraction was assumed to be the full drought year entitlements. Simulation results showed that declines in water level would occur if full normal or full drought year allocations are pumped. Severe drawdowns in excess of 100 m would occur in several areas causing local mining of the aquifer. It was very likely that this level of pumpage would result in some wells drying up and risk damage to the aquifer. For drought year allocation, severity is greater because of the reduced recharge. It appeared that earlier groundwater allocation based on property area was too high in several areas and in excess of the recharge. This deficit of groundwater would continue if the farmers were to use their full entitlements. Based on the modeling results, reallocation of groundwater was undertaken targeting the renewable resource.

Case History 2

This case involves optimizing groundwater pumping rates to lower the water table in Wakool Irrigation District (WID) in southern New South Wales, Australia, covering approximately 210,000 ha. Water table levels have risen significantly as a consequence of irrigation activity, wet winters and inadequate drainage. This is exacerbated by the discharge of regional groundwater from lower aquifers, flat terrain, and meandering prior streams. In 1983, it was estimated that approximately 30,000 ha had a shallow water table and 3000–4000 ha were visibly salinized. The region had experienced rapid salinization with piezometric levels rising to within 0.5 m of the ground surface. Consequently, production losses of up to 50% in rice-growing areas were observed.

In order to lower the piezometric levels below critical depths, a subsurface drainage scheme with forty-eight pumps and two evaporation basins was installed in two stages. The evaporation basins are approximately 2000 ha in size. The pumps extract saline groundwater from 110 tube wells dewatering the upper Shepparton aquifer to protect approximately 47,000 ha of irrigated land. The pumps operated continuosly at non-optimal levels, removing groundwater in excess of the requirement to maintain the piezometric levels at or below critical depths. This resulted in higher operating costs as well as excessive loading of the evaporation basins. Since pumping commenced in 1982, an average volume of 15×10^6 m^3 per year has been pumped to evaporation basins. Annual operating costs have been approximately A$0.3 million of which the electricity charge was the major component. If an optimal pumping schedule could be devised, the piezometric levels could be maintained at or below the critical depth while minimizing the operational costs and effluent load on the evaporation basins. A study was conducted for this purpose. The methodology used and the approach followed for the estimation of net recharge to the aquifer are well documented (Punthakey et al., 1992, 1994). For brevity, only the determination of optimal pumping rates is discussed here.

A linear programming model was developed to determine the optimal pumping rates. The objective function for the problem was defined as (Punthakey et al., 1992, 1994)

$$z = \sum_{i=1}^{I} \sum_{j=1}^{J} d_{ij}^{+} + d_{ij}^{-} \tag{4}$$

where z is the objective variable (m), d^+ is the positive overachievement variable (m), d^- is the positive underachievement variable (m), and I and J are the number of rows and columns in the grid, respectively. Thc achievement variables are defined as follows:

$$h_{pij}^{n+1} + d_{ij}^{+} + d_{ij}^{-} = h_{fij}^{n+1} \tag{5}$$

where h_{pij}^{n+1} and h_{fij}^{n+1} are the predicted and target piezometric levels (m), respectively, at time n+1. The optimal pumping rates is achieved when the piezometric level in a cell reaches h_{fij}. This was defined as

$$h_{fij}^{n+1} = h_{sij}^{n+1} \qquad \text{for } S_{ij} - h_{sij} > dd_{ij} \tag{6}$$

and

$$h_{fij}^{n+1} = S_{ij} - dd_{ij} \qquad \text{for } S_{ij} - h_{sij} \leq dd_{ij} \tag{7}$$

where S_{ij} is the surface elevation (m), h_{sij}^{n+1} is the simulated piezometric level at time n+1, and dd_{ij} is the desired critical depth from the surface (m). Equation (7) implies that the final piezometric levels, h_{fij}^{n+1}, will be equal to the simulated piezometric levels, h_{sij}^{n+1}, when h_{sij}^{n+1} is below the desired critical depth. When h_{sij}^{n+1}, is within the desired critical depth, the optimized pumping rates will result in a water table at the desired critical depth.

With this approach, the range of optimal pumping rates was determined for forty-eight operating pumps. This in turn would improve the efficiency of the subsurface drainage scheme. Results of various scenarios showed that, except for five cells, the capacity of the existing pumps is adequate to provide water table control in the district. The benefits obtainable by operating the pumps within the optimal range include a 25% reduction in disposable saline groundwater to evaporation basin and approximately a reduction of A$75,000 per year in operating costs.

CONSTRAINTS TO SUSTAINABLE ALLOCATION

Social and Financial

In order to sustain water development projects, at least in theory, the cost of all resource development and management must be recovered and society must support and be willing to pay for the services provided (Winpenny, 1994). There can be little sustainable development without attention to overcoming social and financial constraints. In countries where groundwater development costs are high, it is unlikely that the growth of development will continue. For instance, the demand exceeds the recharge in most aquifers where irrigation farming is dependent on groundwater. Moreover, irrigated soils with high salinity levels lead to lower crop yields and farm incomes. Reallocating relatively low-value irrigation allocations to high-value uses requires strong financial commitment, which is often lacking, especially in developing countries. This, together with the lack of adequate marketing opportunities and infrastructure, may cause irrigated lands to go out of production or limit their use to the basic survival of farming communities.

Water pricing is also linked to the sociocultural situation where the ownership of water rights is based on principles of water sharing inherent in traditional customs, with acknowledgment of the right of prior appropriation. Most of the countries of the western Asian region have enacted laws that explicitly specify water resources as public property. Some countries, however, continue to be governed by *Sharia* principles or a combination of *Sharia* law and the *Magalla* code (ESCWA, 1996a). *Sharia* principles governing water legislation hold that water is a gift from God and belongs to the community as a whole, with priority given to human and animal requirements. This belief has caused the concerned authorities to delay or avoid implementing regulations and water pricing schemes that may be considered socially unacceptable (ESCWA, 1996b). For sustainable management, full participation of water users is essential (Akkad, 1989). In its absence, the introduction of any allocation policy may not be successful. In this regard, Ball (1996) stressed the need for the participation of local people in order to change the attitudes in such a way that they feel it is for their own good, they are equal partners, and have confidence to take the responsibility.

Technical

Hydrogeology is often described as "uncertain" since the availability of suitable aquifers and groundwater cannot be predicted with certainty. Kinzelbach and Li (1997) illustrated this fact by means of head measurements at observation wells using a limited number of points. Based on these point measurements, contour lines of equal head were drawn using mathematical interpolation algorithms that do not take into account the principles of conservation of mass or Darcy's law. By fitting a groundwater flow model to the data, these laws are automatically respected and the interpolation is combined with an interpretation. A major drawback of the procedure, however, is that the interpretation found is often not unique since both fluxes and transmissivities are unknown.

Heijde et al. (1986) identified the following problem areas in the use of models in groundwater management: accessibility of models to users, communications between managers (decision makers) and technical personnel, and inadequacies of data and modeling. Difficulties in gaining access to appropriate models was cited as the most serious constraint. Factors affecting accessibility include lack of adequate model documentation, lack of service for the systematic collection and dissemination of information on groundwater models, and lack of training for the professionals who use the models. Other constraints include a lack of confidence on the part of managers in using models as decision-making tools; lack of interest by managers to learn about modeling and failure of technical experts to clearly convey to them model results; the range of applicability; and the limitations and the degree of reliability of models. Furthermore, it is not always easy to formulate management problems precisely for model applications. For example, it may be difficult to state objectives explicitly and quantitatively. Inadequate or insufficient data, high costs, and the length of time it takes to generate data are limiting factors in application of models.

Political

Water allocation often results in disputes among users, resource managers, and policy makers because of the social, economic, and environmental importance of water at all levels and sectors. Rising demand, supply deficit, and a deteriorating environment are the reasons for forcing water onto the political agenda (Morris, 1996). Unresolved disputes may lead to unmanaged water resources. Of those tackling the disputes, capacity building focusing on formal institutional development at the regional and national levels is a necessity. According to McKinney (1989), a collaborative and consensus-building approach to water resources planning and dispute resolution offers many benefits: It avoids the costly and time-consuming outcomes of unmanaged conflicts such as administration appeals, litigation and political actions.

Fragmented water legislation may also lead to mismanagement of water resources, largely due to a lack of enforcement (Caponera, 1992). Effective enforcement, on the other hand, is contingent upon authorities monitoring groundwater withdrawals and suspending or revoking the licenses of violators. Strong political commitment is needed to develop comprehensive water legislation and enforcement mechanisms. Other major areas where political will is needed are the "closure of water resources" at optimal development and water pricing. Once the water resource is closed, water rights can only be acquired through purchase as new allocations are no longer created. This may be difficult politically as decision makers prefer to focus on achievements, perhaps in terms of number of wells drilled or yields of water obtained, rather than on intrinsic policy issues. In this regard, Allan (1996) rightly argued that even economically and environmentally sound, and technically effective, allocation policies will be subordinate to political imperatives.

CONCLUSION

Past efforts to meet water demand have been directed towards developing scarce water resources. While development programs have been successful in stretching available supply, few feasible development options remain. Furthermore, heavy water utilization has affected water quality and overdraft of aquifers. It seems likely that water management will play an even more significant role in the allocation of water resources in the future. Any comprehensive program in water resources allocation and planning in arid countries must take into account the spatial and seasonal distribution of water as well as the cyclical variability of water resources.

Measures to control water demands are becoming increasingly important to water resources planners and decision makers. Physical methods of demand reduction, economic incentives, integrated water resources planning, and implementation of sociological parameters are among the various ways to conserve water resources and achieve sustainability. Since groundwater systems react slowly, predictive capabilities are required for future allocation or planning. In this regard, there is no alternative to modeling. Management options have to be validated in advance to determine whether they comply with sustainability constraints. Application of simulation/optimization techniques allows arriving at reasonable decisions. Finally, there must be a commitment to implementation and the chosen method must be appropriate and effective. These are essentially political issues. Commitment depends on the dominant political ideology. Thus, issues of water management need a strong political will that embraces regulations and direct actions, as well as institutional support capabilities.

REFERENCES

Abrams, L. J. 1996. Policy development in the water sector—The South African experience. In: *Water Policy: Allocation and Management in Practice.* E & FN SPON, London, pp. 21–30.

Akkad, A. A. 1989. Water conservation in arid and semi-arid regions. *Desalinization* 72(1):185–205.

Al-Ibrahim, A. A. 1990. Water use in Saudi Arabia: Problems and policy implementations. *Journal of Water Resources Planning and Management (ASCE)* 116(3):375–388.

Allan, J. A. 1996. Policy responses to the closure of water resources: Regional and global issues. In: *Water Policy: Allocation and Management in Practice.* E & FN SPON, London, pp. 1–12.

Ball, D. M. 1996. Allowing local people to cope with change and take responsibility for sustaining their groundwater resources. In: *Water Policy: Allocation and Management in Practice.* E & FN SPON, London, pp. 362–370.

Caponera, D. A. 1978. Water laws in moslem countries. FAO Paper 20/2.

Caponera, D. A. 1992. *Principles of Water Law and Administration—National and International Level.* A. A. Balkema, Rotterdam.

Caponera, D. A. 1996. Requirements for drafting a modern water code and water legislation in the ESCWA region. Economic and Social Commission for Western Asia.

Chau, T. S. 1989. Analysis of sustained groundwater withdrawals by the combined simulation-optimization approach. *Groundwater* 26(4):454–463.

Conley, A. H. and D. C. Midgley. 1988. National water management in the Republic of South Africa: Towards a consultative partnership with diverse users in a semi-arid country. *Journal of Hydrology* 100 (1):473–487.

De Jong, R. L. 1996. Water sector management legislation and enforcement in the ESCWA region. Expert Group Meeting on Water Legislation in the ESCWA Region, Amman, Jordan.

DWR. 1991. Revised groundwater allocation guidelines for the groundwater resources of the Murray geological basin, New South Wales. Department of Water Resources, Technical Report TS 91.023.

ESCWA. 1996a. Water resources agreements and practices in selected watersheds in the ESCWA region. Economic and Social Commission for Western Asia.

ESCWA. 1996b. Water legislation in selected ESCWA countries. Economic and Social Commission for Western Asia.

Gorantiwar, S. D. and I. K. Smout. 1995. Water resources management of irrigation schemes in semi-arid and arid regions. *WRM'95, Proceedings of Regional Conference on Water Resources Management,* Isfahan University of Technology, Conference Secretariat, pp. 413–422.

Gorelick, S. M. 1983. A review of distributed parameter groundwater management modeling methods. *Water Resources Research* 19(2):305–319.

Heijde, P., Y. Bachmat, J. Bredehoeft, B. Andrews, D. Holtz, and S. Sebastian. 1986. Groundwater management: the use of numerical models. American Geophysical Union, Washington, D.C.

Hutchins, W. A. 1974. Water rights laws in the nineteen western states, Vol. 2. Misc. Publication No. 1906. United States Department of Agriculture, Washington, DC.

Kinzelbach, W. and G. M. Li. 1997. Use of groundwater models in sustainable management of groundwater resources. Keynote Address, Third Gulf Water Conference, Muscat, Sultanate of Oman.

McKinney, M. J. 1989. Role of planning in resolving water policy disputes. Water: laws and management. American Water Resources Association. P2B17-29.

Morris, J. 1996. Water policy: Economic theory and political reality. In: *Water Policy: Allocation and Management in Practice.* E & FN SPON, London, pp. 228–234.

Pigram, J. J. 1991. Water allocation for the environment. Centre for Water Policy Research, University of New England, Australia.

Pigram, J. J., R. J. Delforce, M. L. Coelli, V. Noris, G. Antony, R. L. Anderson, and W. F. Musgrave. 1992. Transferable water entitlements in Australia. Centre for Water Policy Research, University of New England, Australia.

Punthakey, J. F., S. A. Prathapar, and N. M. Somaratne. 1992. Pumpman: Optimizing pumping rates to control shallow watertables in the Wakool irrigation district. Department of Water Resources, Technical Report TS 92.088.

Punthakey, J. F., S. A. Prathapar, and D. Hoey. 1994. Optimizing pumping rates to control piezometric levels: A case study. *Agricultural Water Management* 26:93–106.

Somaratne, N. M, J .F. Punthakey, D. Wooley, R. M. Williams, and J. B. Ross. 1994. Lower Gwydir Valley groundwater model. Department of Water Resources, Technical Report TS 94.031.

Theodossiou, N. and D. Tolikas. 1995. Simulation and optimization modeling in aquifer management. In: *Water Resources Management under Drought or Water Shortage Conditions.* N. X. Tsiourtis (Ed.) A. A. Balkema, Rotterdam.

Tsiourtis, N. X. 1995. Optimal water resources allocation under water shortage conditions via linear programming—simulation models. In: *Water Resources Management under Drought or Water Shortage Conditions.* N. X. Tsiourtis (Ed.) A. A. Balkema, Rotterdam.

UNESCO. 1994. Water resources management: Focussing sustainability. P. Louks (Ed.) IHP-iv Project M-4–3, Unesco.

Utton, A. E. 1987. Selected issues of groundwater management. *Water Resources Policy for Asia.* A.A. Balkema, Rotterdam, Netherlands, pp. 285–311.

Winpenny, J. 1994. *Managing Water as an Economic Resource.* Routledge/ODI, London.

6

Technologies for Domestic Wastewater Treatment and Reuse

S. VIGNESWARAN AND H. H. NGO

Faculty of Engineering
University of Technology, Sydney
PO Box 123
Broadway, NSW 2007, Australia

INTRODUCTION

THE URBAN WATER system is undergoing profound change due to the limitation on available water resources. As a result, the price of water has risen. This can be offset by the introduction of innovative technologies for water reuse. Only a small part of the water supplied to a house is used for drinking, so there are possibilities for a dual supply of water—a high quality supply for drinking and washing and a lower quality supply for toilet flushing and garden watering. Many cities are currently using dual water supplies or are in the stage of developing such systems. Water reuse will reduce the amount of water discharged into the environment.

Water treated at a central facility for reuse may encounter a number of operation difficulties. For example, the reticulation system used to convey the water back to the household is expensive. Water treated on-site can be reused for several purposes such as toilet flushing, garden watering and floor washing. A number of technical alternatives are available to treat the sewage to reusable quality. Physicochemical processes such as deep-bed filtration and membrane processes play major roles among the technologies for water reuse. Deep-bed filtration in modified form finds an important place in water reuse due to its simplicity. Membrane processes have significant advantages over the other processes, such as high quality effluent, minimum sludge treatment requirement, and no or little disinfectant requirement.

The first part of this chapter highlights two membrane hybrid systems tested, namely a household membrane bioreactor (hollow fiber membranes immersed into an aeration tank) and a cross-flow microfiltration (CFMF) system with in-line flocculant/adsorbent addition. The second part of the chapter discusses some simple filtration technologies that can be used in water reuse. Experimental studies conducted with both the systems are highlighted.

MEMBRANE PROCESSES

Wastewater treatment and reuse are important parts of our aqueous environmental management. They provide great opportunities for the membrane community. For example, over the next two decades the projected annual expenditure worldwide for water treatment and tertiary sewage technologies is more than $10 billion/yr (Fane, 1994).

Membrane Bioreactors

Domestic wastewater is generally treated by centralized facilities involving biological processes, most often by an activated sludge process. Conventional activated sludge comprises an aeration tank and a settling tank. In this process, the sludge should be settleable enough to avoid an overflow of sludge into the treated water. This limits the concentration of mixed liquor suspended solids (MLSS) to about 6000 mg/L. Higher concentrations have poorer settling properties. Further, if the return (recycle) sludge is not concentrated enough, MLSS in the aeration tank will be reduced, which results in deterioration of effluent quality.

Substitution of the sedimentation tank by membrane filtration eliminates the above-mentioned problems. There are currently a number of systems where membrane filtration is used as the solid-liquid separation unit. This arrangement leads to many advantages such as high quality effluent, high biomass (MLSS) concentration thus smaller aeration tank requirement, and low food/mass ratio thus minimum excess sludge treatment. The biological oxidation tank coupled with a membrane unit is termed a bioreactor.

Membrane bioreactors were initially developed for biotechnology applications. In the field of wastewater treatment it can be considered as an emerging technology. However, its growth has been quite rapid due to the numerous advantages it offers:

- possibility of very compact units. It is possible to operate the biological treatment at short residence time due to the high biomass concentration (20–30 g/L)

- significant decrease in sludge production due to its operating conditions (high sludge age, low food-to-microorganisms ratio)
- high quality of treated water, which enables reuse: very low turbidity, high level of disinfection, lower BOD and COD than in conventional processes

Two main kinds of membrane bioreactors now existing at (small) industrial scale consist of:

- immersed membranes as initially proposed by Chiemchaisri et al. (1993) and developed at industrial scale by Mitsubishi, Kubota and OTV (BIOSEP process)
- external circulation of the sludge from the activated sludge tank to the UF/MF unit

MEMBRANES IMMERSED IN THE BIOLOGICAL TREATMENT UNIT

This concept looks attractive since no recirculation is needed, air bubbles supplied for aeration of the mixed liquor (biological oxidation) facilitate continuous declogging of membranes, and membranes can be operated at very low transmembrane pressure (as low as 10 kPa) below critical flow condition. This minimizes fouling and thus periodic chemical cleaning, which leads to increased membrane life.

Despite the relative large membrane area requirement, the capital and replacement costs are reduced due to the simplification of membrane equipment (no cartridge) and increased membrane life, respectively.

Design Used at Asian Institute of Technology (Vigneswaran, 1991; Chiemchaisri et al., 1993)

The performance of a hollow fiber membrane module in the activated sludge process (aerobic reactor) supplied with diurnally varying feed was investigated at pilot scale. Two hollow fiber membrane modules were immersed in an aeration tank (working volume 1.5 m^3) of activated sludge system (Figure 1). Microporous hollow fiber membranes of 0.03 μm pore size with 9 m^2 surface area (Mitsubishi Rayon Co., Japan) were used in the study. A centrifugal suction pump was used at 10 minutes "on" and 10 minutes "off" intermittent operation to extract the permeate through the membrane. The effluent flow rate and transmembrane pressure were measured once every two days by rotameter and mercury manometer, respectively. An air flow meter was used to control

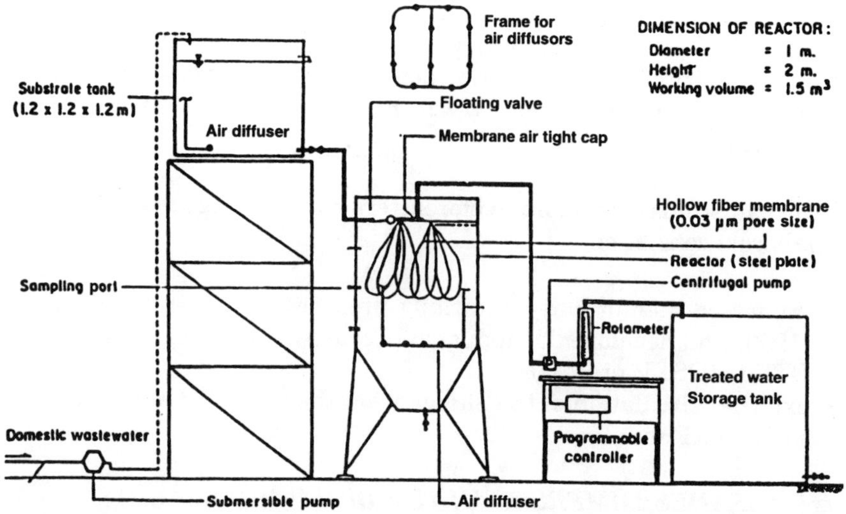

Figure 1. Schematic of pilot-scale membrane bioreactor (Chiemchaisri et al., 1993).

and measure the air flow rate. Domestic wastewater from the Asian Institute of Technology, Bangkok campus, was used as substrate. The seeding microorganisms were obtained from the return sludge at Huay Kwang activated sludge treatment plant, Thailand.

The mean hydraulic retention time (HRT) was determined after permeate flux reached steady state. This mean permeate flux was used as the mean influent flow rate in order to maintain constant volume of mixed liquor in the bioreactor. After steady state was reached, the influent flow was varied corresponding to the actual diurnal change of wastewater flow rate. A solenoid valve and parallel pumps coordinated with a programmable controller were used to control the change of influent. The diurnal variation of wastewater loading was sinusoidal with a mean value of about 1 m^3/d.

In this sixty-day long study, COD, turbidity, and pH of influent and effluent were measured every two days. TKN, NO_2-N and NO_3-N of the samples from influent, effluent and reactor were analyzed once in four days. MLSS, MLVSS and pH of influent, effluent and reactor were measured every two days.

The membrane bioreactor was operated at a volume of 1 m^3. During this period, the average flux tended to decrease slightly but was maintained at a constant value after a month of operation. The mean flux at steady state was found to be around 0.11 m^3/m^2.d (4.6 L/m^2.h) corresponding to an HRT of one day. After the achievement of steady state, the bioreactor loading was varied to represent the diurnal variation and operated for another month to test the performance.

The results showed that the mean flux decreased slightly, but was kept above 0.1 m^3/m^2.d. The transmembrane pressure increased slightly during the experiment. From the results, it was observed that diurnal variation in loading did not alter the stability of the process in terms of the trend of the transmembrane pressure variation. The overall increase in transmembrane pressure was about 20–26 kPa for two months of operation. Of this, about 10–13 kPa increase was during the last ten days of the run.

The effluent turbidity was measured in order to evaluate the effect of fluctuations in the influent turbidity. The diurnal change in influent turbidity fluctuation in the range of 15–35 NTU showed no significant effect in the effluent. The effluent turbidity was well within 0.5 NTU. It was almost constant throughout the run.

The results suggested that throughout the experiment, effluent COD concentration maintained a very low value, less than 10 mg/L. It was independent of the variation in the influent COD concentration (60–200 mg/L) and wastewater loading rate. The COD removal efficiency was in the range of 80–98%.

During the first month, before the start of diurnal variation, more than 90% of TKN was converted to NO_3-N. When the loading was varied diurnally, a fluctuation in the nitrification process was observed. However, a total nitrification of more than 80% was observed during the experiment; it was in the order of 95% after ten days from the start of operation. Therefore it can be concluded that diurnal variation in loading played a minor role in the nitrification process with a satisfactory nitrification rate.

Since the operating condition in this process was aerobic, sufficient air was supplied (also to keep the solids in suspension) through air diffusers at an optimum rate of 7.5 L/min. It was found that although the supplied air was sufficient to achieve saturated DO concentration, it could not suspend all of the solids in the reactor. A portion of the solids was found settled at the bottom of the reactor, under the diffusers, where air was not supplied. This created anaerobic conditions. As a result, the reactor was observed to have two separation zones. An aerobic zone was present, which represented the zone above the diffusers. In this zone DO was 7–10 mg/L and organic matter was biologically degraded. Second an anaerobic zone was also present, which represented the zone below the diffusers. DO was almost zero in this zone. The air flow rate affected the accumulation of the sludge layer and a portion of it was observed to resuspend into the aerobic zone. Although this system produced consistently good effluent quality irrespective of diurnal variation of influent, no nitrogen removal was achieved. Further, permeate flux was low, requiring larger membrane area. These problems have been eliminated in the design adopted by Chiemchaisri (1993) described below.

System Developed at University of Tokyo (Chiemchaisri, 1993)

The experimental set-up used in this study is shown in Figure 2. The system was divided into two parts: a main bioreactor and a separation unit. The separation unit, with a 10 L volume, was immersed into the main bioreactor of 62 L volume. Paddles, driven by a motor, provided shear flow of mixed liquor in the separation unit across the membrane surface at a speed of 290 rpm. Two hollow fiber membrane modules were used for solid-liquid separation inside the separation unit. Diffusers for jet aeration, connected to an air pump, were also installed inside the membrane modules to provide high-speed aeration (40–50 L/min) once in 90 minutes of operational cycle for 1 minute. This application was

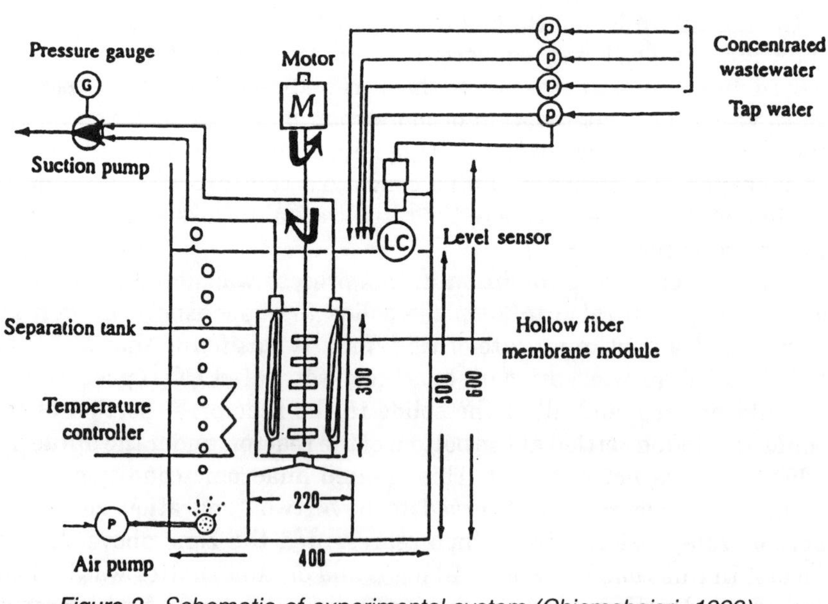

Figure 2. Schematic of experimental system (Chiemchaisri, 1993).

expected to occasionally help in detaching accumulated sludge from the membrane module. Permeate was extracted by a suction pump either under continuous or intermittent operation with different cycle time intervals.

Aeration was supplied through diffusers to the bioreactor at a flow rate of 15 L/min. The pH of mixed liquor was controlled in the range of 6.8–7.5 throughout the study. Whenever a drop in pH took place, adjustment was done by addition of sodium bicarbonate ($NaHCO_3$). The temperature of mixed liquor was controlled by a temperature controller; it varied in the range of 5–25°C.

Polyethylene microporous hollow fiber membrane modules (Sterapore-S filter units, Mitsubishi Rayon Co., Ltd.) of 0.03 and 0.1 μm nominal pore size (STN 314 and 327) were used. Their inner and outer diameters are 270 and 380 μm, respectively. Each membrane module comprised 480 fibers in a loop of 640 mm (320 mm, length of membrane module), making a total surface area of 0.3 m^2 (based on the average fiber diameter).

Synthetic wastewater with the composition shown in Table 1 was used to simulate domestic wastewater. In addition to soluble synthetic wastewater, toilet paper was applied to the system as insoluble waste. Chemical oxygen demand (COD) and total nitrogen (TN) attributed by the toilet paper addition accounted for about 10 and 0.3% of daily COD and nitrogen loading, respectively.

Direct solid-liquid separation in the membrane separation bioreactor was improved by introduction of highly turbulent conditions using paddle mixing inside the separation zone and jet aeration inside the spread membrane module. Long-term observation over 280 days showed a

Table 1. Composition of synthetic substrate (stock solution).

1st day—73rd day		73rd day	
Glucose	10 g/L	$CH_3COONa.3H_2O$	22 g/L
Peptone	1 g/L	Peptone	10 g/L
Yeast extract	1 g/L	Yeast extract	1 g/L
K_2HPO_4	3.2	K_2HPO	3.2 g/L
$(NH_4)2SO_4$	16 g/L	$(NH_4)2SO_4$	16 g/L
$MgSO_4.4H_2O$	4 g/L	$MgSO_4.7H_2O$	4 g/L
$FeCl_26H_2O$	0.02 g/L	$MnSO_4.4H_2O$	0.4 g/L
$CaCl_2.2H_2O$	0.4 g/L	$FeCl_2.6H_2O$	0.02 g/L
		$CaCl_2.2H_2O$	0.4 g/L
		Toilet paper	0.165 m^2/day

promising permeate flux of 0.2 $m^3/m^2 \cdot d$ by intermittent suction of 5-minute on and off cycles (equivalent to 0.1 m/d in continuous operation). The system was operated continuously without any chemical membrane regeneration over that period. Nevertheless, permeate flux was influenced by mixed liquor temperature. A temperature decrease affected the permeate flux not only by increasing the viscosity of the mixed liquor but apparently also by changing the properties of the cake layer (i.e., cake layer thickness and/or porosity of cake layer).

A high degree of organic oxidation was obtained in the system at a moderate temperature of 25–29°C regardless of the aeration mode by operation without sludge wastage. Both soluble synthetic waste and toilet paper were almost completely degraded and oxidized in the system showing a low average effluent COD of 20.8 and 16.5 mg/L under continuous and intermittent aeration, respectively. Long sludge age conditions provided a high degree of organic oxidation as long as sufficient oxygen was supplied to the system. The degree of nitrification depended upon the DO concentration of the mixed liquor. By maintaining a DO level above 1 mg/L in the continuous aeration mode, complete nitrification was achieved. Introduction of intermittent aeration with 180-minute cycles (i.e., 90 minutes of aeration on and off) yielded a nitrogen removal up to 80% or more (DO in the aeration period at 1.5–2 mg/L). An increase in DO level to 4–5 mg/L improved the percentage of nitrification and, therefore, nitrogen removal to more than 90% (Vigneswaran et al., 1996).

During a stepwise short-term temperature decrease from 25 to 5°C every two weeks, organic carbon removal did not deteriorate, showing a stable effluent quality in terms of COD and TOC. Nitrogen removal was achieved by introducing a nonaeration period after nitrification was obtained. Average nitrogen removal at 10 and 5°C was 90 and 85%, respectively. An increase in biomass concentration was observed at low temperatures, which could be attributed to a lower sludge decay rate. This increase in biomass concentration should enhance the nitrification by providing surface area of bioflocs for the oxygen transfer and, thus, more sites for nitrification. An insignificant difference in maximum volumetric nitrification rate at 25 and 5°C suggested that there was not much variation in oxygen availability for nitrifying bacteria at either temperature once high and stable nitrogen removal was achieved. Sufficient oxygen transfer conditions could be maintained in the membrane separation bioreactor even when the temperature was as low as 5°C. Comparison of temperature coefficients among various treatment processes also suggested less temperature dependence on both the organic oxidation and the nitrification rate in the membrane separation bioreactor than in the other processes. Organic oxidation and nitrogen removal was enhanced in the membrane separation bioreactor by providing complete biomass retention and by allowing an increase in biomass concentration at low temperature to compensate for the decrease in biomass activities and oxygen transfer rate.

MEMBRANE UNIT PLACED EXTERNAL TO THE BIOLOGICAL OXIDATION TANK

Ceramic Membrane Bioreactor

A typical example is the design proposed by Lyonnaise des Eaux. They have already commissioned units up to 150 m³/day and a unit of 1500 m³/day is being proposed. In this case, tubular ceramic mf membranes are used with cross-flow velocities greater than 3 m/s.

The performance of a ceramic membrane bioreactor (MBR) in Cronula, Australia wastewater treatment is shown in Table 2 (Trouve et al., 1996). The permeate flux was kept at 150 L/m^2h^{-1}. The membranes were located in an open loop. Aeration and mixing in the activated sludge tank were performed using a hydro-ejector. The plant was automated, thus the only labor cost was due to sludge treatment. The frequency of sludge withdrawal was very low. This system did not give rise to any odor problem. MBR is still more expensive than conventional processes such as activated sludge, and trickling filter. However, permeate flux improvement and process simplification makes the MBR a good competitor for small plants (< 5000 people). It produces high quality treated water suitable for reuse and involves fewer processes than conventional processes. Conventional processes are also more difficult to operate, probably less reliable, and require larger land areas.

Table 2. Treatment efficiency.

		BOD_5	COD	TOC	TKN	Total N	Total P[a]
Raw water (average)	($mg\ L^{-1}$)	154	335	99	51.7	0	8.25
Treated water (average)	($mg\ L^{-1}$)	2.5	27	10	2.1	NO_3-N = 6.2	0.45
Total removed	(%)	98.5	92	90	96	84[b]	95

[a]ferric chloride addition; [b]$100\ [1 - ((TKN + N\text{-}NO_3)_{treated\ water}/TKN_{raw\ water})]$.

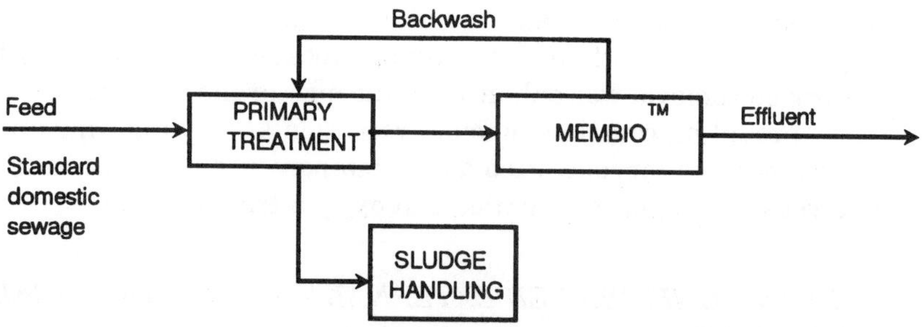

Figure 3(a). MEMBIO™ process.

This result is based on a seven-month analysis. A sequenced aeration mode was adopted with a hydraulic detention time of twelve hours.

MEMBIO™ Process

The MEMBIO™ process is a combination of an aerobic bioreactor and a microfiltration system developed by MemTec. It is integrated into the waste treatment facility [Figure 3(a)].

A simplified schematic of the MEMBIO™ treatment process is shown in Figure 3(b). Wastewater is fed to the bioreactor. After treatment in the bioreactor, the effluent goes to the continuous microfilter (CMF) where the solids, parasites, bacteria and viruses are removed by filtration through a microporous membrane.

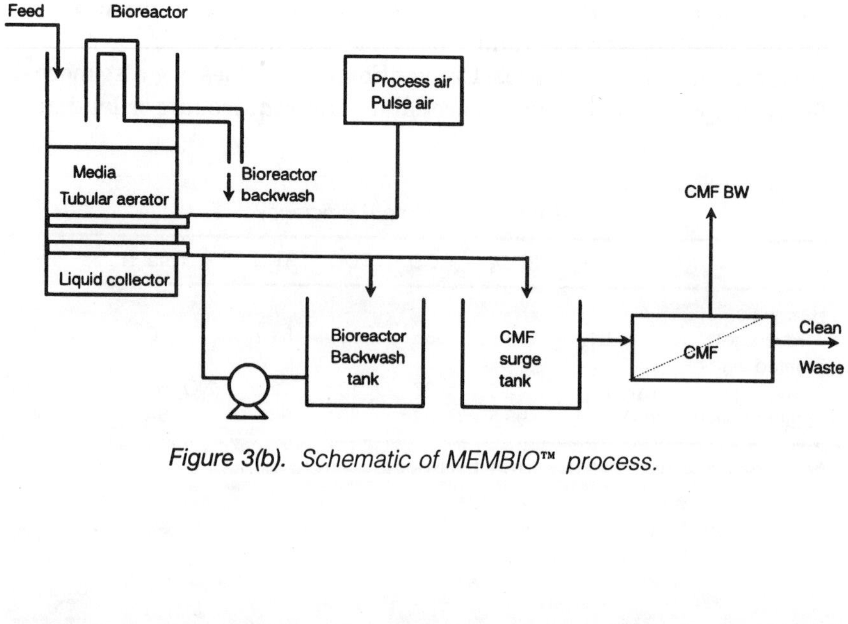

Figure 3(b). Schematic of MEMBIO™ process.

The key difference between the MEMBIO™ bioreactor and other biologically aerated filters is the separation of the process into two distinct components. First the bioreactor converts soluble material into insoluble biomass. This biomass is then removed by either backwashing the bioreactor or by capture on the membrane surface. The MEMBIO™ process has been operated successfully on a number of trade and sewage streams ranging from light domestic sewage at 100 mg/L BOD_5 and trade waste at up to 3700 mg/L BOD_5. Trials have been carried out with a loading rate of up to 8kg BOD_5 per m^3 of bed per day although the feed stream had a low soluble percentage BOD_5. Plants are sized on 3kg soluble BOD_5 per m^3 of bed per day.

A detailed long-term study, which was conducted at a 1 m^3 trial plant with primary settled effluent from Richmond sewage treatment works in Australia, indicated:

- the suspended solids of the filtrate till October 1992 was less than 1 mg/L with an average feed of 100 mg/L
- the average BOD_5 of the filtrate from the CMF was $<$ 5 mg/L. The average feed stream was 130 mg/L BOD_5. The end loading rate was 5.5 kg BOD_5 per m^3 of bed per day

A large demonstration plant, with a capacity in excess of 1.5 mL/day, was trialed on primary settled effluent at a sewage treatment plant (STP) in Sydney, Australia. The bioreactor had two parallel compartments with a total volume of 54 m^3 and a media bed 2.7 m deep. The membrane system (CMF) had 90 modules each with 10 m^2 of hollow fiber microfilters.

The design of the MEMBIO™ process minimizes solid loading on the CMF unit, which has a shell-side feed. In order to maintain throughput, the membranes are regularly (ca 3 × per hour) backwashed with air and clean filtrate. The membranes are chemically cleaned on a daily basis. Results of a trial are summarized in Table 3. The data show that the MEMBIO™ process achieves excellent removals of BOD_5 and suspended solids and a high level of disinfection.

Membrane Bioreactor with Ultrafiltration (UF) System

The combination of a bioreactor with a high level of activated sludge and an ultrafiltration module allows a thorough treatment of wastewater, enabling it to be recycled and used in washing floors and flushing toilets. An industrial use of this system has been developed in Japan where the compact nature of the system and local legislation have allowed it to be installed in office complexes and hotels (Roulet, 1989).

Table 3. Results of MEMBIO™ trial at Cronulla STP.

BOD_5	Input (average)	101 mg/L
	Output	2 to 5 mg/L
	Removal (average)	96%
	Loading (average)	3.0 kg BOD/m^3.day
Suspended solids	Input (average)	56 mg/L
	Output (average)	0.44 mg/L
	Removal	99.2%
Fecal coliform	Input (average)	76×10^5/100 mL
	Output (average)	0.4/100 mL
	Log reduction	ca 7.0
Virus-coliphage	Log reductions	ca 3.0
Average flux = 60 L/M^2 hr		

Mitsui Petrochemical Industries (M.P.C.) have introduced the Rhone-Poulenc system, which combines ultrafiltration and an activated sludge bioreactor. This process has been tested in France for treating wastewater from hospitals and urban power stations. M.P.C. has adopted this system called the "UBIS" system to the treatment of wastewater in office buildings and hotels in Japanese cities. This system has been installed in more than forty buildings and produces more than 5000 m^3/d (Manem and Sanderson, 1996). The "UBIS" system enables reprocessing wastewater from kitchens, wash basins and toilets. The wastewater is transferred to an activated sludge bioreactor, which has a high concentration of sludge (20 grams per liter) and a high level of agitation. These two factors reduce the residence time of the waste to about one hour, compared with twenty-four hours in a traditional system.

The finely dispersed sludge is very efficient but it is difficult to settle in sedimentation tanks. This is where the PLEIADE ultrafiltration module allows for the recovery of purified water from the contents of the bioreactors. Water passing through the membrane will be free of suspended solids, viruses and bacteria (Table 4). This water is stored in a buffer tank where a small amount of sodium hypochlorite is added. It is then reused for flushing toilets.

The design of an ultrafiltration module using turbulence promoters allows very high fluxes over long periods. The flux varies between 100 and 120 liters per square meter per day, over a period of forty-five days.

Normal chemical cleaning is performed every forty-five days, a procedure that takes approximately one to two hours. Electrical consumption is about three kilowatt hours for every cubic meter of treated water.

Table 4. Water quality of the UBIS system (Roulet, 1989).

		Raw Water			Regenerated Water		
Item	Unit	Average	Max.	Min.	Average	Max.	Min.
Temperature	°C	19.5	22.5	15.5	26.3	30.5	21.0
pH	—	6.6	8.5	5.7	6.8	7.9	6.1
BOD	ppm	349.0	577.0	120.0	3.7	5.5	1.8
SS	ppm	96.5	160.0	33.0	~ 0	~ 0	~ 0
n-hexane extract	ppm	11.7	12.4	11.0	< 1.0	< 1.0	< 1.0
MBAS	ppm	6.5	9.9	3.1	0.4	0.4	0.2
Colon bacilli group	pp/cc	18.5×10^4	22×10^4	15×10^4	~ 0	~ 0	~ 0

For instance, to treat 100 cubic meters per day, a plant requires 45 m^2 of floor space and a module with 34 m^2 of membrane area and a 6-cubic meter bioreactor.

The system is very compact compared to traditional processes. For instance, in treating 100 m^3 per day, the floor space saved is equivalent to twenty-five parking places. Further, operation and maintenance of the system is very simple. The system easily withstands the change in load levels. The treated water is of very high quality and can be reused. Energy consumption and running costs are low compared with traditional processes because the treated water can be reused.

Thetford Corporation in Ann Arbor, Michigan, U.S. has developed a membrane bioreactor system known as Cycle-Let-Wastewater Recycling system. Two-phase biological systems are used with tubular organic membranes (Irwin, 1990; Manem and Sanderson, 1996). For color removal and post-disinfection, adsorption and ozonation are used. Thetford Corporation has installed thirty plants across the U.S. (Irwin, 1990).

Microfiltration—Reverse Osmosis Combination

MF has been found to be the most suitable pretreatment for reverse osmosis (RO) application in the production of demineralized water from secondary treated sewage for the use in power plants. For example, the sewage effluent from Dora Creek in New Castle, Australia, wastewater plant after passing through a motorized self-cleaning strainer (0.5 mm aperture) is sent to two Memcor (R) 90M10C CMF units (Figure 4). Each CMF unit comprises ninety filtration modules containing hollow fiber membranes having an average pore size of 0.2 mm. It removes virtually all suspended solids, fecal coliforms and giardia cysts and

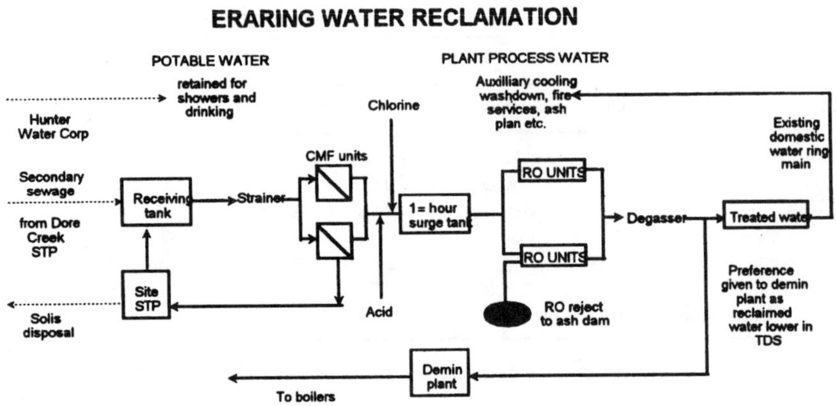

Figure 4. Schematic diagram of Eraring water reclamation plant.

significant amounts of human virus (Reovirus, Enterovirus). When CMF is used as a pretreatment, the silt density index (SDI) of water is consistently reduced to less than 3 allowing RO membranes to be operated at approximately 30% higher flux than is possible with a conventional flocculation/sedimentation/filtration process.

Filtrate from the CMF units is dosed, en route to an intermediate storage tank, with sodium hypochlorite for control of biological growth. Sulphuric acid is also dosed to reduce pH and minimize hydrolysis of RO membranes. Microfiltered water is drawn from the storage tank by centrifugal pumps (15kW), dosed with antiscalant and passed through a 1 micron disposable cartridge guard filter. The RO membranes are cellulose acetate membranes rated at 98% rejection. The water flows tangentially to the membrane surface. Salts and organics are rejected, allowing only water to pass through. Permeate (product water), which is virtually free of all salts and virus, is piped to a degasser tower and stored in a 60 m^3 reclaimed water tank. Treated water, because of its low TDS, is fed preferentially to the demineralizer plant with the balance of the water used for nonpotable water applications around the power plant.

The RO concentrate (reject or brine) passes, via the demineralizer plant wastewater sump, to the station ash dam. Backwash from CMF plant is sent to Pacific Power's on-site wastewater treatment facility and recycled back to the feed receiving tank. This innovative combination of CMF and RO to process secondary treated municipal wastewater achieves near boiler feed quality (Table 5).

Table 5. RO permeate quality (Memtec Technical Bulletin, A2049).

Parameter (mg/L)	Year 0	Year 1	Year 2	Year 3
Chloride	17	20	25	32.0
Sulphate	1.2	1.43	1.8	2.2
Silica	1.4	1.8	2.1	2.6
Sodium	9.5	11.4	14.3	17.8
Magnesium	0.2	0.2	0.2	0.3
Calcium	0.3	0.3	0.4	0.5

Cross-flow Microfiltration (CFMF) with In-Line Flocculation in Wastewater Treatment

In-line flocculation-microfiltration is a method of reducing the degree of internal clogging. The in-line flocculation eliminates the penetration of colloidal particles into membrane pores by the formation of aggregates. This also modifies the deposit characteristics thus facilitating easy membrane declogging. Part of the dissolved matter is also removed by the incorporation of the dissolved solids into the flocs. In-line flocculation-microfiltration is widely practiced in water treatment and will become an important membrane hybrid system in tertiary wastewater treatment and reuse.

The results of a laboratory-scale CFMF with in-line flocculation in sewage treatment were encouraging. The results obtained with flat plate and tubular geometry modules are presented in Tables 6 and 7. Although these results are encouraging, all these studies were conducted with small membrane modules and with no backwashing.

DEEP BED FILTRATION IN TERTIARY WASTEWATER TREATMENT AND WATER REUSE

Deep bed (or granular bed) sand filters have been used as the final clarifying step in municipal potable water treatment since the beginning of the century. Their use is becoming important in wastewater treatment in the tertiary treatment stage as effluent standards are becoming increasingly stringent and as water reuse is envisaged. Most of the wastewater treatment plants incorporate deep bed filtration (with dual media arrangement) in the tertiary treatment step to remove part of the remaining BOD and suspended solids.

Table 6. Performance of CFMF with in-line flocculation arrangement in sewage treatment (cross-flow velocity = 5.14 m/s; transmembrane pressure = 90 kPa; sewage from Asian Institute of Technology, Bangkok; cellulose acetate membrane with flat geometry; temperature = 32.5°C).

Membrane pore size (μm)	Flocculant dose (mg/L)	Influent COD (mg/L)		Effluent (mg/L)	E. Coli (colony/100 mL)		Colour (Hazen)		Permeate flux ($m^3/m^2.h$)	Turbidity (NTU)	
		Total	Filtered		Inf.	Eff.	Inf.	Eff.		Inf.	Eff.
0.45	0	125	60	40	1 × 108	10	200	< 5	0.23	25	0.1
0.45	50	125	60	40	1 × 106	10	200	< 5	0.62	25	0.1
0.20	50	125	60	30	5 × 107	0	200	< 5	0.48	25	0.1

A large number of big buildings in Japan have their own treatment facilities to reuse wastewater as flush water for toilets, air-conditioning water and floor cleaning water. The treatment facilities are usually designed to reduce BOD_5 and suspended solids up to 10 mg/L and 5–10 mg/L, respectively. Figure 5 shows several possible treatment methods applicable for this purpose. Figure 6 presents a complete tertiary treatment process consisting of activated sludge followed by sand filtration together with primary and secondary treatment processes.

If flocculation with alum or iron salts is added prior to filtration, the removal of suspended solids and BOD will be improved while also achieving phosphorus removal. Flocculation can either be achieved in a separate tank prior to filtration (direct filtration, DRF) or in the filter

Table 7. Performance of CFMF with in-line flocculation and powdered activated carbon (PAC) addition (tubular ceramic membranes, pore size = 0.5 μm; cross flow velocity = 3 m/s; P = 69 kN/m²; influent COD = 125 mg/L) (Vigneswaran and Boonthanon, 1993).

Condition	Permeate flux ($m^3/m^2.h$)	Effluent COD (mg/L)
No flocculant	0.10	45
Alum	0.12	30
PAC (60 mg/L)	0.18	34
Alum (120 mg/L) + PAC (60 mg/L)	0.16*	32

*Filtration flux was constant throughout the filter run.

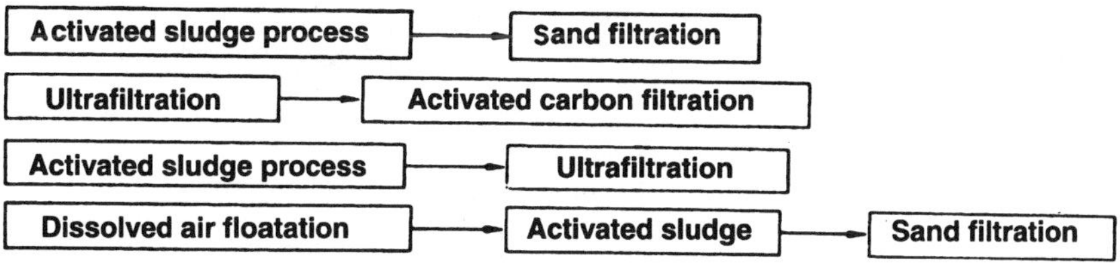

Figure 5. Treatment processes used in big buildings in Japan for the purpose of wastewater reuse (Fujita, 1989).

itself by direct feeding of flocculant. The latter arrangement is known as contact-flocculation-filtration (CFF) where both flocculation and filtration occur within the filter bed.

Direct Filtration (DRF)

Direct filtration refers to the flocculation for a short period followed by filtration without a sedimentation step (Figure 7). Advantages of direct filtration have been attributed to the ability to produce filterable flocs; the wide choice of filter media available; and the use of arrangements that permit higher solids storage capacity in "in depth" filtration without excessive headloss development. In DRF, as the entire solid-liquid separation takes place in the filter itself, the floc size becomes

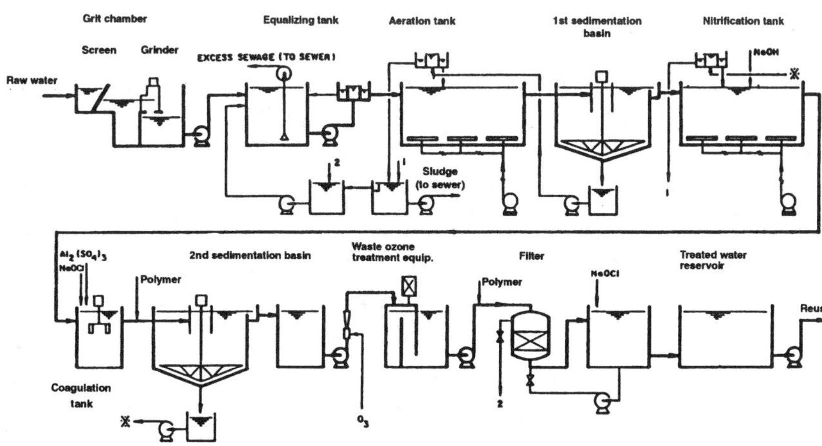

Figure 6. A schematic diagram of a complete treatment process consisting of activated sludge processe followed by sand filtration.

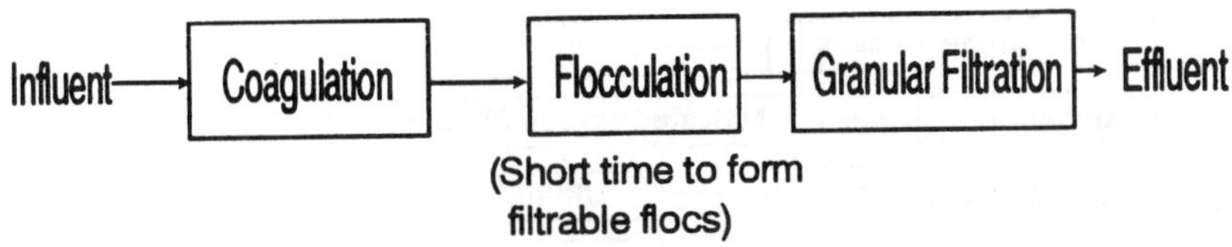

Figure 7. Process diagram of direct filtration.

very important. It is important to verify whether the flocs formed during the flocculation of short duration are appropriate for filtration. In other words, one should choose the appropriate flocculation conditions (flocculation time and mixing intensity or velocity gradient) and flocculant types so that the flocs are of a filtrable size range so as not to clog the filter bed. Information is needed on the effects of flocculant dose, velocity gradient, flocculation time, filter medium type/size and depth, filtration velocity, and particle size and concentration in the influent with respect to floc size, density, aggregate structure and filter run.

In sewage treatment, a majority of the suspended solids and the organics contributing to biochemical oxygen demand are removed in the primary and biological (secondary) treatment stages. The removal of the remaining solids, organics and nutrients (nitrogen and phosphorus), is important if the water is to be reused.

A detailed pilot-scale direct filtration study conducted with secondary effluent from a sewage treatment in NSW, Australia (Vigneswaran et al., 1994), indicated that tertiary filtration with prior flocculation is essential in achieving superior solids and phosphorus removal. A series of filtration runs were made and the best results were obtained with sand (effective size, ES = 0.85–0.95 mm; uniformity coefficient, UC = 1.25–1.50) as filter medium and 4.4 min of preflocculation. Alum at 40 mg/L was found to be the most suitable chemical (Table 8). A dual media filter produced comparable effluent quality (although slightly inferior) while leading to longer filter runs (e.g., eff. SS < 0.8 mg/L, eff. T-P < 0.06 mg/L, eff. fecal streptoccoci = 2 CFU/100 mL, headloss development after 24 h = 16.8 cm). Although fecal coliform removal was higher than 99%, it is important to have post-disinfection.

A series of laboratory-scale DRF studies were carried out on the secondary effluent from a sewage treatment plant (Ngo and Vigneswaran, 1995a, 1995b). The effects of alum dose, flocculation time and velocity gradient were studied. The best filter performance (e.g., PO_4-P < 0.12 mg/L) was found to be at the velocity gradient of $79.6 s^{-1}$, a flocculation time of 2.4 minutes and an alum dose of 40 mg/L. These

Table 8. Filter performance (flocculation time = 4.4 minutes; alum dose = 40 mg/L; filtration time = 24 hours).

Filter media	Headloss after 24 h (cm)	SS (mg/L)		Apparent color (PtCo unit)		Fecal coliform (CFU/100 mL)	
		Influent	Effluent	Influent	Effluent	Influent	Effluent
S/(100 cm)	37.6	9–12.8	1.21	68–95	13	1500–250,000	136
A/(65 cm) and S/(35 cm)	20.7	8–11.2	0.7	56–83	10	2800–12,000	41

Filter media/ (media depth)	Headloss after 24 h (cm)	Fecal streptoccoci (CFU/100 mL)		T-P (mg/L)		TKN (mg/L)	
		Influent	Effluent	Influent	Effluent	Influent	Effluent
S/(100 cm)		7–64	1	0.26–0.45	0.07	0.5	< 0.5
A/(65 cm) and	37.6	36–70	2	0.23–0.34	0.05	0.5–0.86	< 0.74
S/(35 cm)	20.7						

Note: S: sand (ES = 0.85–0.95 mm/U = 1.25–1.50); A: anthracite (ES = 1.8 mm/U = 1.25–1.50).

conditions produced flocs of size 58.7 μm. The floc size in the case of 60 mg/L alum was larger (94.6 mg/L). It led to significantly higher headloss development and resulted in the shorter filter run (Table 9).

Contact-flocculation Filtration (CFF)

In CFF, coagulation is followed immediately by filtration (Figure 8). In this process, both flocculation and solids separation take place directly within the filter media.

Table 9. Filter performance at different alum doses (v = 9.9 $m^3/m^2.h$; G = 79.6 s^{-1}; t_f = 4.8 min; L = 40 cm; sand [ES = 0.85 mm]; filtration time = 240 minutes).

Alum dose (mg/L)	Turbidity (NTU)		Suspended solids (mg/L)		PO_4-P (mg /L)	
	Influent	Effluent	Influent	Effluent	Influent	Effluent
40	12	1.4	26.4	3.6	0.47	0.11
60	31	4.3	34.8	9.0	0.49	0.05

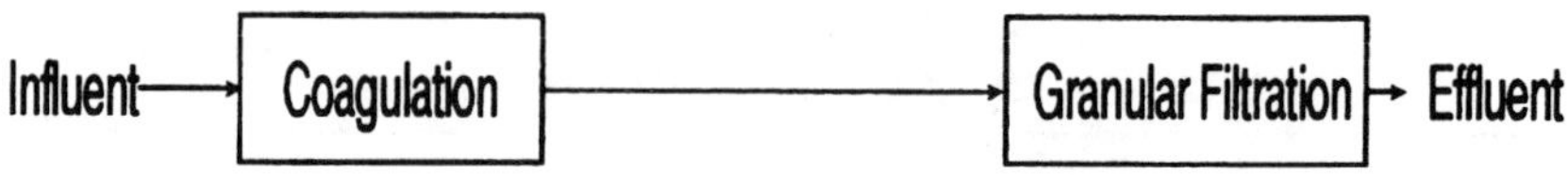

Figure 8. Process diagram of contact-flocculation filtration.

A pilot-scale CFF study was conducted with the effluent from a secondary treatment in Sydney (Vigneswaran et al., 1994). Dual medium (anthracite, ES = 1.8 mm; sand, ES = 0.85–0.95 mm) media filtration was used. The experiments with CFF arrangement, although it gave no reasonable removal efficiency, led to high headloss development and significant decline in flow rate (84% after 24 hours of filtration time). Although effluent phosphorus concentration was within standards, significant deterioration in bacteria and turbidity was observed at the end of 24 hours of filtration (Table 10).

Both DRF and CFF save capital costs since they eliminate most of the unit operations used in a conventional treatment system. However, they suffer from the drawback that the filter becomes clogged rapidly and frequent backwashing is required. To overcome this shortcoming, one could use a mobile bed filter with a CFF arrangement that incorpo-

Table 10. The filter performance of CFF (alum dose = 40 mg/L; filtration time = 24 hours).

Filter media/ (media depth)	Headloss after 24 h (cm)	SS (mg/L)		Apparent color (PtCo unit)		Fecal coliform (CFU/100 mL)	
		Influent	Effluent	Influent	Effluent	Influent	Effluent
A (65 cm) + S (35 cm)	33	5.6–11.6	2.8	59–106	27	2500–16,000	435

Filter media/ (media depth)	Headloss after 24 h (cm)	Fecal streptoccoci (CFU/100 mL)		T-P (mg/L)		TKN (mg/L)	
		Influent	Effluent	Influent	Effluent	Influent	Effluent
A (65 cm) + S (35 cm)	33	30–1000	7	0.16–0.29	0.06	0.5–1.2	< 0.5

Note: S: sand (ES = 0.85–0.95 mm/U = 1.25–1.50); A: anthracite (ES = 1.8 mm/U = 1.25–1.50)

rates continuous washing of sand prefilter, or a prefilter (such as a floating medium filter) with a CFF arrangement.

Mobile Bed Filter

A mobile bed filter normally operates on a continuous basis whereby sand media slowly fall down or move up at an angle, depending on the filter design across which the wastewater passes (Figure 9). Wastewater

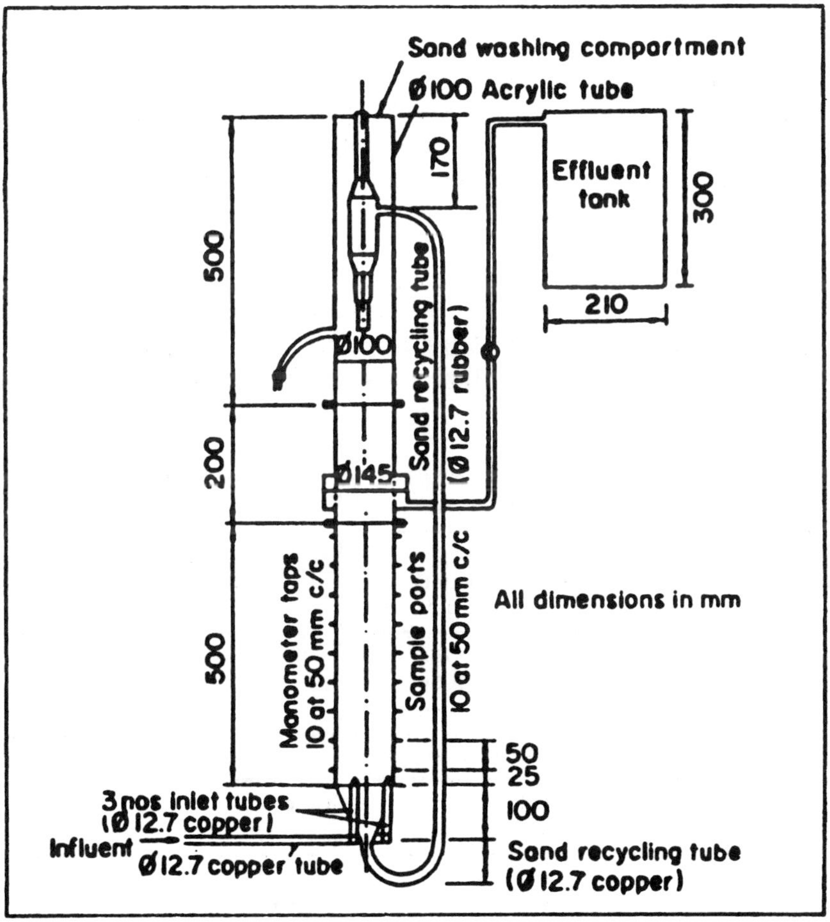

Figure 9. Sectional view of mobile bed filter.

flow direction can be upwards, downwards or radical depending on the design. As the suspended solids are removed, the sand gets dirtier until it reaches the bottom from where it is recycled by air lift or by hydraulic means to a cleaning system at the top of the column. After washing, sand falls to the top to recommence its downward movement. In mobile bed filter, the most important operating parameters are the sand recycling and sludge withdrawal rates, which determine the removal efficiency and economy of operating the filter.

From the mobile bed filter experiments, the sand recycle rate of 190 mL/min and sludge withdrawal rate of 160 mL/min were found to be the most effective, giving the highest removal efficiency and steady performance (Vigneswaran, 1991).

Removal efficiencies in term of COD, T-P and turbidity at different filtration rates are given in Table 11. Headloss was higher for higher flow rates but it was still only 15 cm. There was no headloss development and it remained almost constant throughout the run. Therefore it is economical to operate at a higher filtration rate thus saving capital cost.

A set of experiments was also conducted to find the effect of polyelectrolyte addition on removal efficiency. Results showed that there was not much improvement by adding polyelectrolyte. From the results, it can be noticed that the turbidity and phosphorus removal efficiencies

Table 11. Performance of MBF (influent turbidity = 46 NTU; COD = 54 mg/L; T-P = 2 mg/L).

Exp run	Filtration velocity (m/h)	Sand recycle rate (%/min)[a]	Sludge withdrawal rate (cm^3/min)	Flocculant addition
A	2.5	0.756	160	*
B	5.0	0.756	160	*
C	7.0	0.756	160	*

Average removal efficiency

Exp run	Turbidity effluent (NTU)	COD effluent (mg/L)	T-P effluent (mg/L)
A	2.0	17.7	0.7
B	1.0	16.6	0.5
C	6.9	18.4	0.2

Note: 150 mg/L alum + 0.5 mg/L zetag-32; [a] calculated in terms of % of filter sand volume recycled/min.

were sufficient but that removal of dissolved organics and nitrogen was not sufficient for effluent reuse.

Although the mobile bed filter produces an effluent of good quality with no headloss development, the use of an air compressor to recycle the sand to the top of the layer of filter medium results in an additional energy cost. Based on the results, a sample calculation was made to design a mobile bed filter for a small population of 250 people with per capita sewage production of 250 L/d. This calculation led to a daily energy requirement of 45 kWh. This cost is not high compared to the amount of filtered water necessary for backwash in conventional sand filters (appropriately 3% of daily water production per each backwash). Further, a mobile bed filter is continuous in operation and requires little or no supervision. A number of commercial mobile bed filtration units used in water treatment such as the Dynasand filter, with modification, can be successfully used in wastewater treatment for reuse.

To obtain better quality effluents with respect to nitrogen and dissolved organics, mobile bed filter effluent was posttreated with zeolite ion exchange and activated carbon adsorption, respectively. Table 12 shows the quality of mobile bed filter effluent after posttreatment (Vigneswaran and Bandulahera, 1990).

Floating Medium Filters

In deep bed filtration, the type and size of the filter media are the most important parameters for increasing the retention of particles within the filter bed, for achieving superior effluent quality and for obtaining longer filter run times. In conventional deep bed filters, the filter media usually used are sand (single medium), anthracite and sand (dual media) or anthracite, sand and garnet (multilayer). To fluidize these highly dense media during backwashing, a large quantity of water at a high backwash rate is required. Normally, the water used for backwashing may be up to 5% of the total daily water production.

Table 12. Performance of mobile bed filter with posttreatment.

Post treatment	Turbidity (NTU)	COD (mg/L)	Dissolved COD (mg/L)	T-P (mg/L)	TKN (mg/L)
Zeolite ion exchange	1.0	16.6	—	0.4	0.2
Activated carbon adsorption	0.5	11.8	7.2	0.1	—

Backwash requirements can become very high in the case of direct filtration (more specifically in contact flocculation/filtration) where flocculants are added to achieve flocculation, and the entire solid-liquid separation is within the filter bed itself. With the objective of reducing the backwash requirement, the use of synthetic buoyant filter materials (less dense than water) has been proposed for flocculation and filtration in water and tertiary wastewater treatment (Ben Aim et al., 1993; Ngo and Vigneswaran, 1994, 1995a). The backwashing of filters of this type can be achieved with a small quantity of water at a much smaller backwash velocity. Apart from this advantage, floating filter media are also found to have a high solids retention capacity and low headloss development. Hence the use of floating filter media in direct filters

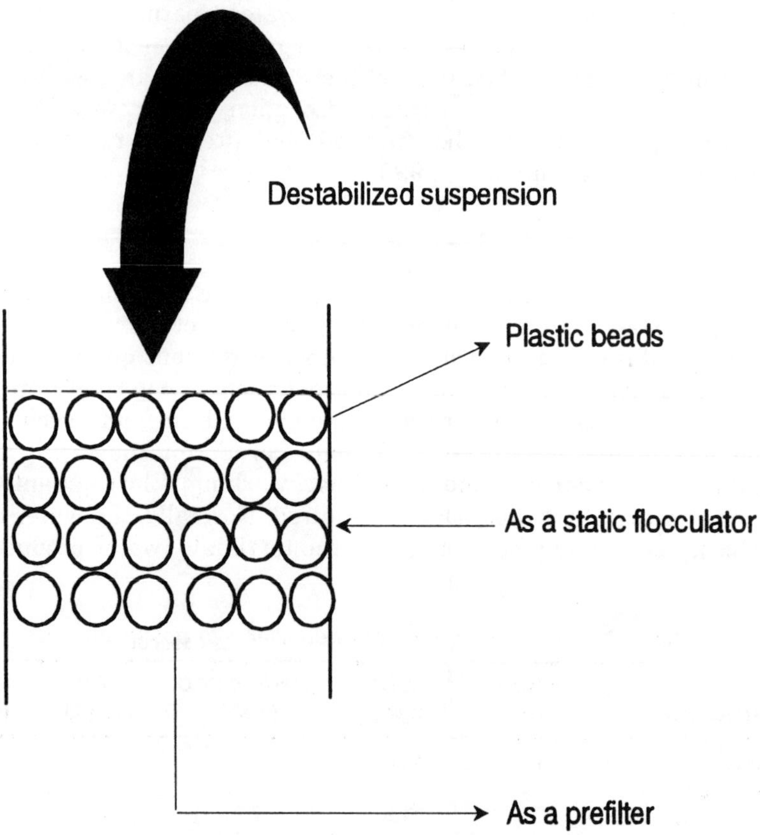

Figure 10. The concept of the DFF system.

Table 13. Performance of the DFF unit at different filtration velocities (polyaluminium chloride dose = 59 mg/L; depth of DFF = 40 cm; filtration time = 240 min).

	Filtration velocity = 5 $m^3/m^2.h$			
Treatment unit	T - P removal (%)	NH3-N removal (%)	Average floc size removal (μm)	Max. headloss development (cm)
DFF	89	80.7	41.6	5.1
	Filtration velocity = 7.5 $m^3/m^2.h$			
Treatment unit	T - P removal (%)	NH3-N removal (%)	Average floc size removal (μm)	Max. headloss development (cm)
DFF	78.5	77.3	39.2	6.0
	Filtration velocity = 9.9 $m^3/m^2.h$			
Treatment unit	T - P removal (%)	NH3-N removal (%)	Average floc size removal (μm)	Max. headloss development (cm)
DFF	61.1	75.3	32	6.5

(where a contact flocculation/filtration arrangement is adopted) is a promising solution. It overcomes the disadvantages of direct filtration, such as limited retention capacity, shorter filter runs (lengths of filtration cycle) and high energy requirement for backwashing.

A detailed study with wastewater using a saturated downflow polypropylene medium flocculator/prefilter showed that this system has a good pollutant removal capacity. Further, it led to very low headloss development and produced uniform microflocs (26–40 μ) suitable for direct filtration (Ngo and Vigneswaran, 1995a,b). The floating medium filter tested is a direct filter that operates in a downflow mode under constant head pressure and incorporates in-line flocculation. Coagulated wastewater enters the filter at the top where it comes into contact with the floating medium. Flocculation occurs in the floating medium due to the promotion of interparticle contacts by the water flow around individual grain of media. This is followed by the separation of particles and flocs by a floating filter medium. Thus, it has a dual function of flocculation and solid-liquid separation (Figure 10). The performance of a floating medium filter with a bed depth of 40 cm at different filtration velocities (Table 13) shows that floating medium filter itself could remove a significant amount of nutrients and solids with negligible headloss development.

The combined system of downflow floating medium and coarse sand filter (DFF-CSF) was developed and tested successfully with the effluent from a secondary treatment unit of a sewage treatment plant. In this system (Figure 11), the floating medium filter was used as a flocculator/prefilter while the sand filter served as a subsequent polishing filter (Ngo and Vigneswaran, 1996).

This system has been assessed in terms of nutrient and bacterial removal in a wastewater treatment plant. The results indicated that due to the ability of DFF to produce uniform microflocs and give a reasonable nutrient removal with low headloss development, the DFF unit can successfully be used as a flocculator/prefilter in nutrient removal and the performance of the DFF-CSF was remarkable in terms

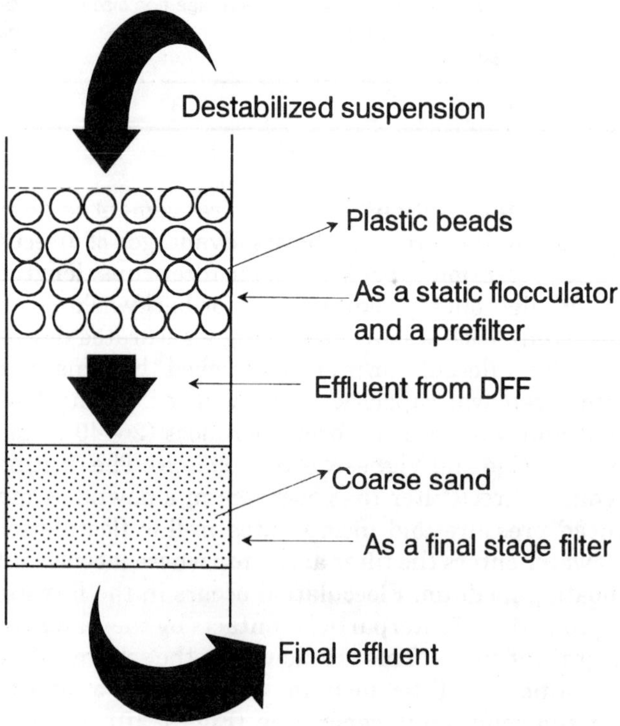

Figure 11. The concept of combined DFF-CSF system.

Table 14. Performance of the DFF-CSF unit at different filtration velocities (polyaluminium chloride dose = 59 mg/L; depth of DFF = 40 cm and depth of CSF = 40 cm; filtration time = 240 min).

	Filtration velocity = 5 $m^3/m^2.h$			
Treatment unit	T - P removal (%)	NH3-N removal (%)	SS removal (%)	Max. headloss development (cm)
DFF + CSF	94.2	87.9	96.6	13.3
	Filtration velocity = 7.5 $m^3/m^2.h$			
Treatment unit	T - P removal (%)	NH3-N removal (%)	SS removal (%)	Max. headloss development (cm)
DFF + CSF	86.1	84	89.1	14.1
	Filtration velocity = 9.9 $m^3/m^2.h$			
Treatment unit	T - P removal (%)	NH3-N removal (%)	SS removal (%)	Max. headloss development (cm)
DFF + CSF	69.5	77.9	77.5	22

of pollutant removal (Table 14). It not only produced superior effluent quality that meets stringent effluent standards but it was also a cost-effective solution for small sewage treatment plants. This system is very compact and requires less energy for backwashing.

The performance of this filter system in terms of bacterial removal was also studied using alum as a flocculant (Peiris et al., 1997). The filter system was effective in removing more than 99% fecal coliforms and streptoccoci when operated at a filtration velocity of 12.5 $m^3/m^2.h$ (Table 15).

CONCLUSIONS

Membrane bioreactors produce a high quality effluent of recycle quality. The process is compact due to high biomass concentration and can successfully be used in wastewater systems in buildings located in areas where water resources are scarce. On the other hand, deep bed filtration finds an important role in small communities and large water

Table 15. Mean percent bacteria removal for various filtration velocities.

Filtration velocity ($m^3/m^2.h$)	Mean % removal (fecal coliforms)	Mean % removal (fecal streptoccoci)
7.5	92.17	91.52
10.0	90.68	90.83
12.5	92.10	90.62

reuse systems. Deep bed filtration is often used with other physico-chemical systems such as flocculation, sedimentation, adsorption and biological oxidation processes. Mobile bed and floating medium filtration systems can be very compact and cost effective as they operate at a high rate and eliminate the use of separate flocculation and sedimentation processes.

REFERENCES

Ben Aim, R., A. Shanoun, C. Visvanathan, and S. Vigneswaran. 1993. New filtration media and their use in water treatment. *Proceedings of the 6th World Filtration Congress,* 18–21 May, Nagoya, pp. 273–276.

Chiemchaisri, C. C. 1993. Enhancement of organic oxidation and nitrogen removal in membrane separation bioreactor for domestic wastewater treatment. Ph.D. Thesis, U. Tokyo, Japan.

Chiemchaisri, C., K. Yamamoto, and S. Vigneswaran. 1993. Household membrane bioreactor in domestic wastewater treatment. *J. Water Sciences and Technology* 27:171–178.

Fane, A. G. 1994) Membrane fouling mechanism in membrane-coupled bioreactor. Korea-Australia Joint Symposium on Sensor and Membrane Tech., Seoul, pp. 229–245.

Fujita, K. 1989. Application of filtration in water and wastewater treatment. In: *Water, Wastewater, and Sludge Filtration.* S. Vigneswaran and R. Ben Aim (Eds.). CRC Press, U.S., pp. 77–100.

Irwin, J. 1990. On site wastewater reclamation and recycling. *J. of Water Environment & Technology* 90–91.

Manem, J. and R. Sanderson. 1996. Membrane bioreactors. *Water Treatment Membrane Processes.* AWWARF/Lyonaise des Eaux/WPR, Chapter 17, pp. 17-1–17-31.

Memtec Technical Bulletin, A 2049, Windsor, Australia.

Ngo, H. H. and S. Vigneswaran. 1994. Application of floating medium filter in organic removal. *J. IAME* 21(3):55–62.

Ngo, H. H. and S. Vigneswaran. 1995a. Application of floating medium filter in water and wastewater treatment with contact-flocculation filtration arrangement. *J. Water Research* 29(9)2211–2213.

Ngo, H. H. and S. Vigneswaran. 1995b. Floating medium downflow flocculator with coarse sand filter—A system for a small community water. *J. of Water,* Aust. Water & Wastewater Association 22(3):34–37.

Ngo, H. H. and S. Vigneswaran. 1996. Application of downflow floating medium flocculator/prefilter (DFF)-coarse sand filter in nutrient removal. *J. of Water Science and Technology* 33(8):63–70.

Peiris, P., J. J. Bailey, H. H. Ngo, and S. Vigneswaran. 1997. Bacterial removal from secondary effluent sewage effluent by a combined downflow floating medium flocculator/prefilter and sand filter. *J. IEMS* 5(4):217–222.

Roulet, R. 1989. The treatment of wastewater using an activated sludge bioreactor coupled with an ultrafiltration module. *Proceedings of Workshop on Selected Topics on Clean Technology,* Asian Institute of Technology, Bangkok, pp. 171–179.

Vigneswaran, S. and R. Bandulahera. 1990. Applicability of mobile bed filtration in domestic wastewater reuse. CIDA-AIT Project Report, Asian Institute of Technology, Bangkok.

Vigneswaran, S. 1991. Application of deep bed filtration in tertiary wastewater treatment. *J. of Water,* Aust. Water & Wastewater Association 19(3):32–34.

Vigneswaran, S. and S. Boonthanon, S. 1993. Cross-flow microfiltration with in-line flocculation. *J. of Water,* Aust. Water & Wastewater Association 19(1):29–31.

Vigneswaran, S., H. H. Ngo, S. Maheswaran, E. Ostarcevis, and A. Potter. 1994. Improved phosphorus and bacterial removal in sewage treatment plant: A pilot-scale study. *J. IAEM* 21(3):38–54.

Vigneswaran, S., H. H. Ngo, C. L. Hwang, C. Chiemchaisri. 1996. Membrane hybrid systems in water reuse. *Watertech 1996,* Sydney, pp. 354–359.

7

Opportunities and Problems of Wastewater Re-use

NARA M. SOMARATNE
Ministry of Water Resources
P.O. Box 2575, Ruwi 112
Sultanate of Oman

MUSHTAQUE AHMED
Department of Soil and Water Sciences
Sultan Qaboos University
P.O. Box 34, Al-Khod 123
Sultanate of Oman

INTRODUCTION

AS THE POPULATION GROWS, the average amount of renewable freshwater available to each person declines, resulting in water stress and scarcity. Although two-thirds of the planet are covered with water, the renewable freshwater is an increasingly scarce commodity (Engelman and LeRoy, 1993). Throughout history, stable social and economic development of numerous cultures and civilizations has essentially resulted from secure access to water. Higher population size and improved standards of living have increased the demand for finite quantities of water and intensified competition and tension among users. If much of the world's urbanization, industrialization, and irrigation is characterized by unsustainable patterns of freshwater use, the present situation will inevitably become worse. The pressures of modern agriculture, industry, and life-styles combined with growing populations and standards of living are causing water tables and river levels to fall and contamination to spread. Water supplies in many cases are becoming unusable and aquatic ecosystems are disappearing. In other instances, water quality is declining or the reliance on nonrenewable water is growing. The need to balance the conflicting water interests of urban residents, farmers, industry, and the natural ecosystems is becoming evident in arid and semi-arid regions and sooner or later will have to be faced worldwide.

With the increase in population, previously dispersed rural/agricultural economies are transforming to concentrated urban economies. Consequently, the established agricultural water rights are being transferred to municipalities to satisfy the growing urban water demand. One of the main solutions to address this problem is the agricultural reuse of wastewater as suggested by Brown (1994). This solution allows municipalities to use agricultural water and then release it after treatment for subsequent agricultural reuse.

In recent years, there has been a growing interest in the use of wastewater for irrigation, especially in arid and semi-arid areas of both developed and developing countries. This shift is attributed mainly to the scarcity of alternate water for irrigation, the high cost of inorganic fertilizers and advanced treatment plants, the low health risk and damage to the soil system if necessary precautions are taken, acceptance by water resources planners, and increased sociocultural acceptance. In addition, general scarcity of water and the high cost of developing other forms of nonconventional water supplies are among other major factors.

Israel has recognized wastewater as a nonconventional source of water supply that can be developed and integrated within the regional water supply system. According to Kanarek and Michail (1994), there are 230 wastewater reuse projects in operation varying size, method of treatment, storage, and mode of operation. Many small farming communities employ waste stabilization ponds to treat their own wastewater and then utilize it for irrigation in their own fields. Others combine their wastewater with that from adjacent townships, thus considerably increasing the volume of wastewater for reuse. In Tunisia, irrigation with reclaimed wastewater is a well-established practice (Bahri, 1996), and citrus and olive orchards are mainly irrigated with treated wastewater. Similarly, Japan and Singapore have been using reclaimed wastewater to flush toilets in commercial and residential buildings for years (Engelman and LeRoy, 1993).

Despite the long history of wastewater reuse and recent technological advancements, the safety of wastewater reuse is still debated in many parts of the world. The presence of pathogenic microorganisms in the effluent wastewater is always a concern for environmental health professionals, due to the risks associated with human exposure to disease-causing organisms. Bacteria, viruses, nematodes, and protozoa are the four major types of pathogens that need to be inactivated or removed before wastewater discharge and reuse. This chapter examines the problems encountered and opportunities available in wastewater reuse, with particular reference to the irrigation application, the biggest consumptive user of diverted water in the world.

CHEMICAL ASPECTS OF IRRIGATING WITH WASTEWATER

One of the important influences on the chemical characteristics of a sewage effluent is the source of freshwater. If significant levels of elements such as calcium, iron, and sodium are present in the source, the wastewater produced after treatment will reflect these characteristics. The same is true for contaminants that find their way into the water source, directly or as a leachate, and those that accumulate during water use. Therefore the main chemical components of domestic sewage effluent are variables that are being affected by local influences. High relative concentration of certain elements in reused effluents such as boron and some heavy metals can cause toxic effects in plants.

Despite their variable nature, nitrogen (in the form of ammonia, nitrite, and nitrate), phosphate, potassium, sulphates, chlorides, sodium, boron, surfacants, and heavy metals are typical constituents of wastewater. The build up of sodium chloride, boron, and toxic metals from industrial wastes poses the greatest chemical limitations on effluent reuse. Salinization can decrease yields, especially for salt-sensitive plants. If organic matter is largely removed by pretreatment (primary and secondary sedimentation), the bulk of the metals, between 44% and 96% of total crude sewage metal load, will also be reduced. This can substantially increase the reuse potential of wastewater.

Nutrients

Although enhanced nutrient concentrations are one of the advantages of reusing wastewater, the chemistry is not simple, and there are several potentially deleterious aspects to consider. This is most significant in the case of nitrate. At low concentrations, nitrate is entirely beneficial. However, high concentrations accelerate the growth beyond the plant's capacity to synthesize fructose. This results in fruits that are large, but have poor flavor or insipid taste. Studies in the U.S. (Bouwer and Idelovitch, 1987) have shown that application rates in excess of 90 mg L^{-1} are detrimental to nitrate-sensitive crops, while concentrations in the range of 60–80 mg L^{-1} are more than sufficient to meet the nitrate demands of most crops. Such a range is commonly below that of raw water, which may also be richer in nitrate than many treated wastewaters. However, wastewater treatment plants of low denitrifying capacity present a significant risk of deterioration in crop quality. In addition, it is environmentally dubious to introduce high nitrate to a groundwater environment by way of irrigation leachate. A chopped straw biofilter (Diab et al., 1993), which is both cheap and efficient, may be appropriate for small-farm scale applications. At the other extreme, some municipal wastewaters are so rigorously treated to reduce nitrogen species that nitrogen fertilizers must be used at the end user stage. In such cases, there is an obvious need for dialogue between the interested parties in order to attain a mutually acceptable level of inorganic nitrogen delivery. As a general rule, a crop nitrate limit of 88 mg L^{-1}, equivalent to 20 mg L^{-1} of total nitrogen, is safe for irrigation water. However, the leachate fraction, irrigation rate, soil drainage characteristics, and individual crop tolerances are related variables, making it inappropriate to set too rigid guidelines. Applied research would be useful to ascertain empirical crop-nitrate effects.

Under optimum conditions, approximately half of the applied nitrate is biologically utilized, a quarter is lost as nitrogen gas following biodenitrification, and the remainder is leached to the groundwater. This fraction is increaased substantially if the applied nitrate is significantly greater than the crop nutrient requirement. For this reason, special attention should be given to the ultimate fate and disposal of irrigation leachate when using treated wastewater.

Unlike nitrate, phosphate is a highly nonconservative solute that is strongly affected by adsorption processes. It is concentrated in raw wastewater and largely adsorbed to the organic substrate of the treatment processes. Nevertheless, some phosphate enrichment is to be expected in treated wastewaters. This can be advantageous provided that the phosphate is not immediately lost by adsorption to the soil surface during flood irrigation. To avoid such loss (and for obvious water conservation reasons), subsurface microirrigation is the preferred irrigation method since it delivers phosphate along with water to the root zone with minimal losses. In addition to the vertical concentration of total phosphate in the upper root zone, there is some evidence that the lateral distribution may be significant for some crops, particularly for wheat since it grows better with a more uniform delivery of nutrients.

Many farmers have been dissuaded from using subsurface microirrigation with treated wastewater due to the high capital expense as well as fouling problems. In arid countries, fouling by iron-metabolizing bacteria may be less serious than elsewhere. However, the high temperatures tend to cause biofilm development, which results in hydraulic constriction and phosphate adsorption. For this reason, the treated wastewater should be chlorinated during the last hour or two of each irrigation cycle if the water is not already chlorinated. Slight potassium enrichment is likely during the primary water use of the wastewater treatment cycle, especially at the lower concentration range. Such enrichment is entirely beneficial to low to medium potassium soils and does not present any known problems.

Heavy Metals

The moderate to heavy alkalinity levels of natural and wastewaters exert strong negative solubility control on heavy metals, resulting in concentrations seldom exceeding the limits. Consequently, unless there is some input of acidic and/or heavy metal-contaminated industrial effluents, the likelihood of heavy metal problem is limited. Nevertheless, it would be prudent to make provisions for heavy metal analyses, on an irregular basis, in areas of higher than usual risk of contamination.

There are two areas where heavy metal contamination is likely to be concentrated. Studies in China have shown a marked enrichment in the root zone (Yediler et al., 1994). Given the rapid immobilization of heavy metals in alkaline soils, it is advisable to monitor a few microirrigation vents where roots are well developed. The second area, important from the consumer's viewpoint, is where heavy metals are concentrated in the seeds and other edible parts of many crops (Barman and Lal, 1994). However, risk of this form of contamination is extremely low.

Organic Chemicals

Phenolic compounds and other soluble oil derivatives (BTEX—Benzene, Toluvyn, Ethylene, and Xylene) have some capacity for transfer to crops, thereby imparting a slight deleterious taste. This is especially true when the finished water is chlorinated. However, adequately treated domestic wastewater is low in such organics while most, if not all, such organic compounds should be below organoleptic detection levels.

Biological Oxygen Demand (BOD)

The values of BOD range from zero in very well-oxygenated systems, to over 1000 mg L^{-1} in poorly treated or grossly overloaded sewage treatment plant (STP) outflows. The normal range of BOD does not have any significant effect on mature crops. High BOD levels, however, may cause poor germination of seeds or inhibit the early growth of seedlings in a variety of crops (Nashikar, 1994).

The BOD is often regarded as the primary index for assessing the quality of treated water. However, the correlation between BOD and some other deleterious parameters is low. Low or even zero BOD is no guarantee that pathogens and refractory organic pollutants have been reduced to safe levels (Juanico, 1993).

BIOLOGICAL ASPECTS

Following the World Health Organization (WHO) experts' meeting in Engelberg in 1985, it was revealed that the risks from irrigation with wastewater were smaller than previously thought. The bacteriological standards used were also found to be unjustifiably restrictive. However, the risks from helminths were highlighted. Health guidelines were discussed further in 1987 when Engelberg Report (1985) recommendations regarding acceptable levels of pathogenic contaminants were endorsed, advocating wider applications of effluent reuse.

Generally, fecal coliform counts are used as the standard indicator of fecal pollution and the biological quality standard for reuse. However, the use of *E. coli* for pollution quality monitoring has been considered inadequate in areas where helminths are present. In these situations, helminth egg numbers in the effluent are now considered to be the most important factor in assessing reuse potential and relative health risks (WHO, 1989). Sewage wastewaters can contain pathogens from four groups: bacteria, viruses, protozoa and helminths. The risks of infection are high for viruses, some salmonella, and helminths with relatively small level of contamination. When considering a wastewater usage for irrigation, knowledge of pathogens' survival time (Table 1) is crucial for safe use.

Shuval et al. (1986) quoted much longer survival times for helminths (many months), illustrating the need for caution when considering guidelines for safe use. Epidemiological studies have shown a direct link between pathogenic infections and unsafe irrigation practices or inappropriate crop use (Shuval et al., 1986; Feachem et al., 1983; WHO, 1989). These include direct contact and waterborne disorders such as diarrhea and dysentery. Also, they include those requiring an intermediate host such as schistosomiasis and malaria spread by water snails and mosquitoes, respectively.

Table 1. Survival times of excreted pathogens on crops at 20–30°C (adopted from WHO, 1989).

Pathogen	Survival Time (Days)
Viruses	
Enteroviruses	< 60 but usually < 15
Bacteria	
Fecal coliforms	< 30 but usually < 15
Salmonella spp	< 30 but usually < 15
Shigella	< 10 but usually < 5
Vibrio cholerae	< 5 but usually < 2
Protozoa	
Entamoeba histolytica cysts	< 10 but usually < 2
Helminths	
Ascaris lumbriocoides eggs	< 60 but usually < 30
Hookworm larvae	< 30 but usually < 10
Taenia saginata eggs	< 60 but usually < 30
Trichuris trichiura eggs	< 60 but usually < 30

Bacteria

Sewage wastewater treatment has traditionally been directed towards suspended sediments, BODs, and the presence of pathogenic organisms, as implied by fecal coliforms. Provided that solid organic residues are filtered out, the risk of "bacterial shielding" from oxidizing agents in finished waters is small. Any bacterial risk is usually associated with the subsequent regrowth of biofilms in the wastewater reticulation system. In such cases, bacterial shielding, and even vigorous regrowth, can occur especially in *Acinetobacter, Aerromonas,* and *Pseudomonas.* Some concern has been expressed about the detection of "viable but nonculturable *Salmonella spp,*" and possibly of *Vibrio cholerae,* in chlorinated wastewaters (Watkins and Cameron, 1991). Such concerns are real and should be investigated as the family enterobacteriaceae can cause killer diseases, such as typhoid and bacillary dysentry, especially in children. Such bacteria live in the enteric tract of mammals and include coliforms, vibrio cholerae, salmonella, and shigella.

Several factors strongly mitigate any bacterial health risk from irrigation reuse of wastewater. First, the high standard of secondary and tertiary treatment very strongly attenuates any residual bacterial population. Second, irrigation of dry surface soils strongly desiccates any exposed *E.Coli* and presumably bacteria in general. This feature has been demonstrated even in temperate climates (Hunter and MacDonald, 1991) and is especially true of hot climate. Finally, any residual bacteria are subjected to both desiccation and solar ultra violet (UV) disinfection in spray irrigation resulting in short survival times (relative to bacterial survival within soils). An example for fecal coliforms, given by Feachem et al. (1983), is "normally less than 15 days, but occasionally up to 30 days."

Viruses

Viruses and bacteria are of prime importance for potable water quality. However, risks of viruses in wastewater used for irrigation are less significant and are subordinate to the risks from helminths and some parasitic protozoans. This does not imply that viruses are not present. Many viruses are effectively immune from low levels of chlorination while smaller viruses may be immune to the filtration treatment. However, most viruses are readily killed by desiccation, high air temperatures, solar UV exposure, and low humidity or soil moisture. An example of the survival time on soil and crop surfaces for enteroviruses at up to 30°C is normally less than fifteen days, but occasionally up to sixty days (Feachem et al., 1983). A few viruses, such as *Hepatitis A*, have much higher survival rates in the environment. No unequivocal viral infection has been reported from agricultural workers using secondary or treated wastewater. This is despite the expectation that agricultural workers are at high risk of infection through spray irrigation (aerosol transmission).

Helminths

Helminth survival and transmission probably present the most significant risks in wastewater use. Helminths are pathogens that primary treatment works are least capable of removing. Moreover, helminth eggs are largely, if not entirely, immune to chlorination and are capable of survival for many months under warm and moist soil conditions. Conventional treatment can remove helminth eggs, but require very efficient filtration or multiple stabilization lagoons. A faster but more complex and expensive alternative is to use a single-stage lime-accelerated sedimentation system (Gambrill et al., 1992). With any such treatment, the risk of intestinal nematode egg infection, even for agricultural workers, is still low. The Engelberg guideline consists of one or less viable intestinal nematode egg (ova) per liter. Compared to raw sewage water, this is a reduction of 5 or more orders of magnitude, and hence may be regarded as a highly conservative limit.

Protozoan Parasites

A particular problem with pathogenic protozoans is their substantial resistance to chlorination and their ability to survive for long periods (in some cases up to a year) in warm moist environments. This is particularly true for *Cryptosporidium spp,* which is killed by dessication, but otherwise robust, with a thermo tolerence of up to 60°C. Hence any absence of index bacteria, such as *E. coli,* is not necessarily an indication of the absence of protozoans. Of the various cases of waterborne *Cryptosporidiosis* that have been reported, all were associated with drinking water (Smith, 1992). However, one would also expect spray irrigation to be a potentially infective pathway. The *Amoebiosis* diseases caused by the species of *Rhizopodes* are very common, especially in warm and humid developing countries. The most important of these is the *Entamoeba histoytica,* which destroys the erythrocytes and causes ulcerations and intestinal hemorrahage and can reach the brain, bones, and spleen. The remaining significant pathogens are *Naegleria fowleri, Entamoeba histolitica,* and *Giardia intestinalis,* none of which has high risk of infective pathways during agricultural use. On the other hand, the survival of some of these protozoans within moist soils could constitute environmental reservoirs from which infective pathways could evolve. This is most likely in the case of flood irrigation. Finally, protozoans can harbor and protect bacteria that would otherwise have been oxidized. Overall, it is particularly important to remove helminths, and to a lesser extent protozoans, from wastewater prior to land applications.

APPROPRIATE TREATMENTS

Table 2 summarizes the data of Feachem et al. (1983) in terms of orders of magnitude of reduction, R_k, of microorganisms. The parameter R_k represents the logarithm of the ratio of input to output concentrations of microorganisms. As discussed above, viruses can be disregarded for irrigation wastewater purposes. Therefore, some combination of treatments must be chosen for the appropriate removal of helminths, bacteria, and protozoans cysts.

Table 2. R_k factors expected from various treatment processes (after Feachem et al. 1983).

Treatment Process	Helminths	R_k Protozoans	Bacteria
Primary free sedimentation	0–2	0–1	0–1
Primary chemical sedimentation	1–3	0–1	1–2
Activated sludge	0–2	0–1	0–2
Biofiltration	0–2	0–1	0–2
Aerated lagoon	1–3	0–1	1–2
Oxidation ditch	0–2	0–1	1–2
Chlorination	0–1	0–3	2–6
Stabilization ponds	1–3	1–4	1–6
Effluent reservoir	1–3	1–4	1–6

Feachem et al. (1983) showed that the internationally acceptable Engelberg guidelines require an R_k of at least 4 for bacteria, and at least 3 for helminth eggs. Comparing that with the information in Table 2 clearly shows that the choice of final treatment for bacteria is between chlorination and wastewater stabilization ponds. Analogous comparisons for protozoans and helminths reveal that stabilization ponds are the preferable treatment for these pathogens.

The context in which treatment takes place is also important. In areas of high population density, for example, the risk of human infection from widespread reuse of water is higher than elsewhere and hence chlorination is the appropriate option. A supporting argument is that land costs in metropolitan areas are very high. Hence, the alternative of stabilization ponds, which requires much larger areas, would not be viable due to the high capital cost.

There is also the question of public expectations. In a rapidly developing country, there is a natural preference for fast, unobtrusive processing that would mitigate against extensive stabilization ponds. On the other hand, not all of the advanced wastewater treatments are necessarily the best for all circumstances. In rural and small urban settings where the availability of space is not a high priority, there are several advantages to the use of cascaded stabilization ponds for various reasons. First, they require little maintenance and are easy to operate. Second, they are cheap to install and do not incur the continued expense of chlorine dosing. Finally, they provide a higher attenuation level than chlorination for agricultural use where helminths and protozoans cysts are the major concern.

LEGISLATION AND STANDARDS

In many parts of the world, standards for water quality criteria have been developed for wastewater reuse. Of particular relevance are the reuse for irrigation; discharge to rivers, streams, and wadis; aquaculture and reservoirs; and groundwater recharge. The specified standards for treated wastewater may vary according to the planned end use. However, the adverse effects on human health, the soil-plant system, and other environmental considerations must be evaluated. The degree of treatment required varies according to the specific reuse application and associated water quality requirements. The simplest treatment systems involve solid-liquid separation processes and disinfection. More complex treatment systems involve a combination of physical, chemical and biological processes. Accordingly, the standards applicable to treated wastewater may also vary.

Summaries of selected parameters from some of the international microbiological standards are presented in Table 3. Table 4 gives the chemical standards of Oman, Jordan, and US EPA/FAO for wastewater reuse applications.

As can be seen from Table 3, microbiological criteria vary considerably. However, the recommended standards are generally less stringent than those of the U.S. Regulations provide a legal framework for reclaimed wastewater reuse in many countries. To be effective, however, regulations require close coordination among the authorities responsible for implementation of reuse projects and relevant authorities responsible for wastewater collection, treatment, and disposal. This responsibility includes operating the system, monitoring the application of water laws, regulations, and other policies related to wastewater reuse. These laws prohibit the use of raw water in agriculture and irrigation with reclaimed wastewater of any vegetables to be eaten raw.

CONSTRAINTS IN WASTEWATER REUSE

Apart from the sociocultural background of some of the Islamic countries, the most severe constraint regarding wastewater reuse is the financial cost of wastewater reuse projects. Periodic monitoring of the physical, chemical, and biological quality of wastewater is expensive and requires considerable input of resources. Common wastewater reuse categories and their potential constraints are listed in Table 5.

Application of unjustifiably strict microbiological standards for wastewater is another constraint. It may lead to the standards not

Table 3. Microbiological standards applied for wastewater reuse.

Country/State/ Organization	Reuse for	FC/100 mL	TC/100 mL	Helminth Egg/L	E.Coli/100 mL	FS/100 mL	Salmonella	Giardia Cysts	Entero Virus
WHO	Irrigation (A)	< 1000							
	Irrigation (B)								
U.S. guidelines	Agricultural reuse	< 200							
	Landscape	< 1							
California	Agricultural (unrestricted access)		< 2.2 and max < 23						
	Agricultural (restricted access)		< 23						
Arizona	Agricultural (unrestricted access)		< 1					< 1 per 40L	< 1 per 40L
	Agricultural (restricted access)		< 75%					< 1 per 40L	< 125 per 40L

Table 3 (continued).

Country/State/ Organization	Reuse for	FC/100 mL	TC/100 mL	Helminth Egg/L	E.Coli/100 mL	FS/100 mL	Salmonella	Giardia Cysts	Entero Virus
Florida	Agricultural reuse	< 200							
	Restricted access	< 1							
	Agricultural reuse, fire protection, flush	75% and max < 25							
Australia	Irrigation and discharge	< 200							
New South Wales	Irrigation (A)	< 300 and							
	Irrigation (B)	max < 2000							
	Residential and urban reuse	< 750 and max < 5000	< 10						
		< 1							< 2 per 50L
France	Irrigation (A)	< 1000	< 1						
	Irrigation (B)		< 1						
Cyprus	Irrigation (A)	< 50–80%							
	Irrigation (B)	max							
		< 100							
		< 3000 and max < 10000							

Continued

Table 3 (continued).

Country/State/ Organization	Reuse for	FC/100 mL	TC/100 mL	Helminth Egg/L	E.Coli/100 mL	FS/100 mL	Salmonella	Giardia Cysts	Entero Virus
Jordan	Irrigation (A)	< 20		< 1					
	Irrigation (B)	< 1000		< 1					
Tunisia	Ocean discharge								
	Agricultural		< 2000			< 1000	0 per 5L		
	reuse		< 2000	< 1		< 1000	0 per 5L		
South Africa	Irrigation (A)								
	Irrigation (B)	< 1000							
Kuwait	Irrigation (A)		< 100						
	Irrigation (B)		< 10000						
Israel	Irrigation (A)		< 12, 80%						
	Irrigation (B)		< 2.2, 50%						
			< 250						
Mexico	Restricted		< 10,000						
	Unrestricted	< 2000	< 1000						
Saudi Arabia	Irrigation (A)			< 2.2					
	Irrigation (B)			< 2.2					

Irrigation (A): Irrigation of crops likely to be eaten uncooked, sports fields, and public parks.
Irrigation (B): Irrigation of cereal crops, industrial crops, fodder crops, pastures and trees.
FC: Fecal coliform.
TC: Total coliform bacteria.
FS: Fecal streptococci.

Table 4. Comparison of existing chemical standards for wastewater used for irrigation in Jordan (Al-Lafi, 1996), Oman (MRME, 1993), and U.S. EPA (1992)/FAO (1979).

Parameter	Omani Standards[a] Class A	Omani Standards[a] Class B	Jordanian Standards[a] (for vegetables eaten cooked)	U.S. EPA[a]/FAO[a] Standards Long-term	U.S. EPA[a]/FAO[a] Standards Short-term[b]
BOD_5	15	20	150	—	—
COD	150	200	500	—	—
Phenol	0.001	0.002	0.002	—	—
Al	5.0	5.0	5	5	20
As	0.1	0.1	0.1	0.1	2
Ba	1.0	2.0	0.1	—	—
B	0.5	1.0	1.0	0.75	2.0
Cd	0.01	0.01	0.01	0.01	0.05
Co	0.05	0.05	0.05	0.05	5.0
Cr	0.05	0.05	0.1	0.1	1.0
Cu	0.5	1.0	0.2	0.2	5.0
Hg	0.001	0.001	0.001	—	—
Li	0.07	0.07	2.5	0.075	0.075
Mo	0.01	0.05	0.01	0.01	0.05
Ni	0.1	0.1	0.2	0.2	2.0
Pb	0.1	0.2	5.0	5.0	0.075
Va	0.1	0.1	0.1	0.1	1.0
Zn	5.0	5.0	2.0	2.0	10.0
SO_4^{-3}	400	400	1000	—	—
Se	0.02	0.02	0.02	0.02	0.02
F	1.0	2.0	1.0	1.0	15.0
PO_4-P	30	30	—	—	—
NO_3-N	50	50	50	—	—
NH_4-N	5	10	—	—	—
Na	200	300	230	—	—
CN	0.05	0.1	0.1	—	—
TDS	1500	2000	2000	500	2000

[a]Maximum permissible concentrations in mg/L.
[b]Less than 20 years of continuous cropping.
Class A: Vegetables likely to be eaten raw, fruits likely to be eaten raw and within two weeks of any irrigation, public parks, hotel lawns, recreational areas, areas with public access, lakes with public contact (except places that may be used for praying and hand washing).
Class B: Vegetables to be cooked or processed, fruit if no irrigation within two weeks of cropping, fodder, cereal and seed crops, pastures, and areas with no public access.

being enforced due to the strictness, the public health being jeopardized by the unregulated use of raw wastewater, and the unnecessarily high cost of treatment.

In order to provide a basis for establishing water quality guidelines for uses in agriculture, industry, and environmental protection, experimental

Table 5. Wastewater reuse categories and potential constraints.

Reuse Category	Potential Constraints
Agricultural irrigation	Effects of water quality, particularly salts, on soils and crops, as well as health risk.
Landscape irrigation	Public health concerns related to pathogens, surface, and groundwater pollution.
Industrial reuse	Reclaimed wastewater constituents related to scaling, corrosion, biological growth, and fouling. Public health concerns, particularly aerosol transmission of organics and pathogens in cooling water.
Groundwater recharge	Organic chemicals in reclaimed wastewater and their toxicological effects. Total dissolved solids, metals, and pathogens in reclaimed wastewater.
Recreational/environmental use	Health concerns of bacteria and viruses, eutrophication due to N and P.
Nonpotable urban uses	Public health concerns over pathogens transmitted by aerosols. Effects of water quality on scaling corrosion, biological growth, and fouling.
Potable reuse	Organic chemicals in reclaimed wastewater and their toxicological effects. Esthetics, public acceptance, and public health concerns on pathogens transmission, including viruses.

programs and research are needed to determine the effects of reclaimed wastewater use in localities. As shown in Table 5, it is often necessary to find answers to the following questions: What are the crops that might be produced under reclaimed wastewater? What are the effects of the use of reclaimed wastewater on soil, crop and groundwater and what are the chemical and microbiological levels of contaminations? What recommendations can be formulated for agricultural use of wastewater considering the results acquired? Is it possible to blend with existing groundwater?

CONCLUSION

In many arid areas, wastewater can be regarded as a valuable resource under certain conditions. The use of treated wastewater is contingent upon protecting against pollution and affording maximum safeguard for the environment and the health and welfare of the people against waterborne diseases. In many arid countries, recycling and reuse of wastewater is probably the only significant and sustainable source of water yet to be fully utilized. In areas with sewerage and treatment facilities, the potential of using treated wastewater for reuse at a minimum additional cost is significant. However, only large urban arid areas in the more advanced countries have the necessary facilities. New infrastructure including pumping stations, treatment facilities, storage installations, and large networks of pipelines needs to be built before large volumes of treated wastewater can be supplied.

Multiple benefits result from wastewater recycling. Apart from increasing the supply of water, protection of the environment and providing nutrients for the soil are significant benefits. The safety concerns of wastewater reuse are of immense importance from the standpoint of public health. The potential risks include pathogenic contamination of drinking water sources, human exposure to disease-causing organisms, and pollution of groundwater. Microbiological and chemical standards are currently applied in several countries. However, conservative standards remain a barrier to rapid growth of wastewater reuse because the building of treatment plants that meet these strict standards requires very expensive, technologically advanced equipment, and high operational costs.

Concerted efforts are required to further the philosophy of wastewater reuse. Efforts will bear fruit if the risks associated with wastewater reuse are minimized to such levels that the public will have full confidence in the process. Economic considerations play an important role in whether such efforts are made as it is technically possible to produce water of any desired quality from any water source. Country-specific experimental programs and research will help achieve better understanding and instill a greater level of confidence in the problem solving and risk minimization of wastewater re-use resulting in better design, management, and operation of individual plants.

REFERENCES

Al-Lafi, M. 1996. Legislations, regulations and standards of wastewater re-use in Jordan. Workshop on Wastewater Treatment and Re-use in Jordan, Aqaba, Jordan.

Bahri, A. 1996. Reclaimed wastewater re-use in Tunisia: Achievements and constraints. Workshop on Wastewater Treatment and Re-use, Aqaba, Jordan.

Barman, S. C. and M. M. Lal. 1994. Accumulation of heavy metals in soils and cultivated vegetables and weeds grown in industrially polluted fields. *J. of Environmental Biology* 15(2):107–115.

Bouwer, H. and E. Idelovitch. 1987. Quality requirements for irrigation with sewage water. *J. of Irrigation and Drainage Engineering* 113(4):516–535.

Brown, D. P. 1994. Integrated watershed management in the South Platte basin: Status and practical implementation. *Proceedings of the 1994 South Platte basin,* Greeley, Colorado.

Diab, S., M. Kochba, and Y. Avnimelech. 1993. Development of a biofilter for turbid and nitrogen rich irrigation water. *Bioresource Technology* 44(2):137–140.

Engelberg Report. 1985. Health aspects of wastewater and excreta use in agriculture and aquaculture. IRCWD News No: 23, Dübendorf, Switzerland.

Engelman, R. and P. LeRoy. 1993. Sustaining water: Population and future of renewable water supplies. Population and Environmental Program. Population Action International.

FAO. 1979. Water quality for irrigation. Irrigation and Drainage Paper No: 29. Food and Agriculture Organization, Rome, Italy.

Feachem, R. G., D. J. Bradley, H. Garslick, and D. D. Mara. 1983. *Sanitation and Disease: Health Aspects of Excreta and Wastewater Management.* John Wiley and Sons, New York, U.S.

Gambrill, M. P., D. D. Mara, and A. A. Silva. 1992. Physicochemical treatment of tropical wastewaters: Production of microbiologically safe effluents for unrestricted crop irrigation. *Water Science and Technology* 26(7–8):1449–1458.

Hunter, C. and A. MacDonald. 1991. The occurrence of coliform bacteria in the surface soils of two catchment areas in the Yorkshire Dales. *J. of Irrigation Water Engineering and Management* 5:534–538.

Juanico, M. 1993. Alternative schemes for municipal sewage treatment and disposal: A case study. *Ecological Engineering* 2(2):101–118.

Kanarek, A. and M. Michail. 1994. *Water Supply* 12(1-2):1R9-17–IR9-19.

MRME. 1993. Regulations for wastewater re-use and discharge. Ministry of Regional Municipalities and Environment, 145/93, Sultanate of Oman.

Nashikar, V. J. 1994. Effects of domestic wastewaters with various BOD levels on the germination and early seedling growth of some crops. *Environmental Ecology* 12(3):507–509.

Shuval, H. I., A. Adin, B. Fattal, E. Rawitz, and P. Yekutiel. 1986. Wastewater irrigation in developing countries: Health effects and technological solutions. World Bank Technical Paper No: 51, The World Bank, Washinton, D.C.

Smith, H. V. 1992. Cryptosporidium and water: A review. *J. of Institution of Water Engineering and Management* 6 (Aug):443–450.

US EPA. 1992. Guidelines for water re-use. United States Environmental Protection Agency EPA/625/R-92/004, Cincinnati, Ohio.

Watkins, J. and S. A. Cameron. 1991. Recently recognised concerns in drinking water microbiology. *J. of Institution of Water Engineering and Management* 5 (Dec):624–628.

WHO. 1989. Scientific group on the health aspects of use of treated wastewater in agriculture. WHO Technical Report Series, Geneva, Switzerland.

Yediler, A., P. Grill, T. Sun, and A. Kettrup. 1994. Fate of heavy metals in a land treatment system irrigated with municipal wastewater. *Chemosphere* 28(2):375–381.

8

Wastewater Disinfection: Assessment of Available Technologies for Water Reclamation

VALENTINA LAZAROVA

Center of International Research for Water and Environment (CIRSEE)
Lyonnaise des Eaux
38, rue du Président Wilson
78230 Le Pecq, France

INTRODUCTION

MUNICIPAL WASTEWATER DISINFECTION guarantees environmental and public health protection. Such treatment is generally used for a selective elimination, inactivation or destruction of disease-causing organisms such as bacteria, viruses and parasites (protozoan cysts and helminth eggs). Nondisinfected effluent or effluent from some stabilization pond systems may be suitable for certain limited nonpotable uses. However, for all reuse applications where direct or indirect human contact with reclaimed water is possible, an appropriate disinfection step is essential.

Although conventional wastewater treatment plants (predominantly activated sludge systems) remove up to 3 log units of pathogenic bacteria and viruses and up to 0.5 log units of protozoa cysts, the effluent's microbial contamination remains high (Table 1). Consequently, some surface waters and coastal areas are subject to microbial contamination and pose periodic public health and epidemiological risks. Microbiological parameters are not systematically included in wastewater treatment standards. However, legislative restrictions exist in many countries for wastewater discharge into bathing zones or shellfish farming areas. Additional rules and regulations apply to microbiological quality before wastewater reuse, reclamation and recycling.

Table 1. Microorganism concentration in raw wastewater and typical percentage removal by biological treatment systems (adapted from Crook et al., 1994).

Organism	Concentration (number/100 mL)	Secondary Treatment, % removal Primary	Activated Sludge	Trickling Filter
Fecal coliforms	10^4–10^9	< 10	0–99	85–99
Fecal streptococci	10^4–10^6	< 10	0–99	85–99
Salmonella	10^2–10^4	0–15	70–99	85–99
Shigella	1–10^3	< 15	80–90	85–99
Helminth ova	1–10^3	50–98	limited	60–75
Cysts and parasites	0–10^4	< 50	limited	limited
Enteric virus	10^2–10^5	limited	75–99	0–85

Wastewater disinfection levels are determined by standards or recommendations, which are specific to each country or region. For example, regarding fecal coliforms (FC) for bathing water standards, France requires 2000FC/100 mL (WHO standards), while 200FC/100 mL are required in Australia and recommended in the U.S. (EPA). For shellfish farming water, the values vary between 300FC/100 mL in France and 14FC/100 mL in the U.S. Some U.S. states have even more stringent regulations.

Municipal wastewater reuse standards for irrigation are generally also country- and crop-specific. Table 2 lists some of the more stringent standards for irrigation and urban applications (Levine et al., 1997). New water quality parameters have appeared (e.g., for viruses and

Table 2. List of some wastewater reuse standards (Levine et al., 1997).

Country/Region/State	Reuse Application	Indicators	Guide Value
WHO (France, Tunisia,	Irrigation (Category A)	Fecal coliforms/100 mL	< 1000
Argentina, etc.)	(Category B)	Nematodes/L	< 1
Australia	Irrigation	Fecal coliforms/100 mL	< 300
	Urban applications	Fecal coliforms/100 mL	< 1
		Viruses/50 L	< 2
		Parasites (cysts)/50 L	< 1
Japan	Urban applications	Total coliforms/100 mL	< 1
U.S., EPA	Irrigation I	Fecal coliforms/100 mL	< 200
	Irrigation II*	Fecal coliforms/100 mL	< 1
Florida, U.S.	Urban applications	Fecal coliforms/100 mL	< 1

*Irrigation of parks, gardens or crops that can be eaten raw.

cysts, Australia). In order to directly or indirectly reuse wastewater effluents while complying with these standards, wastewater disinfection becomes necessary.

This chapter provides a classification of existing disinfection technologies, with an emphasis on the development of modern disinfection practices based on advanced treatment processes. It also covers the factors influencing disinfection efficiency. Some design parameters are also provided. Finally, a cost evaluation and feasibility assessment of advanced disinfection technologies, based on various French projects, are discussed and compared with chlorination/dechlorination, the most widespread technology.

CLASSIFICATION OF DISINFECTION PROCESSES

Throughout the world, various disinfection technologies are used for municipal wastewater disinfection (Figure 1). Three main groups of processes can be distinguished: extensive treatment technologies such as tertiary lagooning (maturation ponds), infiltration-percolation and soil-aquifer treatment; and intensive technologies such as ozonation, UV irradiation, low-pressure membrane filtration or other processes;

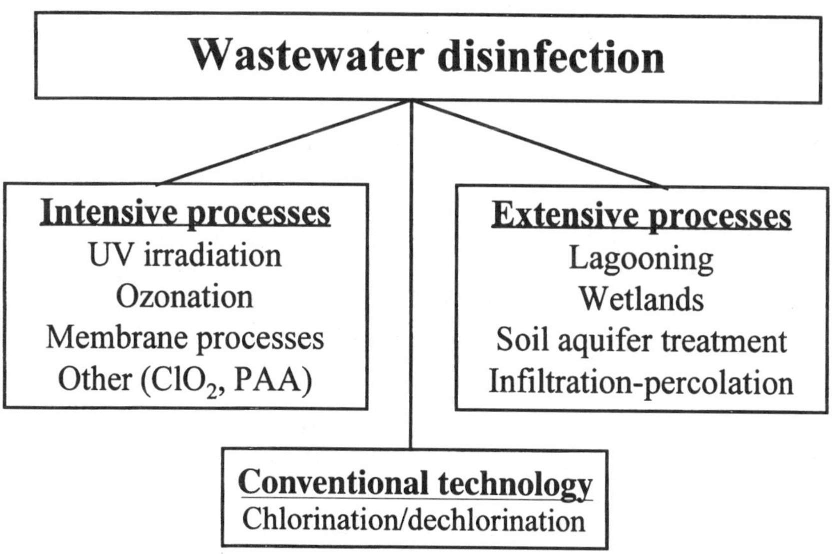

Figure 1. Classification of wastewater disinfection processes.

Table 3. Recommended treatment schemes in relation to wastewater reuse applications (adapted from Shelef and Azov, 1996).

Type of Reuse Application	Extensive Treatment	Intensive Treatment
1. Irrigation of restricted crops	E.1. Stabilization ponds in series or aerated lagoons; wetlands; infiltration-percolation	I.1. Secondary treatment by activated sludge or trickling filters with or without disinfection
2. Irrigation of unrestricted crops, vegetables eaten raw	E.2. Idem as E.1. with polishing steps and reservoirs	I.2. Idem as I.1 with filtration and disinfection
3. Urban uses for irrigation of parks, sport fields, golf courses	E.3. Idem as E.2	I.3. Idem as I.2
4. Groundwater recharge	E.4. Idem as E.2, or soil aquifer treatment	I.4. Idem as I.2 with nutrient removal (when necessary)
5. Dual distribution for toilet flushing		I.5. Idem as I.3 with activated carbon (when necessary), or MBR and disinfection
6. Indirect and direct potable use		I.6. Secondary, tertiary and quaternary treatment, including activated carbon, membrane filtration and advanced disinfection

and the conventional process of chlorination/dechlorination. The choice of the appropriate disinfection process depends on wastewater quality after secondary treatment, category of reuse application, existing water reclamation standards, and local, urban, economic and geographical conditions. Table 3 shows several recommended treatment schemes for the main groups of reuse applications. The most common type of reuse in arid and semi-arid regions is still the irrigation of agricultural crops and urban areas (parks, gardens, golf courses).

EXTENSIVE DISINFECTION PROCESSES: PRINCIPLE, DESIGN PARAMETERS AND FIELD OF APPLICATION

Wastewater stabilization pond treatment, also known as lagooning, is a low-technology system that has been practiced for over 3000 years. It reproduces, in a controlled environment, the natural purification and disinfection processes found in lakes and streams. Sunlight is the principal germicide factor. Different pond types have been developed to remove pathogens, organic matter, nutrients, color and heavy metals.

Over the last decade, an increased number of studies conducted in different countries (Shereif et al., 1995; El Hamouri et al., 1996; Grimason et al., 1996; Rångeby et al., 1996) have shown that stabilization pond systems in series can produce effluent with microbiological water quality suitable for unrestricted irrigation (Table 2, WHO guidelines category A, < 1000 FC/100 mL and < 1 helm. egg/L). The hydraulic residence time varies between 20 and 90 days depending on climate conditions. The optimal lagoon depth is 1.2–1.5 m. For disinfection purposes, two types of maturation ponds can be used: aerated or nonaerated lagoons (Figure 2) depending on space constraints. Under optimal operating conditions, the disinfection efficiency is 3–5 log removal, with maximum values of up to 5–6 log removal for fecal coliforms. According to Grimason et al. (1996), in the case of nematode parasite removal alone (WHO guidelines Category B), the retention time may vary from 10 to 30 days depending on local conditions.

Stabilization ponds are a common type of treatment systems in the Mediterranean countries (France, Portugal, Spain; Renaud et al., 1997), North Africa (Kenya; Grimason et al., 1996; Tanzania; Mayo, 1996; Tunisia; Ghrabi et al., 1989; etc.), South America and Argentina (Renaud et al., 1997). This technology is recommended for towns with between 2000 and 5000 inhabitants (Ghrabi et al., 1989). About 13% of existing stabilization ponds are tertiary disinfection lagooning (maturation ponds) after secondary treatment (activated sludge or trickling filter). These are used in Mediterranean countries, Namibia and South Africa (AGHTM, 1996).

Among the largest pond systems reported are those in Nairobi, Kenya (80,000 m^3/d, peak capacity 240,000 m^3/d; Pearson et al., 1996), and Mendoza, Argentina (160,000 m^3/d, up to 290 ha; Kotlis, 1997).

One of the main disadvantages of this technology is its restricted operating flexibility, especially with regard to flow and seasonal variations. A recent study performed in South Africa (Jagals and Lues, 1996) demonstrated that a series of stabilization ponds were not completely efficient in removing pathogens from irrigation water. Reliable and efficient disinfection with elimination of fecal coliforms, viruses and helminth eggs may be guaranteed by coupling an activated sludge unit with tertiary treatment ponds. Another important disadvantage is the amount of space required; for example, 5 m^2/p.e. for European conditions (AGHTM, 1996).

(a) nonaerated tertiary lagoon after activated sludge treatment (40,000 p.e.).

(b) aerated tertiary lagoon after activated sludge treatment (80,000 p.e.).

Figure 2. View of tertiary lagooning for wastewater disinfection in France.

Wetlands offer another environmentally sound solution. They are used in rural areas, sometimes in natural wetlands (marshes, bogs, swamps) but more frequently in constructed wetlands. The four general classes of constructed wetland are: (1) surface flow marshes, very popular in the U.S., with a mean surface area of 2–5 ha and a maximum of 400 ha (Kadlec, 1995); (2) vegetated subsurface flow beds, which are found acceptable throughout Europe, Australia and South Africa (Wood, 1995); (3) submerged aquatic beds, less frequently used; and (4) floating aquatics. The literature suggests that gravel-filled subsurface flow wetlands successfully remove protozoan and helminth parasites. If they are properly designed and operated, the WHO guidelines for unrestricted irrigation (Category A) may be reached. However, several field studies performed in constructed wetlands for secondary treatment in Egypt (Williams et al., 1995) and Uganda (Denny, 1997) show that pathogen reduction (2–3 log reduction of fecal coliforms and coliphages) is not sufficient to meet the relevant water quality standards for reuse.

Soil aquifer treatment (SAT) provides wastewater purification during flow through unsaturated soils and the aquifer itself to recovery wells. This method ensures treatment of lower quality effluents than is possible with direct injection into the aquifer, without microbial contamination of the freshwater. This technology is used in arid and semi-arid regions, in the U.S. and around the world. One of the largest water reclamation plants is in the Dan region, Israel, serving a total population of about 1.3 million with an average wastewater flow of 270,000 m^3/d (Kanarek and Michail, 1996). The removal efficiency for different pollutants (suspended solids, nutrients, pathogens, trace metals and organics) varies widely with soil type, loading rates and temperature. The hydraulic residence time may vary from several months to one year.

Infiltration-percolation is an efficient extensive wastewater treatment process that ensures high treatment and disinfection efficiency with lower residence time (only a few hours or days). This treatment process acts as an aerobic biological filter and removes suspended solids, organic matter, nutrients and microorganisms from the influent. It is based on flooding-drying cycles of deep sand infiltration basins. Disinfection efficiency depends on media, temperature, pH and, most important, saturation level and wastewater residence time (Brissaud and Salgot, 1994). Enhanced disinfection of 3–4 log removal can be achieved if the filter bed remains unsaturated for a residence time of at least thirty hours (i.e., minimum time for a secondary effluent to pass through a 4 m sand dune filtering bed at 0.5 m/day).

Figure 3. View of the infiltration-percolation plant for tertiary disinfection in Pierra, Spain (200 m^3/d, 432 m^2).

In France and Spain, the size of infiltration-percolation plants is, on average, 600 m^3/day for secondary treatment plants and 3600 m^3/day for tertiary treatment plants (Figure 3). Larger plants exist in Israel and in the United States. There are also small treatment plants in Morocco (Ouazzani et al., 1996).

The main advantage of this technology is the simplicity of operation and the relatively low capital and operating costs. It requires less space than tertiary lagooning or soil aquifer treatment, about 0.3 m^2/p.e. (AGHTM, 1996). The most important disadvantage is the sensitivity of this process to variations in flow and water quality, especially suspended solid concentration.

INTENSIVE DISINFECTION PROCESSES

Chlorination/Dechlorination

Chlorination has become a very common wastewater disinfection technology to satisfy regulations, despite remaining problems such as toxic by-products and questionable efficiency on ammonia-rich effluents. According to a recent WEF study (Isaac, 1996), in 1990, 78% of U.S. publicly owned treatment works were disinfecting treated wastewater with chlorine or hypochlorite and 28% of these practiced dechlorination. Use of dechlorination avoids by-product toxic effect, but increases disinfection costs by up to 20–30%. Many wastewater reclamation projects in Europe and around the world now include chlorination as a final treatment step (Figure 4).

Over the years, a number of theories have been proposed to explain the disinfection effect of chlorine and its related compounds, including oxidation of protoplasm, protein precipitation, modification of cell wall permeability and hydrolysis. While all these effects may be operative, the predominant mechanism depends on the type of organism, its physiological state, and specific conditions such as wastewater quality and environmental conditions (pH, t°C).

Chlorine doses reported in the literature vary considerably depending on wastewater quality and are typically about 5–20 mg/L and 30–60 min contact time. According to some U.S. recommendations (EPA, Title 22), the required contact time can be up to 90–120 min. Under these conditions, a good inactivation of total and fecal coliforms occurs (4–6 log removal), and also of viruses (Mandra et al., 1996; Warriner et al., 1985). Lower disinfection doses and contact time are characteristic of good quality effluent; for example, filtered-activated sludge effluent. Organic and humic compounds or nitrite increase chlorine demand. Suspended solids also affect chlorine disinfection efficiency, because microorganisms are shielded by or embedded in particles.

It is also important to stress that the chlorine dose required for disinfection varies significantly between the different categories of microorganism, as well as within a given category. For example, young bacteria are more sensitive to chlorine than old bacteria and high free chlorine concentration (beyond the breakpoint) is required to inactivate many cysts and viruses of concern.

The disinfection efficiency and the possibility of maintaining residual chlorine are the main advantages of chlorine disinfection. However, important disadvantages have been reported, such as (1) production of by-products that have been identified as potential human carcinogens and harmful for the environment at very low concentration, less than 0.1 mg/L (Abarnou and Miossec, 1992; Szal et al., 1991); (2) poor inactivation of spores, cysts and some viruses at the low dosages used for coliform removal; and (3) need for stricter safety regulations leading to high investments in scrubbing systems and other safety equipment.

(a) chlorine contact tank

(b) chlorine storage

Figure 4. Chlorine disinfection of wastewater after tertiary treatment by multimedia filter and activated carbon filtration in Taif, Saudi Arabia, before reuse for irrigation (67,000 m³/d, 270,000 p.e.).

UV Irradiation

UV irradiation has emerged as a viable alternative to chlorination with a comparable and often more effective disinfection efficiency for control of viruses and bacteria (Baron et al., 1995; EPA, 1992; Lazarova et al., 1997). When new treatment plants have to be built, UV systems are more cost-effective than chlorination. Furthermore, UV-irradiated effluents exhibit no toxicity. These characteristics and the recent change in wastewater treatment regulations in North America and Europe have prompted a number of facilities to select this technology for disinfection. Over 600 UV disinfection units are in place in the U.S. alone, the great majority installed since 1986 (EPA, 1992). Numerous pilot and full-scale studies have demonstrated that UV disinfection can consistently achieve the objective of 200FC/100 mL (Braunstein et al., 1996; Lazarova et al., 1997; Whitby et al., 1984; Whitby and Palmateer, 1993).

The germicidal effects of UV light involve photochemical damage to RNA and DNA, leading to effective cell inactivation. If damaged cells replicate, they produce mutant daughter cells that are unable to replicate (EPA, 1986). However, photoreactivation of damaged pathogens can occur and has been reported in numerous studies (Whitby and Palmateer, 1993).

The lack of standardization in determining UV dosage in UV systems makes it difficult to directly compare results from different studies. Figure 5 illustrates the difference between UV design parameters obtained with a collimated beam in distilled water and wastewater and pilot-scale experiments. Moreover, the variability of wastewater characteristics and reactor hydrodynamics leads to greater inaccuracy in predictions of large-scale UV plant performance. Some empirical (Loge et al., 1996; Darby et al., 1995) and numerical (Blatchley et al., 1997; Do-Quang et al., 1997) models have been developed to avoid these constraints by taking into account the influence of water quality and hydrodynamics.

Disinfection data based on various types of wastewater showed that 30–45 mW s/cm^2 of UV radiation was sufficient to ensure 3–5 log removal of total and fecal coliforms and fecal streptococci (AGHTM, 1996; Lazarova et al., 1997). The required doses could be higher (up to 112 mW s/cm^2) in the case of more stringent standards; for example, 2.2TC/100 mL based on the 7-day median sample (Braunstein et al., 1996). The required UV doses for a given pathogen log removal are significantly influenced by wastewater quality, and especially UV transmittance and

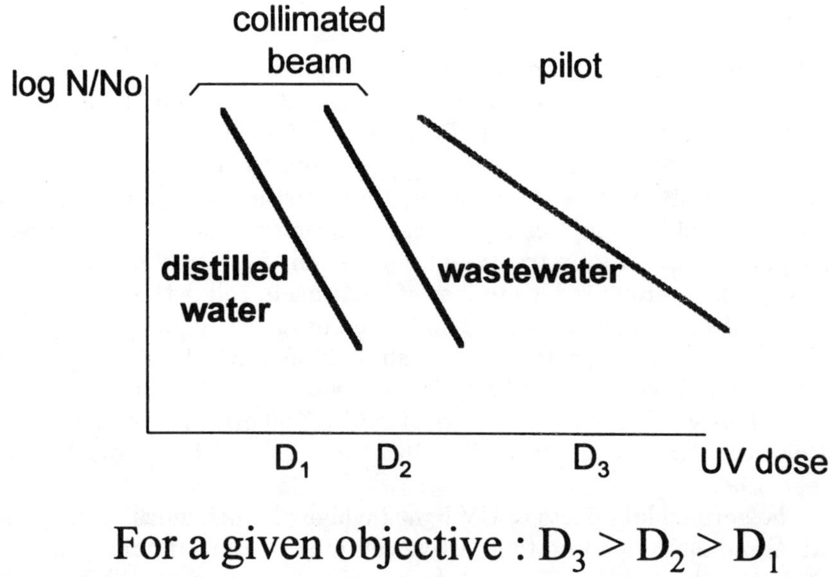

Figure 5. Effect of raw water and experimental pilot-scale experiments on UV disinfection dose.

suspended solids. Water transmittance is affected by certain inorganic (iron, manganese) and organic (humic substances, dyes) compounds, as well as by suspended solids. To minimize this phenomenon, water transmittance must be above 50–60%.

The most important effects of suspended particles on UV disinfection are microorganism shading and UV light scattering. In a number of studies (EPA, 1992; Whitby and Palmateer, 1993; Mandra et al., 1996), pathogen inactivation has been found to decrease significantly as the total suspended solids concentration of the effluent increases. Figure 6 illustrates this effect for total coliform inactivation in different filtered and nonfiltered wastewater effluents. The effect of filtration before disinfection was drastic: The UV dose required for a given log inactivation could be more than halved.

UV radiation virucidal efficiency varies with virus type (Figure 7). A low UV dose (30 mWs/cm^2) enables the removal of 6 log of the bacteriophage φX174, whereas a high dose (100 mWs/cm^2) inactivates less than 5 log of the highly resistant bacteriophage MS2. These results are consistent with previous data showing a 4-log unit MS2 removal with a UV dose of 75 mWs/cm^2 (Oppenheimer et al., 1993) to 100–200

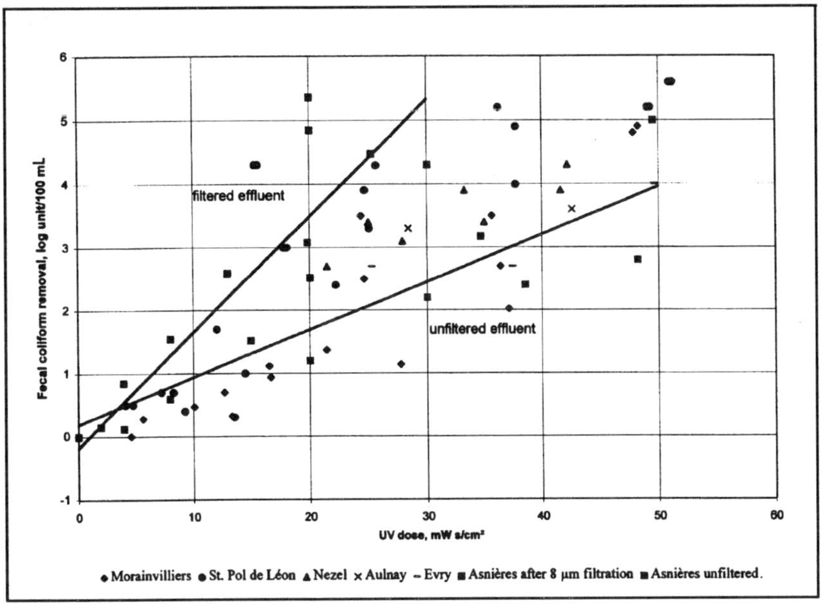

Figure 6. Influence of UV dose and filtration on fecal coliform inactivation for different WWTP effluents in France.

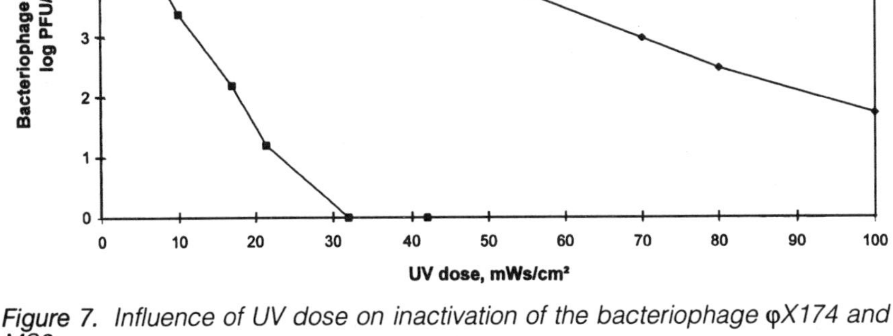

Figure 7. Influence of UV dose on inactivation of the bacteriophage φX174 and MS2.

Table 4. Estimated relative effectiveness of UV radiation and chlorine for representative microorganisms of concern in wastewater (adapted from Darby et al., 1995 and AGHTM report, 1996).

	Organism	Relative Resistance UV Irradiation	Chlorine
Bacteria	Fecal coliforms	1.0	1.0
	Salmonella typhosa	0.9–1.0	1.0
	Staphylococcus	1.0–1.5	2.5
	Clostridium botulinum	1.6	NR
	Bacillus (spore)	6.6	NR
Virus	Bacteriophage *E. Coli*	1.0–2.0	5–6
	Adenovirus	0.6–0.8	0.5
	Poliovirus type 1	1.0–3.0	6–7
	Coxsackie A-2	0.8–1.0	6–7
Fungi	Aspergillus niger	50	NR
	Penicillium chrysogenus	8	NR

mWs/cm^2 (Braunstein et al., 1996; Havelaar et al., 1990; Nieuwstad et al., 1991).

According to Table 4, for three types of viruses reported, as well as for inactivation of Staphylococcus bacterial strains, UV irradiation is more effective than chlorination.

At present, low-pressure mercury arc lamps are the principal means of generating UV energy for disinfection. Recently, medium-pressure lamps have been tested, but little information about their efficiency, clogging problems and energy consumption is available. Most large-scale UV disinfection systems in use today consist of open channels with two common lamp configurations: horizontal systems (lamps oriented parallel to the flow) or vertical systems (Figure 8).

Manufacturers use calculation methods to evaluate average doses delivered by their system according to flow rate. However, the dose effectively received by the water during its residence time in such a reactor cannot be accurately described by a single average value, as would be the case in collimated beam experiments where hydrodynamic conditions are totally controlled. The flow pattern of the water and the intensity field result in a distribution of doses over a cross-section. When a reduction of more than 2 log units is sought, which is usually the case in wastewater disinfection, the presence of weaker irradiation zones within the reactor (due to low intensity, high velocity or both combined) has a strong effect on the global disinfection performance (Schoenen, 1996).

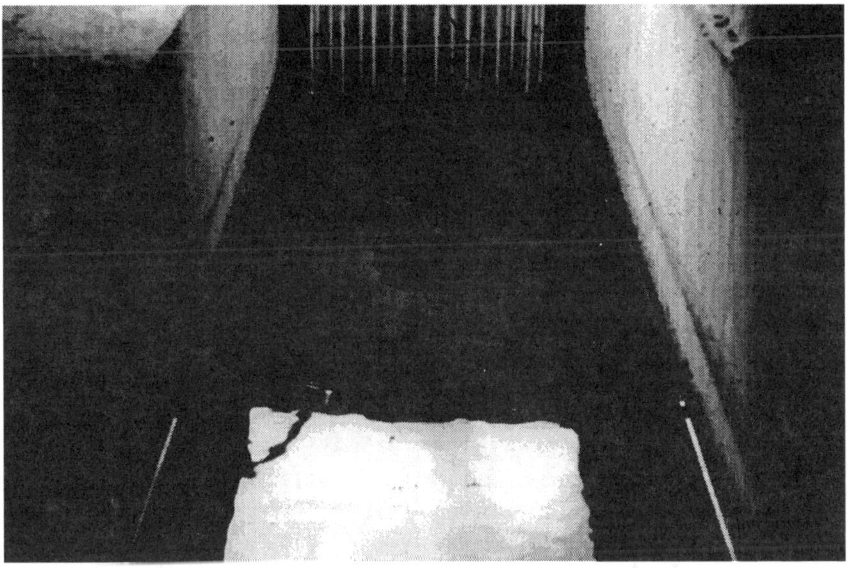

Figure 8. View of the vertical system of disinfection by UV irradiation in open channels.

The numerical modeling approach (Figure 9) provides a better understanding of local flow-mixing mechanisms in UV reactors. Used as a tool for evaluating alternative reactor configurations, it could achieve significant improvements in disinfection efficacy (Do-Quang et al., 1997). Using such models, it has been suggested that internal baffles be introduced into these high-velocity regions to reduce shortcircuiting

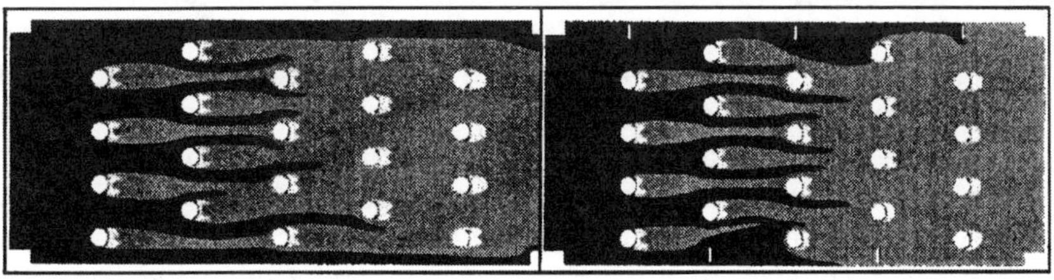

a) b)

Figure 9. Numerical simulation of temperature distribution in UV reactors: (a) unmodified and (b) modified reactor with baffles (source: Do-Quang et al., 1997).

and improve transverse mixing. Comparing the temperature field in the modified system [(Figure 9(b)] with that in the unmodified one [(Figure 9(a)] it was observed that a more homogeneous temperature distribution was obtained with baffles, resulting in a better overall heat transfer yield (> 10%). Assuming that the radiation itself behaves as an additional heat source, the same tendency can be expected for UV intensity evolution in relation to water flow rate.

Major advantages of UV disinfection include its simplicity, minimal space requirement and absence of toxic by-products. As with chlorine, amoebic and protozoan cysts are the most difficult to inactivate. Another disadvantage is microorganism photoreactivation and regrowth.

Ozonation

Ozone is a quick-acting and highly effective disinfectant against bacteria, viruses and protozoa. In addition, ozone by-products are mainly biodegradable compounds. Probably the most dangerous by-products reported in drinking water disinfection are bromates. No toxic compounds have been found in ozonated secondary effluent (Langlais et al., 1992). Several studies on wastewater ozonation reported enhanced water quality with up to 20% of BOD and COD reduction (AGHTM, 1996). Most mutagenicity studies show that ozonation reduces or removes mutagenicity in water. The germicidal effect of ozone consists of modifying the cell wall structure, resulting in microorganism kill.

Wastewater ozonation has been used in the U.S. for more than fifteen years. There are also ozonation treatment works in the Middle East, South Africa, France and Spain (Figure 10). A review of the U.S. experience in this area has been provided by Robson and Rice (1990).

As with chlorine and UV irradiation, ozone doses are influenced by water quality and can be reduced by preliminary filtration of the effluent (Stover and Jarnis, 1981). The typical ozone demand of secondary treated effluent is in the range of 1–3 mg/L (Rakness et al., 1993). It is important to underscore that with small applied ozone concentrations of up to 7–10 mg/L, corresponding to small residual ozone concentrations of 0.1 mg/L, more than 3 to 4 log removal can be obtained for total coliforms, fecal coliforms and fecal streptococci within a relatively low contact time of 10 min (Figure 11), thereby ensuring less than 200FC/100 mL (Robson and Rice, 1990; Langlais et al., 1992; Mandra et al., 1996). Higher doses of up to 40–50 mg/L may be required to achieve the most stringent Title 22 requirements of 2.2TC/100 mL (Legan, 1982; Robson and Rice, 1990).

Figure 10. View of the ozone generator for disinfection in a wastewater treatment plant in Spain.

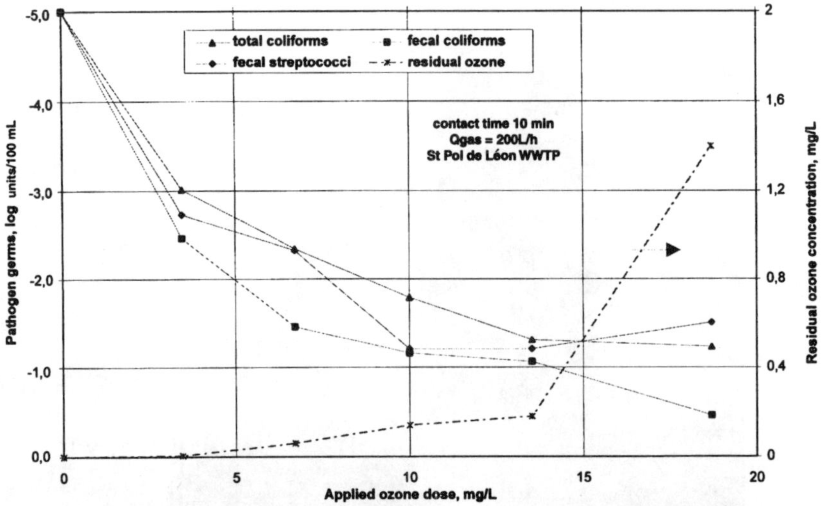

Figure 11. Example of disinfection of fecal contamination test germs by ozone in relation to applied dose (source: CIRSEE; Lazarova et al., 1997).

Another recognized particularity of ozone is that indicator bacteria are more resistant than viruses. In general, lower ozone concentrations, in the range of 1–5 mg/L, and lower contact times are necessary for 3–5 log virus removal, including the highly resistant MS2 strain (Mandra et al., 1996; Warriner et al., 1985). Ozone is also more effective on parasites: *Giardia* and *Cryptosporidium* cysts and *Acanthamoeba* and *Naegleria* amoebae (Perrine, 1990).

The design parameters found in pilot plant experiments can be used to forecast disinfection facilities. However, great care has to be taken to include the ozone demand of the real effluent, which may reach a few mg/L, and the performance characteristics of the ozone contactor (short-circuiting, back-mixing).

The main advantage of ozonation is its very high disinfection efficiency for all pathogens of concern in wastewater (bacteria, viruses and parasites). Ozonation also enhances water quality and is an acceptable solution for large treatment plants.

Membrane Filtration

Microfiltration and ultrafiltration are being intensively studied in France and the U.S. for wastewater disinfection and removal of colloids

Table 5. Efficiency of wastewater disinfection by hollow fiber membranes UF and MF (source: CIRSEE; Lazarova et al., 1997).

Indicator	Influent Concentration	Permeate Concentration Ultrafiltration	Microfiltration
Total coliforms	10^5–10^7/100 mL	0/100 mL	0/100 mL
E. coli phages	10^4–10^5 PFU/100 mL	0/L	10–10^3/100 mL
Enterovirus	0–20 MPN/L	0/L	ND*
Cryptosporidium	10–10^2/L	0/L	0/L
Giardia cysts	10–10^2/L	0/L	0/L

*ND not determined.

and larger molecular-weight organics. These technologies offer a completely different disinfection principle, that of separating the pathogen from the treated wastewater by a physical barrier (i.e., membranes).

Membrane filtration is used in Australia, Japan, Spain and the U.S. for water reuse applications requiring high water quality, such as groundwater recharge, greywater recycling and industrial wastewater recycling.

Suspended solids, bacteria and parasites are readily removed by both UF and MF (Table 5). Ultrafiltration (0.01 μm cutoff) also ensures total virus removal, while viruses that had passed through the more porous MF membrane (0.2 μm cutoff), 1 to 3 log PFU/100 mL coliphages, were found in the microfiltered permeate.

One of the most important advantages of membrane disinfection is the absence of bacterial regrowth and residual toxicity. Membrane disinfection is also less sensitive to wastewater quality than the physicochemical disinfection technologies discussed. However, to ensure reliable membrane operation and minimize operating costs, a pretreatment step (tertiary filtration, for example) is recommended.

UF disinfection may also be successfully combined with intensive biological treatment in membrane bioreactors, MBR® (Manem and Sanderson, 1996). This technology is advantageously used for complete wastewater treatment in small compact units for greywater recycling in France and Japan, and for industrial recycling in France (Figure 12).

Another new application under investigation for low-pressure membranes is their use as a pretreatment to reverse osmosis in California, U.S., for production of high quality water such as would be required for indirect potable reuse.

The major advantages of membrane disinfection include the excellent water quality and the absence of toxicity and regrowth problems. The

Figure 12. Industrial wastewater treatment in France by membrane bioreactor before reuse for irrigation (150 m^3/d, cosmetics manufacture).

obstacle to wider application is its cost: capital and operating costs for energy, cleaning and membrane replacement.

Other Disinfection Processes

Chlorine dioxide is a more effective disinfectant than chlorine, both for bacteria and viruses in a broad pH range (AGHTM, 1996; Junli et al., 1997; Warriner et al., 1985). This process has been tested and applied for wastewater disinfection before reuse in China, France, Israel and the U.S. (Narkis et al., 1988). The required pathogen removal (up to 4–5 log fecal test germs) has been achieved with ClO_2 doses of 2–5 mg/L and contact time of 5–15 min. These design parameters (4.8 mg/L for 15 min) have been confirmed on a large scale in France, WWTP of Deauville, 115,000 p.e. (AGHTM, 1996). Junli et al. (1997) report higher disinfection efficiency of ClO_2 in the inactivation of highly resistant viruses such as *Adenovirus, Coxsackie, Poliovirus,* and *Herpes simplex.*

In recent years, another new biocide compound studied for wastewater disinfection is peracetic acid (PAA), produced by the reaction of hydrogen peroxide with acetic acid (Lefevre et al., 1992). Several laboratory and pilot studies have shown the removal of 3 log of fecal test germs by PAA doses in the range of 5–10 mg/L within 10-min contact time. The effect of PAA on virus removal depends on the type of virus considered: Higher doses and contact time are necessary to remove highly resistant viruses (AGHTM, 1996; Mandra et al., 1996). One of the major disadvantages of PAA disinfection is the increased organic content of the effluent.

COST EVALUATION OF WASTEWATER DISINFECTION

Numerous economic studies of wastewater disinfection have been described. It is difficult to compare the results directly, as there are several sources of uncertainty. For example, unit prices of civil engineering work, energy, manpower, chemicals, UV lamps, UF membranes and process equipment vary widely from one market or country to another. Moreover, wastewater quality also varies widely, from relatively diluted effluent in the U.S. to very concentrated sewage in arid areas and on islands.

Figure 13 compares, on a ratio basis, the total annual cost per cubic meter treated effluent (including civil engineering work depreciation, equipment depreciation and operating costs) calculated for three different disinfection technologies and three plant sizes, based on French treatment plants (Lazarova et al., 1997). As expected, the specific disinfection cost decreases as plant size increases. For the reported plant sizes, the cost reduction varied by a factor of 2 to 7, depending on the disinfection technology used. Operating costs are about 45% of total annual costs for UV irradiation. For ozonation, this figure increases from 47% for a WWTP of 20,000 p.e. to 61% for a WWTP of 500,000 p.e. Operating costs are proportionately highest for chlorination, from

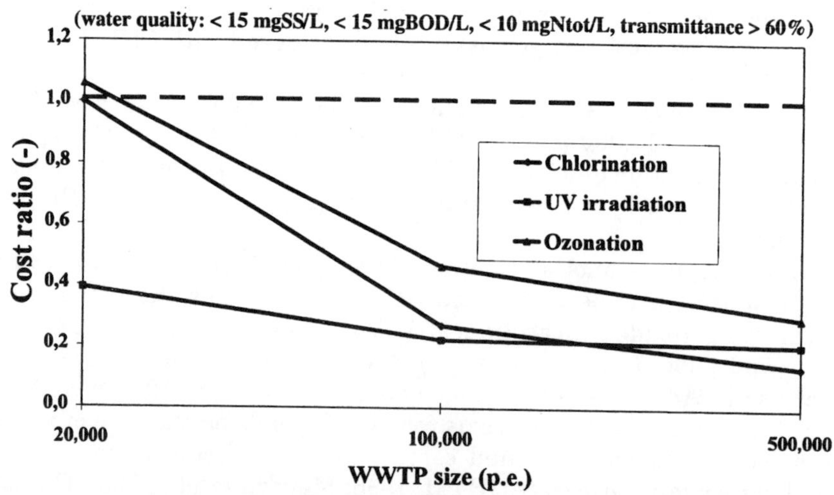

Figure 13. Cost evaluation of different disinfection technologies on the basis of cost ratios (the unit of reference is chlorination for a WWTP of 20,000 p.e.).

51% to 78%. In the case of poor wastewater quality, overall disinfection costs may double.

The main operating and maintenance cost incurred by chlorination and ozonation is the cost of the chemicals. Operating costs for the UV systems include mostly lamp replacement and cleaning. Energy costs are about 2–5% of operating costs for chlorination and 12–36% for UV irradiation and ozonation, depending on plant size.

The range of cost ratios shown in Figure 13 indicates that UV disinfection is substantially less expensive than chlorination/dechlorination. These results are consistent with the conclusions of the WERF expert group (Darby et al., 1995) and other authors (Blatchley et al., 1996). Moreover, WWTP upgrading with UV irradiation is easier and requires few on-site modifications. Ozonation also seems to be a competitive alternative to chlorination, especially in the case of large plants (up to twice as expensive as UV radiation). Moreover, residual ozone could be recycled in the biological treatment step. Other authors reported higher differences in total annual costs between ozone and UV: a factor of 7 (Legan et al., 1982). The major problems in ozone plant operation in the past involved the ozone generator systems. Today, these have been replaced by new, more efficient devices.

Membrane filtration is more expensive than the physicochemical disinfection processes. However, the effluent quality obtained is excellent from the microbiological and physicochemical points of view. This technology will be the only acceptable solution for advanced reuse applications with stringent water quality requirements (indirect potable reuse or urban uses). In fact, for the operating conditions assumed, ozonation and UF systems met higher water quality standards, including total virus removal.

COMPARISON OF INTENSIVE DISINFECTION PROCESSES

Critical analysis of the reported performance, advantages and disadvantages of intensive disinfection processes shows that virus removal requires a significant increase in the chlorine, PAA and UV doses compared to indicator fecal bacteria reduction alone. Ozonation and ultrafiltration possess high bacteria, parasite, and virus removal capacities under the same operating conditions. Moreover, in addition to its disinfecting action, ultrafiltration improves the physicochemical characteristics of the effluent. The resulting excellent UF permeate quality makes it particularly favorable for groundwater recharge and specific urban applications. Finally, peracetic acid is not considered an appropriated technology for wastewater disinfection because of the high doses required for virus removal and the considerable increase in organic pollution of the effluent.

Table 6 summarizes the main technical and economic characteristics of advanced disinfection processes compared with chlorination/dechlorination. It includes the major criteria that have to be taken into account: technical performances regarding regulatory issues, environmental impact and cost considerations.

Table 6. Main technical-economic characteristics of advanced disinfection technologies compared with chlorination.

Characteristics	Chlorine	PAA	Ozone	UV	MF	UF
Bactericidal action	++	++	+++	++	+++	+++
Virucidal action	++	+	+++	+	none	+++
Bacterial regrowth potential	+	+++	++	+	none	none
Residual toxicity	+++	++	+	none	none	none
By-products	+++	none	+	none	none	none
Operating costs	+	++	++	++	+++	+++
Investment costs	+	+	+++	++	+++	+++

"+" low; "++" middle; "+++" high.

CONCLUSIONS

Assessment of advanced technologies for wastewater disinfection and reuse shows the advantages of UV irradiation and ozonation in producing a highly disinfected effluent without toxic by-products. UV disinfection is substantially less expensive than chlorination/dechlorination for new disinfection facilities. Chlorine is very effective for bacteria inactivation, but the requirement of higher doses for removal of some chlorine-resistant bacteria and viruses and the potential toxicity of chlorination by-products make this process less attractive. These constraints can be avoided by using chlorine dioxide.

UV virucidal effect depends on virus type and higher doses must be applied for total virus removal.

Peracetic acid is considered to be an inappropriate technology for wastewater disinfection because of the high doses required for virus removal and the considerable increase in organic pollution in the effluent.

Ozonation offers the advantage of having a strong effect on all types of viruses and protozoa cysts, even at low treatment doses and short contact times.

Membrane disinfection ensures excellent water quality (no residual toxicity, no bacterial regrowth) and high reliability of operation. The main disadvantage is its relatively high cost. However, membrane processes are the only acceptable technology that guarantees total wastewater effluent disinfection combined with the high physicochemical water quality required for advanced reuse applications such as indirect potable reuse.

Extensive disinfection technologies are well suited to local conditions in regions and developing countries where there are no space constraints. Although good levels of disinfection efficiency have been reported for some existing extensive treatments, there is no guarantee of total disinfection because the processes are highly sensitive to variations in flow and water quality, and to environmental conditions.

REFERENCES

Abarnou, A. and L. Miossec. 1992. Chlorinated waters discharged to the marine environment chemistry and environmental impact. An overview. *The Science of Total Environment* 126:173–197.

AGHTM report 1996. Principaux procédés de desinfection complémentaire. TSM (2):107–118.

Baron, J., P. Joret, J. Lesavre, and J. Y. Perrot. 1995. Désinfection par ultraviolets des poliovirus et des bactériophages FRNA spécifiques. *Tech. Sci. Méthodes* Décembre, (12):931–936.

Blatchley, E. R., K. C.Bastian, R. K. Duggirala, J. E. Alleman, M. Moore, and P. Schuerch. 1996. Ultraviolet irradiation and chlorination/dechlorination for municipal wastewater disinfection: assessment of performance limitations. *Wat. Envir. Res.* 68(2):194–204.

Blatchley III, E. R., Z. Do-Quang, J. M. Laîné. 1997. Process modelling of ultraviolet disinfection. *WEF Conference,* Chicago, 18–22 Oct.

Braunstein, J. L., F. J. Loge, G. Tchobanoglous, and J.L. Darby. 1996. Ultraviolet disinfection of filtered activated sludge effluent for reuse applications. Wat Environ Res. 68(2):152–161.

Brissaud, F. and M. Salgot. 1994. Infiltration percilation as a tertiary treatment. *Colloque Sci. Tech. Int.,* "Mieux gérer l'Eai," Marsielle, France, 2:391–398.

Crook, J., D. A. Okun, and A. B. Pincince. 1994. Water reuse. WERF Project 92-WRE-1.

Darby, J., M. Heath, J. Jacangelo, F. Loge, P. Swaim, and G. Tchobanoglous. 1995. Comparison of UV irradiation to chlorination: guidance for achieving optimal UV performance. WERF Project 91-WWD-1, Alexandria, VA.

Denny, P. 1997. Implementation of constructed wetlands in developing countries. *Wat. Sci. Tech.* 35(5):27–34.

Do-Quang, Z., R. Djebbar, E. R. Blatchley III, J. M. Laîné. 1997. Computational fluid dynamics (CFD) modeling of UV disinfection reactor performances: Optimization of the flow in vertical lamps open channel. Conference of the American and Canadian Chemical Society, 25 July, Edmonton.

El Hamouri, B., A. Handouf, M. Mekrane, M. Touzani, A. Khana, K. Khallayoune, and T. Banchokroun. 1996. Use of wastewater for crop production under arid and saline conditions: Yield and hygienic quality of the crop and soil contamination. *Wat. Sci. Tech.* 33(10-11):327–334.

EPA. 1986. Design manual: Municipal wastewater disinfection. EPA/625/1–86/021. U.S. EPA, Washington, D.C.

EPA. 1992. Ultraviolet disinfection technology assessment. EPA/832/R-92/004. U.S. EPA, Washington, D.C.

Ghrabi, A., M. Ferchichi, and M. Drakidès. 1989. Treatment of wastewater by stabilization ponds application to Tunisian conditions. *Wat. Sci. Tech.* 28(10):193–199.

Grimason, A. M., H. V. Smith, G. Young, and W. N. Thitai. 1996. Occurrence and removal of *Ascaris sp.* ova by waste stabilisation ponds in Kenya. *Wat. Sci. Tech.* 33(7):75–82.

Havelaar, A. H., C. C. Meulemans, W. M. Pot-Hogeboom, and J. Koster. 1990. Inactivation of bacteriophage MS2 in wastewater effluent with monochromatic and polychromatic ultraviolet light. *Wat. Res.* 24:1387–1392.

Isaac, R. A. 1996. Disinfection dialogue: The evolving perception of need, benefits and detriments. *Wat. Env. Tech.* May, pp. 67–72.

Jagals, P. and J. F. R. Lues, 1996. The efficiency of a combined waste stabilization pond/maturation pond system to sanitise waste water intended for recreational reuse. *Wat. Sci. Tech.* 33(7):117–124.

Junli, H., W. Li, L. Nenqi, L. X. Li, S. R. Fun, and Y. Guanle. 1997. Disinfection effect of chlorine dioxide on viruses, algas and animal planktons in water. *Wat. Res.* 31(3):455–460.

Kadlec, R. H. 1995. Overview: Surface flow constructed wetlands. *Wat. Sci. Tech.* 32(3):1–12.

Kanarek, A. and M. Michail. 1996. Groundwater recharge with municipal effluent: Dan region reclamation project, Israel. *Wat. Sci. Tech.* 34(11):227–233.

Kotlis, L. 1997. Water reuse. National Report Argentina. *Proceedings of the IWSA World Congress,* Madrid, IR8–9/IR8–11.

Langlais, B., B. Legube, H. Beuffe, and M. Doré. 1992. Study of the nature of the by-products formed and the risks of toxicity when disinfecting a secondary effluent with ozone. *Wat. Sci. Tech.* 25(12):135–143.

Lazarova, V., J. C. Bourdelot, M. L. Janex, J. Manem, and J. M. Laîné. 1997. Wastewater reuse: Technical and economic evaluation of disinfection processes. *Proceedings of the Conference of the Benefical Reuse of Water and Solids,* Marbella, Spain, 6–9 April, 8–3/8–15.

Lefevre, F., J. M. Audic, and F. Ferrand, 1992. Peracetic acid disinfection of secondary effluents discharged off coastal seawater. *Wat. Sci. Tech.* 25(12):155–164.

Legan, R.W. 1982. Alternative disinfection methods: A comparison of UV and ozone. *Ind. Wat. Eng.* March/April: 12–21.

Levine, B., V. Lazarova, I. H. Suffet, and J. Manem. 1997. Wastewater reuse standards: Goals, status and guide-lines. *Proceedings of the Conference of the Benefical Reuse of Water and Solids,* Marbella, Spain, April: 13/59–13/71.

Loge, F.J., W. E. Robert, M. Heath, J. Jacangelo, G. Tchoganoblous, and J. D. Darby. 1996. Ultraviolet disinfection of secondary wastewater effluents: Prediction of performance and design. *Water Environment Research,* 68(5):900–916.

Mandra, V., V. Lazarova, N. Dumoutier, and J. M. Audic. 1996. Étude comparative de la désinfection des eaux résiduaires urbaines par l'acide paracétique, l'irradiation UV et l'ozone. *Proc. Colloque Journées Information Eaux,* 18–20 Septembre, Poitiers, (2) 61/1.

Manem, J. and R. Sanderson. 1996. Membrane bioreactors. In: *Water Treatment Membrane Processes.* J. Mallevialle, P. Odendaal, and M. R. Wiesner (Eds.). McGraw-Hill, AWWA, p. 17.1.

Mayo, A.W. 1996. BOD_5 removal in facultative ponds: Experience in Tanzania. *Wat. Sci. Tech.* 34(11):107–117.

Narkis, N., R. Offer, and N. Betzer. 1988. Chlorine dioxide as a disinfectant at each step of the advanced treatment of effluents intended for reuse. *Proceedings of the 61st Annual Conference of WPCF,* Dallas, Texas, U.S.

Nieuwstad, Th. J., A. H. Havelaar, and M. Van Olphen. 1991. Hydraulic and microbiological characterization of reactors for ultraviolet disinfection of secondary wastewater effluent. *Wat. Res.* 25:775–782.

Oppenheimer, J.A., J. E. Hoagland, J. M. Laîné, J. G. Jacangelo, and A. Bhamrah. 1993. Microbial inactivation and characterization of toxicity and by-products occurring in reclaimed wastewater disinfected with UV radiation. *Proceedings of the WEF Speciality Conference Series,* TT043,13.

Ouazzani, N., K. Bousselhaj, and Y. Abbas. 1996. Reuse of wastewater treated by infiltration percolation. *Wat. Sci. Tech.* 33(10–11):401–408.

Pearson, H. W., S. T. Avery, S. W. Mills, P. Njaggah, and P. Odiambo. 1996. Performance of the phase II Dandora waste stabilisation ponds, the largest in Africa: The case for anaerobic ponds. *Wat. Sci. Tech.* 33(7):91–98.

Perrine, D. 1990. Cinétiques d'action du chlore et de l'ozone sur les trophozoïtes et les kysts d'amibes libres de germes *Naegleria* et *Acanthamoeba*. *J. Français d'Hydrol.* 21(1):113–121.

Rakness, K.L., K. M. Corsaro, G. Hale, and B. D. Blank. 1993. Wastewater disinfection with ozone: Process control and operating. *Proceedings of the Conference Ozone in Water and Wastewater Treatment,* San Francisco, 1:S-7–16/S-7–32.

Rångeby, M., P. Johanson, and M. Pernrup. 1996. Removal of faecal coliforms in wastewater stabilisation pond system in Mindelo, Cape Verde. *Wat. Sci. Tech.* 34(11):149–157.

Renaud, P., V. Lazarova, B. Levine, and J. Manem. 1997. Water reuse. international report. *Proceedings IWSA World Congress,* Madrid, IR8–1/IR8–7.

Robson, C. M. and R. G. Rice. 1991. Wastewater ozonation in the USA: History and current status—1989. *Ozone Science and Engineering* 13(1):23–40.

Schoenen, D. 1996. The influence of inhomogeneous irradiation in UV disinfection—experimental findings and theoretical considerations. *J. Water SRT—Aqua* 45(3):120–129.

Shelef, G. and Y. Azov. 1996. The coming era of intensive wastewater reuse in the Mediterranean region. *Wat. Sci. Tech.* 33(10-11):115–125.

Shereif M. M., M. El-S. Easa, M. I. El-Samra, and K. H. Mancy. 1995. A demonstration of wastewater treatment for reuse applications in fish production and irrigation in Suez, Egypt. *Wat. Sci. Tech.* 32(11):137–144.

Stover, E. L. and R. W. Jarnis. 1981. Obtaining high-level wastewater disinfection with ozone. *JWPCF* 53(11):1637–1647.

Szal, G. M., P. M. Nola, L. E. Kennedy, C. P. Barr, and M. D. Bilger. 1991. The toxicity of chlorinated wastewater: Instream and laboratory case studies. *Res. J. WPCF* 63(6):910–920.

Warriner, R., K. D. Kostenbader, D. O. Cliver, and W. C. Ku. 1985. Disinfection of advanced wastewater treatment effluent by chlorine, chlorine dioxide and ozone. Experiments using seeded poliovirus. *Water Res.* 19(12):1515–1526.

Whitby, G. E. and G. Palmateer. 1993. The effect of UV transmission, suspended solids and photoreactivation on microorganisms in wastewater treated with UV light. *Wat. Sci. Tech.* 27(3-4):379–386.

Whitby, G. E., G. Palmateer, W. G. Cook, J. Maarschalkerweerd, D. Huber, and K. Flood. 1984. Ultraviolet disinfection of secondary effluent. *JWPCF* 56(7):844–851.

Williams, J., M. Bahgat, E. May, M. Ford, and J. Butler. 1995. Minaralisation and pathogen removal in gravel bed hydroponic constructed wetlands for wastewater treatment. *Wat. Sci. Tech.* 32(3):49–58.

Wood, A. 1995. Constructed wetlands in water pollution control: Fundamentals to their understanding. *Wat. Sci. Tech.* 32(3):21–29.

9

Aridity and Water Resources of the Arabian Peninsula

MAMDOUH SHAHIN
Water Resources Engineering Consultant
Flaviusstraat 11
2275 VG Voorburg
The Netherlands

INTRODUCTION

THIS CHAPTER DEALS with the aridity and water resources of the Arabian Peninsula. In view of the broad scope of this topic, the underlying basic components are briefly reviewed. The chapter begins by highlighting the geographical and physiographical features of the peninsula. This is followed by a general description of the climate. (For the purpose of illustration, data from four meteorological stations are included in Appendixes at the end of the chapter.)

The next section presents a variety of definitions and methods commonly used for classifying the climate and the level of aridity. These definitions and methods are based on precipitation, temperature and/or evapotranspiration. Since the rainfall in arid zones is widely variable both in time and space, it has been found necessary to analyze statistically the annual rainfall at a reasonable number of stations. In some cases the analysis has been extended to monthly, daily and short-duration rainfall data. Potential evapotranspiration at several stations is also presented and discussed.

The Arabian Peninsula is probably the largest surface on earth that is not traversed by any perennial stream whether springing from inside or outside the peninsula. Part of the rain water is carried by the wadis, which flow depending on the prevailing rain characteristics and/or infiltrate into the ground and recharge the groundwater aquifers. Certain sections of the chapter have, therefore, been devoted to wadi flow and groundwater resources. Environmental impacts of large-scale groundwater abstraction in the peninsula are reviewed.

What used to be called nonconventional resources are no longer nonconventional. For example, desalination of seawater in the Arab Peninsula has expanded immensely since the late 1950s. About two decades later the use of treated wastewater for industrial and agricultural purposes started to catch up.

The chapter ends with conclusive remarks, references and appendixes.

GEOGRAPHY AND PHYSIOGRAPHY OF THE ARABIAN PENINSULA

The Arabian Peninsula is a vast terrain characterized by aridity of climate and scarcity of water resources. The peninsula as defined here is located approximately between the latitudes 32°N and 12°30′N, and between longitudinals 35°E and 60°E. The surface covered by the peninsula is about 3×10^6 km^2. This surface is shared between seven Arab states, all of them sovereign and independent. These are the states of Bahrain, Kuwait, Qatar, Sultanate of Oman, Kingdom of Saudi Arabia, the United Arab Emirates and the Republic of Yemen (see the map, Figure 1). The seven states are members of the United Nations Organization, the Arab League and the World Council of Islamic Countries.

The surface area occupied by the peninsula is far from equally divided between these seven countries. Quite the opposite. Saudi Arabia occupies the largest part, 72.7% of the area, while Bahrain occupies the smallest part, 0.021%.

The geographical boundaries, and so the specification of the countries sharing the Arabian peninsula as given in this chapter, may be different from those given in other references. As an example, Petrov (1976) in his treatise of the deserts of the world included what is called the eastern subregion of the Arab region or Al-Mashreque Al-Arabi in the total surface of the peninsula. This subregion includes Iraq, Syria, Lebanon, Jordan and Palestine.

The central part of the Arabian Peninsula is a vast desert plateau (Najd) covered with sand and pebbles filling many flat depressions, commonly known as playas. This plateau is situated in Saudi Arabia and surrounded by sand deserts: the Great Nufud from the north and Rub Al-Khali from the south. These two deserts are connected from the easternmost edge by a narrow sand belt known as the Ad-Dahna desert.

The Najd plateau slopes to the east to join the low-lying Al-Hasa desert. The same plateau is bounded from the west by the slopes of the Hejaz mountains, which run parallel to the Red Sea coast and rise up to 2000 meters above mean sea level (m amsl).

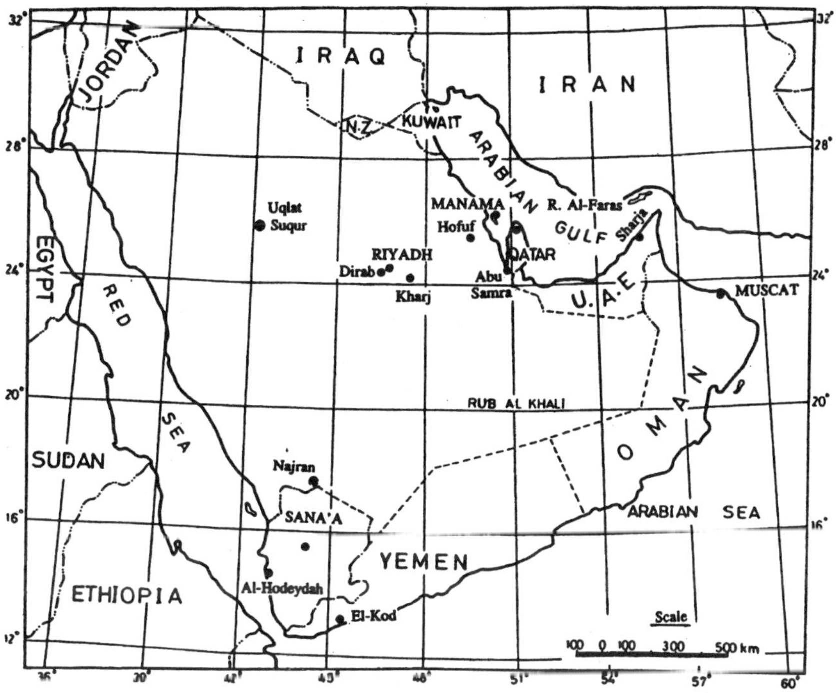

Figure 1. Map of the Arabian Peninsula.

The southwestern part of the peninsula is occupied by the Tihama plain. This plain also runs parallel to the coast of the Red Sea covering a surface 440 km long and 45 km average width. The highest elevation is about 400 m amsl. The plain is a flat to rolling terrain intersected by almost east-west running wadis, which during and shortly after heavy rain storms can carry enormous floods from the mountain slopes to the Red Sea.

The Tihama plain is bounded by the slopes of the western Yemeni mountains, which slope upwards from 400 m to 2000 m amsl. This is known as the midlands of Yemen. It is a rugged terrain, where wadis have cut deep canyons with almost vertical walls. These wadis drain their flow westwards to the Red Sea. To the east of the midlands one finds the central highlands of Yemen. The elevation of the mountains ranges from 2000 m to 3000 m, with peaks exceeding 3500 m. The surface of this physiographic unit is less rugged than the western slopes.

The eastern slopes are gentle and terminate in the interior region of the high Yemeni plateau. The interior mountain plains are almost flat varying in altitude from 2200 m to 2700 m amsl. The surface is mostly covered with lava flows underlain by thick alluvial deposits. They slope eastwards forming a large flat desert with elevations varying from about 1000 to 2000 m. The drainage of these slopes is about east towards Ar-Rub Al-Khali desert.

The south Arabian mountains, commonly referred to as Hadramaut plateau, rise to about 2500 m amsl. This plateau is also traversed by a large number of wadis. The Omani mountains occupy the southeastern part of the peninsula and reach the highest elevation, about 3000 m amsl, at the peaks of Al-Akhdar mountains in northern Oman. These mountains are separated from the Gulf of Oman in the east by the coastal plain of Al-Batinah and are bounded from the west by the central plateau, which is formed of a number of plains covered by pebbles and sand deserts. The western and northern fringes of the central plateau are protruding into the large desert of Ar-Rub Al-Khali. The region of Dhofar, Oman, is separated from the mountains in the north by 200 km long sand deserts. It also separates the central plateau in Oman from the Hadramaut plateau in the southern part of Yemen.

The remaining ground in the eastern part of the Arabian Peninsula slopes gently towards the Gulf, with elevations varying between 200 m and 0 m amsl. The only exceptions from this are the central and northern mountain ranges in the United Arab Emirates, with peak heights of about 2000 m in the north and about 1200 m in the south.

The Arabian Peninsula has a coastal line about 7400 km long. The length of coast line in km per country is: 120 for Bahrain, 300 for Kuwait, 1700 km for Oman, 420 km for Qatar, 2250 km for Saudi Arabia, 600 km for the U.A. Emirates and 2000 km for Yemen.

The existence of innumerable wadis is another feature of the Arabian peninsula. Some of them have cut their sections in deep, narrow canyons, the rest are shallow and broad. Depending on the nature of the ground surface, like slope and land use, some of the wadis are not able to carry the flow very far. In such cases water disappears rapidly by seepage from the beds and walls. A good example is the case study reported by FAO (1981) on one of the wadis in Saudi Arabia.

CLIMATE OF THE ARABIAN PENINSULA

General

The geographical position and relief of the Arabian Peninsula are the main factors affecting its climate. The region is to a large extent a transitional area between the equatorial and mid-latitude climates. A characteristic feature of all subtropical latitudes, of which the Peninsula occupies a fairly large area, is the prevalence of aridity.

The average daily global radiation in the peninsula is high, in the order of 500 cal cm^{-2} d^{-1} (7700 MJ m^{-2} y^{-1}) in the zone between the 20°N and 30°N latitudes, falling down to the north and south. The figures for radiation in the summer are somewhat larger than the winter figures. The high radiation figures are coupled with a high relative sunshine, 70 to 90%. The annual number of sunshine hours varies from 3500 in the interior to 2800 along the southern shores of the Red Sea and the Gulf of Aden. The cloudiness in the peninsula is generally moderate to scarce depending on the season and the location. Very likely, the interior of Saudi Arabia is the location with the least cloudiness all year round. Generally speaking, the total daytime hours per year are in the order of 4000. The shortest day length is about 10 h and the longest about 14 h depending on the season of the year.

According to Martyn (1992), the weather and climate of the region are formed under the influence of a variety of seasonal combinations of high- and low-pressure centers superimposed on the solar conditions. The Mediterranean depressions make the weather in the cool half year. Most of the cyclones, mainly in the winter and spring, move eastwards across Mesopotamia to the Persian Gulf and the Gulf of Oman. From time to time in spring, depressions from northern Egypt reach the Arabian Peninsula where they affect the weather along the Red Sea as far as the Kamaran Island and possibly even as far as Bahrain. In the cool half-year maritime polar air transformed over the Mediterranean Sea moves in towards the Red Sea and the Syrian Desert. The warm half-year is characterized by masses of maritime tropical air from the Indian Ocean moving across the Gulf of Oman and the Persian Gulf. Over Mesopotamia and the center of the Arabian Peninsula these air masses are transformed into continental air and as such can reach the eastern Mediterranean. In summer, the south of the peninsula comes under the influence of equatorial air, which arrives at the Arabian Sea coasts from the Somalia basin, and at the south of Yemen and Aden at great heights from the Indian Ocean. For more information about the general circulation of the atmosphere and the factors controlling it, the reader is referred to, among others, Lockwood (1974).

The coldest month of the year all over the peninsula is January. The maximum temperature along the Red and Arabian Sea coasts varies in this month from 20 and 25°C. The corresponding temperature along the Persian Gulf and the Gulf of Oman is somewhat cooler as it varies between 15 and 20°C. The nighttime temperature is low in the north of Saudi Arabia and the Yemeni highlands where it falls to below 5°C, temperatures as low as −7°C can be reached. Both on the Red and the Arabian Sea coasts and in Ar Rub Al-Khali desert the minimum is above 10–15°C. The southern end of the Red Sea and the Gulf of Aden have nighttime temperatures of above 20°C. The warmest months on the other hand are July-August. In these months mean maximum temperatures of 35°C and absolute maximum of 45°C or more are reached in several parts of the Peninsula. Along the coasts of the Red Sea and the Gulfs of Aden and Oman the minimum temperature exceeds 25°C, and between Kuwait and Qatar 30°C.

The relative humidity of the atmospheric air varies depending on the distance from the sea coast as well as the elevation. For the same location the humidity varies with the time of year. The summer season is the driest period of the year, varying from 20–30% in the interior to 50–65% along the seashores. These figures increase by at least 20% in the winter season, according to Martyn (1992).

The north of the Arabian Peninsula experiences dry continental northeasterly winds in the winter season. These winds become westerly and northwesterly in the summer. Easterly winds are the rule in the Gulf of Oman and the Persian Gulf. The south coast of the peninsula is exposed to easterly winds; at altitudes of 3000 m the winds are southwesterly. Certain areas are typified by local winds; for example, Al-Kharif (autumn), which blows in the Gulf of Aden, and the Shamal (north) in the north of the Persian Gulf.

The tables in Appendix 1 include the monthly and annual averages of the air temperature, T°C; mean relative humidity, RH%; sunshine duration in h d^{-1}, S; global radiation in cal cm^{-2} d^{-1}, R; wind speed in m s^{-1} at height of 2 m, u_2; precipitation in mm mo^{-1}, p; and free water evaporation in mm d^{-1}, E_0.

Precipitation

The principal mechanisms responsible for the rainfall are the cold frontal troughs, many of which originate and develop over the north Atlantic Ocean and the Mediterranean Sea, and the tropical cyclones which originate over the Arabian Sea. The cyclone tracks over the peninsula can be seen from Figure 2.

The dry season in the peninsula south of latitude 30°N lasts from April-May to October-November. In this period the monthly precipitation is generally less than 10 mm mo^{-1}. The Asir and the Yemeni

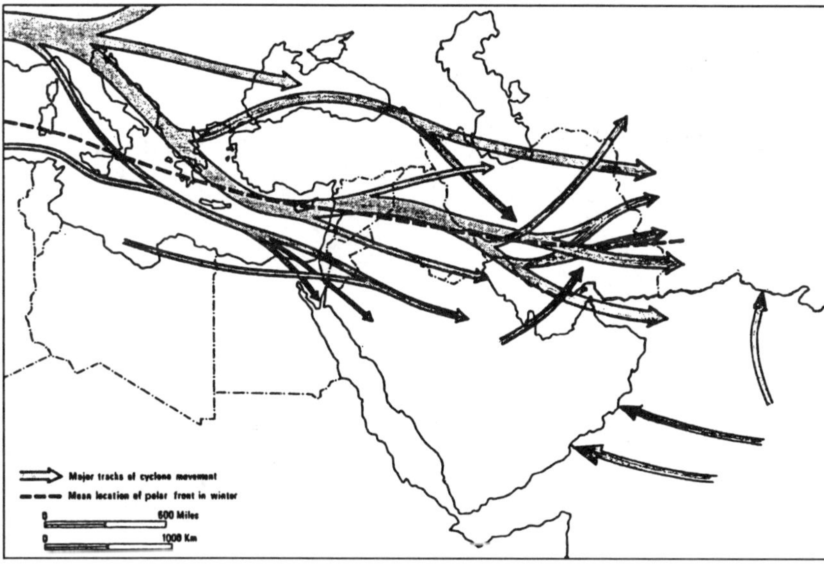

Figure 2. Cyclone tracks over the Arabian Peninsula (Beaumont et al., *1976; Ministry of Water Resources of Oman, 1995).*

mountains are exceptions as they receive in March and August, respectively, more than 50 mm mo^{-1}. The rainfall on the latter is essentially caused by the southwest monsoon (Monsoonal Intertropical Convergence Zone), which upon deflection to the west rises up high and blows across the mountains of Yemen. The winter rainfall on these mountains is caused by the jet winds, which blow northwestwards or southeastwards and meet at around 20°N (Red Sea Convergence Zone) to form a regional front bringing rainfall to several areas.

The annual rainfall is generally limited. With the exception of a maximum rain depth of about 800 mm y^{-1} falling to the west of the main watershed west of Sana'a, the annual rainfall in the largest part of the peninsula falls below 100 mm. This can be seen from Figure 3 (UNDP, 1992). The annual rainfall has been estimated at about 219 × 10^9 m^3 corresponding to an overall average of 72 mm. The distribution of these figures is given in Table 1 (Shahin, 1989).

Shahin (1996) gave tables including the mean annual rainfall depths at about 180 locations in the Arab Region. Thirty-five of them are distributed among six countries in the Arabian Peninsula: 1 in Bahrain, 6 in Kuwait, 2 in Oman, 10 in Qatar, 6 in Saudi Arabia and 10 in Yemen. The tables also give the coordinates and the number of years

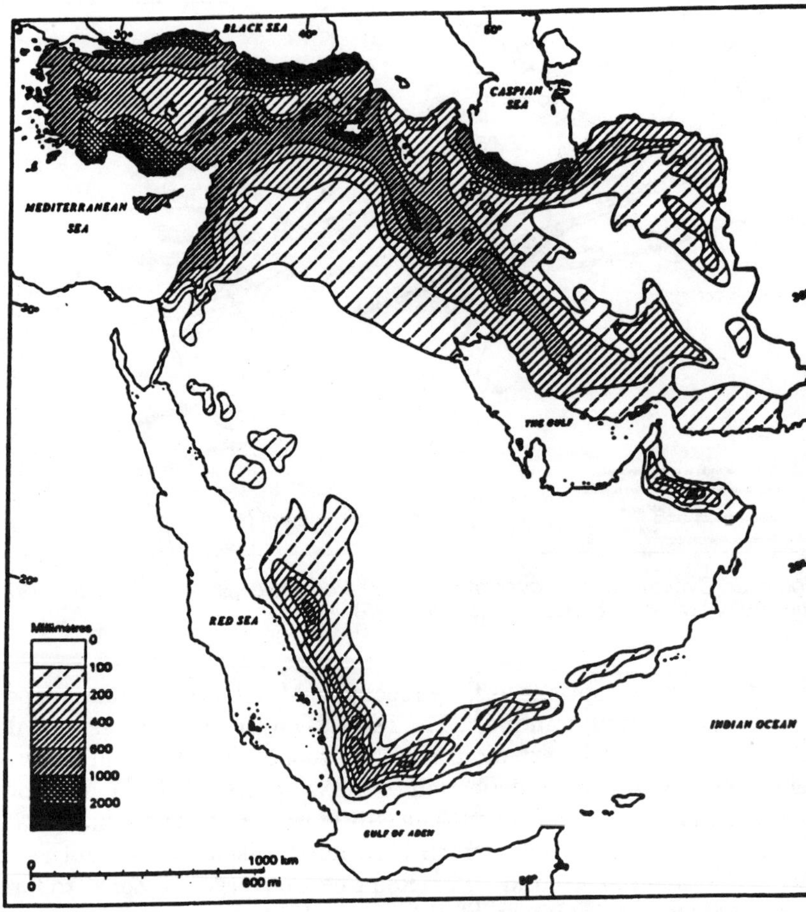

Figure 3. Annual rainfall on the Arabian Peninsula (UNDP, 1992).

of record for each station. They also include the coefficient of variation, CV, and the relative interannual variability, $IAV_{rel.}$, in percentages. These parameters can be expressed as follows:

$$CV = n\,[\Sigma\,(P_i - P_m)^2/(n-1)]^{1/2}/\Sigma\,P_i \quad (1)$$

where,

P_i = the annual depth of rain for year i, i = 1,2, . . . , n
P_m = mean annual rainfall = $\Sigma\,P_i/n$
Σ = summation of terms for i from 1 to n

Table 1. Annual rainfall per country in the Arabian Peninsula (Shahin, 1989).

Country	Annual Rainfall mm	Annual Rainfall 10^9 m^3	Country	Annual Rainfall mm	Annual Rainfall 10^9 m^3
Bahrain	80	0.05	Saudi Arabia	59	126.2
Kuwait	135	2.4	U.A. Emirates	89	7.4
Oman	71	15.0	Yemen	139	67.15
Qatar	73	0.8			

and

$$IAV_{rel.} = 100 \, \Sigma \, | \, P_{i-1} - P_i \, | \, / \, P \, (n-1), \, \% \tag{2}$$

where,

$$\Sigma = \text{summation of terms for i from 2 to n}$$

Due to the limited range of the annual mean rainfall for the thirty-five investigated stations and the wide variability of each CV and $IAV_{rel.}$ for about the same P_m, any attempt to draw a single relationship between either CV or $IAV_{rel.}$ and P_m does not make much sense. All that can be said is that there is a tendency for each CV and $IAV_{rel.}$ to become smaller with increasing P_m.

It is unfortunate that one can hardly find in the published literature any adequate data for the purpose of analyzing annual rainfall series. However, analyses of two rainfall series have been found in literature of limited distribution.

The Aden rain observations started in 1870 at Steamer Point and continued till 1936, with seven missing years in the period 1871–1886. There is a continuous record from 1940–1989 for the nearby Aden Khormakser station. Upon testing the two series it has been concluded that both series can be considered as samples of the same population, even at a high significance level. This means that the two series can be combined into one homogeneous series, with a mean of 50.8 mm and CV of 0.97. Further analysis has led to a number of conclusions, most important of which is that there is no significant trend nor well-defined cycles of wet and dry years. Instead, short periods of above-average rainfall (1–3 years) alternate with irregular, but usually longer periods of below-average rainfall (Ministry of Oil and Mineral Resources of Yemen and TNO Institute of Applied Geosciences, 1995).

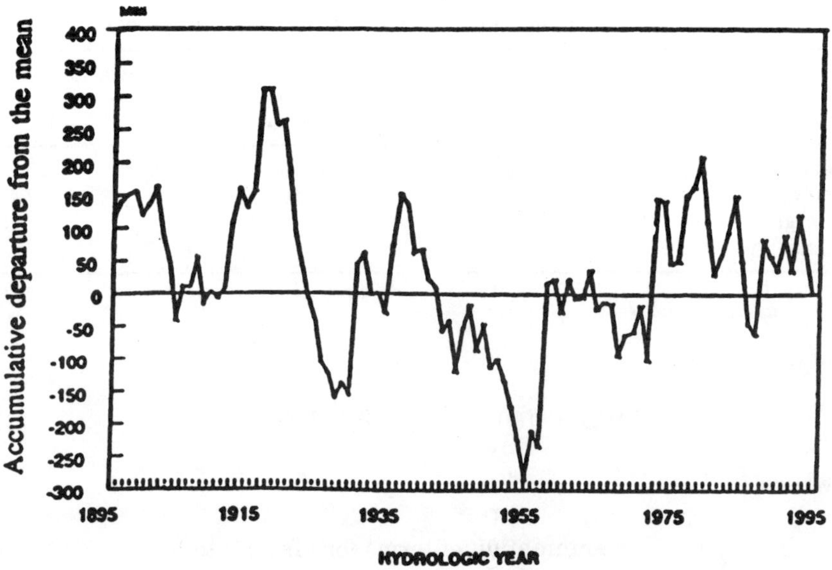

Figure 4. Cumulative departure of the annual rainfall from the long-term mean, Muscat (Ministry of Water Resources of Oman, 1995).

The second case is that of the cumulative departure from the long-term mean plot of the annual rainfall at Muscat. Figure 4 shows sustained periods of above-average and below-average rain-fall (Ministry of Water Resources of Oman, 1995). Quite remarkable is the nine-year period, 1919–1927, of below-average rainfall. The persistence of such a long dry sequence is a major challenge to effective water resources management in Oman.

The tables in Appendix 1 include the mean monthly rainfall totals at four stations in Qatar, Oman, Saudi Arabia and Yemen. Recently, Farquharson et al. (1996), in a paper on rainfall and runoff in Yemen, included a table of monthly and annual rainfall at six additional stations in Yemen. The monthly distribution of the annual rainfall at each station depends on the location. Nevertheless, the months of maximum rain-fall remain, as mentioned earlier, April-May and August-September. Mean monthly totals of above 100 mm have been observed at Taiz and Dhalla. In the same paper, the authors suggested that the average fall per rainday is nearly constant throughout Yemen; the average value of storm rains of over 5 mm depth was found to be 17 mm. Therefore, it has been concluded that storm cells are able to produce the same magnitude of rainfall across the country and that the difference in monthly or annual rainfall between locations is due to differences in the number of storms.

In a paper on the Northern Oman flood study, Wheater and Bell (1983) gave a summary of Muscat rainfall data for the period 1893–1959. The annual mean for this period was 106 mm, or about 15% larger than the mean for the period mentioned in Appendix 1. The maximum monthly rainfall reached 171.2 mm in December and 143.0 mm in January. The corresponding maxima for Al-Manama, Bahrain, for the period 1903–1987 (1913–1928 missing) were 97 and 136 mm, respectively. For the same period the mean was 14.4 mm for December and 17.1 mm for January.

The maximum rainfall in each month of the year at nine stations, three stations in each of Kuwait, Saudi Arabia and Oman, is included in Appendix 2.

Wan (1976) studied point rainfall characteristics of the western part of Saudi Arabia using four groups of rain-gauging stations, representing annual rainfall zones of 90–140, 140–200, 222–325 and 325–450 mm. The results made it possible to establish depth-duration ratios for durations of 5, 15, 30, 120 min. and 24 hours. The ratios of the depths for the various durations to the one-hour rainfall of the same return period are included in Table 2. From that investigation it was concluded that the ratios had practically constant values within each duration class, indicating that they are independent of the return period. In addition, the results for rainfall durations of two hours or less, except those for the stations in the lowest rainfall zone, agreed remarkably with the corresponding ratios derived from the U.S., Australia and the U.S.S.R.

In a somewhat similar study, Wheater and Bell (1983) developed an intensity-duration-frequency relationship for short-duration rainfall in northern Oman, based on the station-year method. The derived rela-

Table 2. Depth-duration ratios for rainfall in Saudi Arabia (Wan, 1976).

Station Group	Mean Annual Rainfall, mm	Ratio of Rainfall of a Given Duration to 1 h Rainfall				
		5 min	15 min	30 min	120 min	24 h
A	109	0.37	0.65	0.85	1.20	1.67
B	176	0.31	0.57	0.78	1.19	1.07
C	278	0.28	0.53	0.75	1.21	2.28
D	388	0.29	0.57	0.81	1.21	2.08

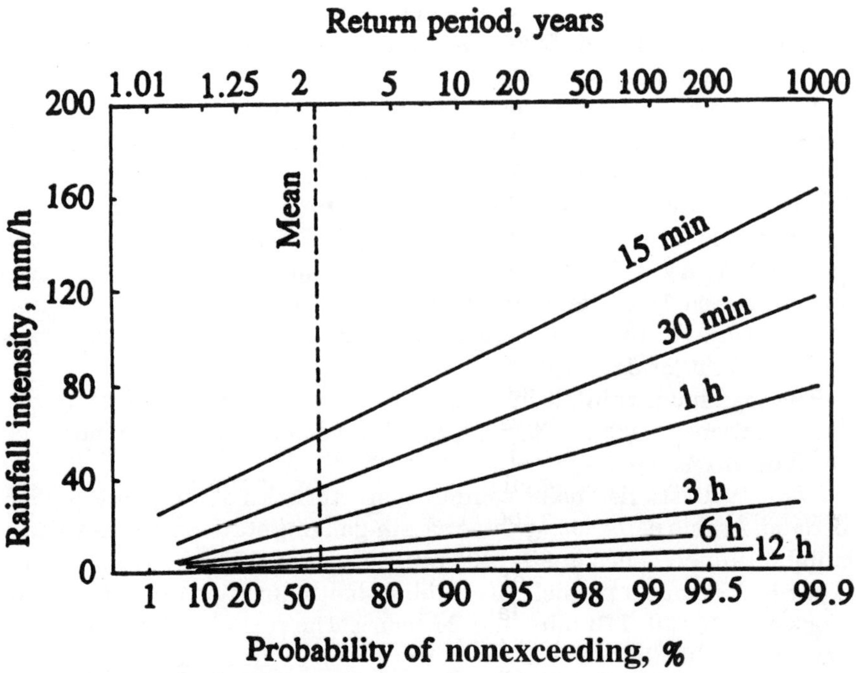

Figure 5. Intensity-duration-frequency relationship for short duration of rainfall in northern Oman (Wheater and Bell, 1983).

tionship is shown graphically in Figure 5. It was also found that the ratios obtained for the least rainfall zone, zone A (90–140 mm y^{-1}), in Saudi Arabia agree well with the ratios developed for northern Oman. For example, the ratio of 30 min to 24 h rainfall in Saudi Arabia is 0.51 for zone A (0.32–0.49 for other zones), whereas the value for Oman is 0.56.

Evaporation and Evapotranspiration

Accurate information about the loss of water by evaporation from free water surfaces and by evapotranspiration from vegetated or cultivated surfaces is necessary for drawing up water and moisture balances, respectively, as well as for classifying the climate.

Evaporation estimates are usually obtained from hydrologically and/or climatologically based methods. The water balance is often regarded as the best method for estimating evaporation. Other estimation methods include the heat balance, the bulk aerodynamic (Dalton) and the combined approach (Penman).

The latter approach requires data about extraterrestrial radiation, temperature, relative humidity, cloudiness, horizontal wind speed and albedo. Measurement of the evaporative capacity of air is taken by means of the Piche tube and evaporation pans, especially the United States Weather Bureau (USWB) Class A pan.

The Penman-Monteith equation, used frequently for estimating the potential or reference evapotranspiration, requires additionally parameters related to atmospheric resistance as well as bulk stomata resistance.

The studies on Penman's evaporation at Wadi Najran and Uqulat Suqur (FAO, 1981), the measurements obtained from the Al-Hassa area and later the estimates of evapotranspiration at the observation stations of Ar-Riyadh, Karaj and Dirab (Salih and Sendil, 1984; Sha'lan and Salih, 1987), all in Saudi Arabia, are probably the most informative published materials on evaporation and evapotranspiration in the Arabian Peninsula.

Estimation of the free water evaporation using the Penman method (Penman, 1948) requires a large number of meteorological factors that are not readily available. As a result, early evaporation studies in Saudi Arabia have been carried out with the aim of reducing these factors. The then available data at the Uqulat Suqur meteorological station were used. Figure 6 shows a graphical plot of predicted Penman, $E_{o(predicted)}$, using a reduced number of factors, versus the calculated Penman using all factors, E_o, for 6 years (72 months).

The simple linear relationship between these two variables is

$$E_{o(predicted)} = -2.403 + 0.0184\,R_c \tag{3}$$

where R_c is the incoming radiation in cal cm^{-2} d^{-1}.

The data of the USWB Class A pan at Riyadh and other stations in Saudi Arabia in the period 1965–1986 (Sha'lan and Salih, 1987) multiplied by a conversion factor of 0.75–0.85 show that the free water evaporation at those locations falls in the range from 7.0 to 8.2 mm d^{-1}, or 2500–3000 mm y^{-1}.

The data of R. Al-Faras station, Qatar, using the same values of the conversion factor give much less evaporation, in the range from 1750 to 2150 mm y^{-1}. This result is not different from what would be expected as this station represents coastal conditions, whereas Riyadh represents a harsh interior climate. Al-Manama, Bahrain, is similar to

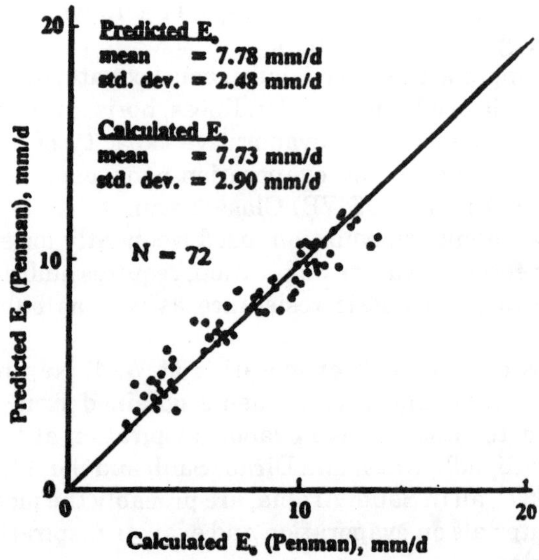

Figure 6. Calculated E_o versus E_o using only R_c, both for Penman evaporation (FAO, 1981).

R. Al-Faras, Qatar, with an average daily free water evaporation of 5.3 mm, or about 1940 mm y^1.

Another location along the sea coast, though in a hotter environment, is Al-Seeb, Oman, where the Piche evaporimeter is still used. Assuming a conversion factor of 0.75 and an average annual piche evaporation of 9.2 mm d^{-1}, the equivalent free water evaporation is in the order of 2500 mm y^{-1}.

Sana'a is characterised by a high elevation, about 2250 m amsl, and a relatively moderate temperature. The estimated average daily evaporation using Penman's method is 5.6 mm. Using a reduction factor of 0.85 yields 4.85 mm and 1750 mm as equivalent figures for the daily and annual free water evaporation.

The crop water use, ET_c, is by definition the maximum evapotranspiration from any crop whereas the grass reference evapotranspiration, ET_o, is the maximum water loss from clipped grass surfaces having 0.12 m height and a bulk surface resistance of 70 cm^{-1} (Allen, 1994).

The evapotranspiration studies carried out in 1970–1971 at Al-Hofuf Agriculture Research Station, Saudi Arabia, are probably the most pioneering in this field in the Arabian Peninsula. They were continued later, 1980–1981, at the Experimental Station of King Faisal University, Saudi Arabia. In these investigations alfalfa was used as a reference crop. Estimates of ET_0 and ET_r for the surrounding locations were based on temperature, relative humidity, radiation, combination and pan methods. ET_r is the maximum evapotranspiration from a soil covered with alfalfa grass 0.5 m high.

This writer has recently reviewed the published results of all those experiments and combined them with estimates obtained for other countries in an attempt to cover a larger surface in the Arabian Peninsula (Shahin, 1998). A total of fourteen stations covering the countries of the peninsula were used. The locations of these stations are indicated on Figure 1. Other particulars of interest are listed in Table 3.

The methods used for estimating the grass reference evapotranspiration, ET_0, were: the FAO Blaney-Criddle and the FAO-Penman as described by Doorenbos and Pruitt (1992) with correction for elevation included, the Hargreaves and Semani (1982, 1985), and the FAO-Penman-Montieth described by Allen (1994).

Table 3. Data on the meteorological stations used in the study of evapotranspiration by Shahin (1998).

Ser. No	Country Place	Location of Station	Latitude N	Longitude E	Altitude m	Period of Record
	Kuwait					
1	Shuwaikh	Coastal	29°20′	47°57′	11	1965-89
	Bahrain					
2	Manama	Coastal	26°12′	50°36′	—	1989–93
	Qatar					
3	R. Faras	Coastal	25°49′	51°20′	14	1972–92
4	A. Samara	Coastal	24°44′	50°50′	3	1976–92
	Saudi Arabia					
5	Al-Hofuf	Interior	25°30′	49°35′	160	1970–71
6	Riyadh	Interior	24°39′	46°43′	564	1970–81
7	Al-Kharj	Interior	24°10′	47°24′	430	1973–81
8	Dirab	Interior	24°25′	46°34′	630	1977–81
9	U. Suqur	Interior	25°50′	42°11′	—	1970–72
10	Najran	Highland	17°38′	44°20′	—	(6 y)
	Oman					
11	Muscat	Coastal	23°36′	58°38′	5	1981–82
	Yemen					
12	El-Kod	Coastal	13°05′	45°20′	5	1982–83
13	Hodeida	Coastal	14°48′	42°57′	10	1983–90
14	Sana'a	Highland	15°22′	44°13′	2240	1982–86

The methods used for estimating the alfalfa reference evapotranspiration, ET_r, are Jensen and Haise (1963) and Jensen (1966). The results are given in Appendixes 3(a) and 3(b).

The grass reference evapotranspiration for coastal areas in the Arabian Peninsula varies from a minimum of about 1700 mm y^{-1} for El-Kod, southern Yemen, to a maximum of about 2200 mmy^{-1} for Muscat, Oman. Since the estimate for El-Kod was based on the meteorological data of one year only, it should be used very cautiously. Other than these two extreme stations (i.e., El-Kod and Muscat), the estimates for the others fall in the range from 1850 to 2150 mm y^{-1}. The effect of high temperature on ET_0 at the coastal zone throughout the year is counterbalanced by high humidity, the small difference between the mean maximum and mean minimum temperatures as well as the cloudiness.

Though all are located in the interior, the climates of Sana'a, Yemen, and the group of Riyadh, Kharaj and Dirab stations in Saudi Arabia are very different, rendering the corrcsponding estimates of ET_0 different too. A difference in elevation of about 1500 m helps to bring the mean annual temperature of Sana'a to at least 8°C below the average of the three Saudi stations. Further, the annual rainfall at Sana'a makes its climate less arid than that of Saudi Arabia. Last, there is a difference of some 20% in the global radiation between the Yemeni station and the Saudi stations. All these factors combined bring the estimate of ET_0 at Saana'a to no more than 75% of the average ET_0 estimate for Riyadh, Kharaj and Dirab.

ARIDITY

The increasing availability, on a global scale, of climatic data on air temperature and precipitation has motivated meteorologists and other scientists to search for criteria for classifying the climate. Actually this search started more than a century ago. Since then several classification systems have been developed. Arid climate is one of the climate classes, semi-arid is another class. The two classes combined, refer to those areas of the earth's surface that receive less precipitation than is needed for a full utilization of heat and light capacities in the synthesis of plant matter. In other words, in plant production the deficiency of water is the most effective controling factor.

In the next subsections we will discuss four approaches to quantifying the extent of aridity of the Arabian Peninsula.

Aridity Indexes

Köppen (1936) suggested that the limits of arid and semi-arid regions in the subtropics may be established in terms of the mean annual precipitation, P_m in mm, and the mean annual temperature, T_m°C, by the relations

$$P_m = 10\,(T_m + 7) \tag{4}$$

and

$$P_m = 20\,(T_m + 7) \tag{5}$$

Köppen adjusted Equation (5) for semi-arid zones to read

$$P_m = 20\,(T_m + 14) \tag{6}$$

and

$$P_m = 20\,T_m \tag{7}$$

for areas with summer and winter rainfall, respectively.

According to the above equations the limits of aridity dictated by the conditions for the four locations listed in Appendix 1 are: $P = 331$ mm for Rahdat al Faras, Qatar; 355 mm for Al-Seeb, Oman; 319 mm for Al-Ryiadh, Saudi Arabia; and 253 mm for Sana'a, Yemen. For these locations, in the same order, the limits of semi-aridity are: 522, 570, 498 and 646 mm, respectively, assuming that the rainfall in the first three stations takes place in the winter time and in the last station takes place in the summer time.

The conclusion one can draw is that the areas these four stations represent are arid lands.

Thornthwaite (1948) suggested using the potential evapotranspiration, ET_p, instead of temperature, with precipitation for classifying the climate. This suggestion was further elaborated by FAO/UNESCO in "International Arid Zones Research Programme" (UNESCO, 1977). Since then boundaries between the different climatic zones have been set accordingly. The limiting values of the index P/ET_p are as follows:

Hyper-arid	$P/ET_p < 0.03$
Arid	$0.03 \le P/ET_p < 0.20$
Semi-arid	$0.20 \le P/ET_p < 0.50$
Subhumid	$0.50 \le P/ET_p \le 0.75$

The potential evapotranspiration, ET_p, as presented and discussed in the previous section varies from across locations in the peninsula. The value of ET_p falls from 1800 to 2500 mm y^{-1}. From the rainfall data in Appendixes 1 and 2 the P/ET_p index for the major part of the peninsula falls in the range of 0.03–0.20. Only in a small part of Yemen, where the annual precipitation equals or exceeds 500 mm, can the climate be described as semi-arid. In several parts of the peninsula, where the annual rainfall is at or less than 75 mm and the potential evapotranspiration is at or more than 2500 mm, can the climate be described as hyper-arid.

A rainfall-temperature index has been developed by Bagnouls and Gaussen (1953) to define the wet season of the year. According to this approach, only months with $p/T'_m > 2$ can be regarded as wet. In this relationship, p is the monthly rainfall in mm and T'_m is the mean monthly temperature in degrees centigrade. Applying this criterion to the four stations listed in Appendix 1, one can conclude that of the forty-eight station months only March and April at Sana'a are wet months, all remaining months are dry. In other words, for a winter month in the peninsula to be wet, the rainfall must be between 35 and 45 mm depending on the location; for a summer month it must be beween 45 and 70 mm.

Budyko (1956) related the net radiation, R_n, available to evaporate water vapor from a wet surface to the heat required to evaporate the mean annual precipitation, P_m. He denoted the ratio R_n/PL, where L is the latent heat of vaporization, by the radiational heat index of dryness. The index values for the different environments are as follows:

Environment	Index
Tundra	$R_n/PL \leq 0.33$
Forest zone	$0.33 < R_n/PL \leq 1$
Steppe	$1 < R_n/PL \leq 2$
Semi-desert	$2 < R_n/PL \leq 3.40$
Desert	$3.40 < R_n/PL$

Assuming that R_n is 40–50 k cal cm^{-2} y^{-1}, P = 100 mm y^{-1} and L is about 60 cal cm^{-2} mm^{-1}, the value of the index R_n/PL can be found equal to 6.7–8.3 (i.e., larger than 3.4). This result leads to the conclusion that the overall environment of the peninsula is desert. Only those areas with annual precipitation of 300 mm or more can be described as semi-desert. The Yemeni mountains receiving annual precipitation of 600 mm or more fall in the steppe class according to the dryness index of Budyko.

The relative interannual variability of rainfall was investigated by Wallén (1968) in relation to the possibility of dry-land farming in the

Middle East. The expression developed by Wallén for IAV_{rel} was given earlier in the discussion of rainfall.

Wallén concluded from his studies on the Middle East that the normal minimum precipitation conditions for dry-land farming is 240 mm of annual rainfall and 37% for the interannual variability from the mean. In the mountainous areas where the relative interannual variability is particularly high, a greater mean annual rainfall total is necessary to ensure dry-land farming than in lowland areas. Since these conditions are hardly met in the largest part of the peninsula, one cannot rely on the possibility of having dry-land farming except in a few limited areas.

In addition, in order to descibe a season as being wet, it is essential to consider not only the mean monthly rainfall but also its variability. For the Middle East, Wallén suggested that the start and end of the reliable rainfall season be considered to correspond to a monthly rainfall of 25 mm and a variability of 100%. For this purpose he developed a parameter q expressed as

$$q = p / IAV_{rel} \tag{8}$$

The value of q = 0.25 has been adopted to establish the dates of the beginning and the end of the wet season.

The results from the above-mentioned methods confirm the arid to hyper-arid nature of more than 90% of the surface covered by the Arabian Peninsula.

RUNOFF

General

The rainfall that reaches the ground either evaporates, infiltrates into the soil, or lies in storage in natural and/or artificial depressions. After the losses arising in these ways, any surplus may run off along or beneath the ground surface to the nearest stream channel. This surplus water is often referred to as the runoff. In the Arabian Peninsula the streams collecting the runoff are known as wadis.

According to Wilson (1969), when the rainfall is particularly intense or prolonged, or both, surplus runoff becomes too large for the stream network to carry. As a result, the streams become filled and overflow and in doing so may cause harm to nature as well as to human activities. This kind of event occurs periodically in certain areas in the peninsula. One of the most devastating effects of these floods is the washing away of the fertile top soil in which crops are grown. In urban areas floods can cause great damage to property, pollution of water supplies, threat to human lives and often total disruption of communications.

Runoff Coefficient

The runoff coefficient can be defined as the ratio of the runoff to the rainfall. It varies from near zero for a small storm to a relatively high value for a major storm and is dependent upon the initial moisture conditions. Since the runoff coefficient varies with the magnitude of the rainfall, at least up to a certain depth, this coefficient, in general, must be small for the wadi basins in the Arabian Peninsula. However, the available observations of wadi flows in Yemen do not support this universally accredited conclusion.

On the question of runoff in Yemen, Farquharson et al. (1996) commented that the intermittent nature of wadi flows and the instability of many sites in the alluvial beds of the wadis make it difficult to measure runoff without continuous calibration of gauging sites. The difficulties of access have meant that reliable long-term records of runoff are available only for the main wadis. At other sites the flow series are too short and the only way by which reasonable lengths of record can be obtained is to extend the measured flows using measured and generated rainfall data. In the report on Water Resources Assessment of Yemen (Ministry of Oil and Mineral Resources of Yemen and TNO Institute of Applied Geosciences, The Netherlands, 1995), it has been shown that the appropriate figure for the runoff coefficient of ungauged wadis is 0.055. Shahin (1996) found that the runoff depth, R, in mm, for the Tihama Plain and the eastern escarpment in Yemen, can be estimated from the approximate relationship

$$R = 0.08\,(P_m - 100) \tag{9}$$

where P_m is the mean annual rainfall in mm.

The tables in Appendix 4 give the average values of the runoff coefficients for a number of gauged and ungauged wadis in some countries in the Arabian Peninsula.

Flood Prediction

Flood hydrology of arid zones has till the present been largely unquantified with the exception of a few flood studies in northern Oman and southwestern Saudi Arabia.

The floods in northern Oman were studied by Wheater and Bell (1983), who determined the peak discharge of Wadi Adai using the slope-area method after adopting a roughness coefficient based on experience with similar terrain. This discharge has been used for calibrating a certain numerical model based on the kinematic wave approximation suggested for estimating the response of the wadi catchment to a well-recorded severe rain storm. The model proved capable of reproducing the magnitude and time of the occurrence of the peak discharge quite satisfactorily. The return period of the peak discharge, 1160 $m^3 s^{-1}$, of that wadi has been reported to correspond to 100–300 years.

Three methods of flood prediction have been investigated by Nouh (1988) using the records of thirty-two gauged stations in the southwest region of Saudi Arabia. In the first method, regional curves were developed to establish the ratio of the peak discharge corresponding to a certain return period given that return period (Figure 7). The mean annual flow was estimated from the characteristics of the wadi basin. The procedure underlying the second method is to estimate the annual maximum rainfall for a given duration and return period and then apply this estimate to a certain model for predicting the annual maximum flood flow that corresponds to the same return period. In the third method, flow duration curves are developed and used together with the mean annual calendar-day flood flow to predict the peak flood flow for a given return-period. This parameter (i.e., the mean annual calendar-day flood flow) is estimated from the characteristics of the drainage basin. The study concluded by recommending that the first method be applied to any site in the region whenever reliable data are available to compute the mean annual flow at that particular site.

The use of frequency analysis of peak wadi discharges for hydrograph design was suggested by Wheater and Brown (1989) as a result of their study of the water resources in southwest Saudi Arabia for the period 1983–1988.

To investigate the possible magnitudes of maximum floods, historical flood records in Saudi Arabia, Yemen and Oman have been collected and, together with estimates of probable maximum floods, plotted against the drainage area of the corresponding wadis as shown in Figure 8. The probable maximum floods as used here are meant to be the highest floods recurring, on the average, once in 10,000 years. In Figure 8 the so-called Creager Envelope curves with C-values of 10, 25, 50 and 100 are drawn.

The enveloping curve of C = 25, supposedly corresponding to 50 years' return period, exceeds all historical floods. On the other hand,

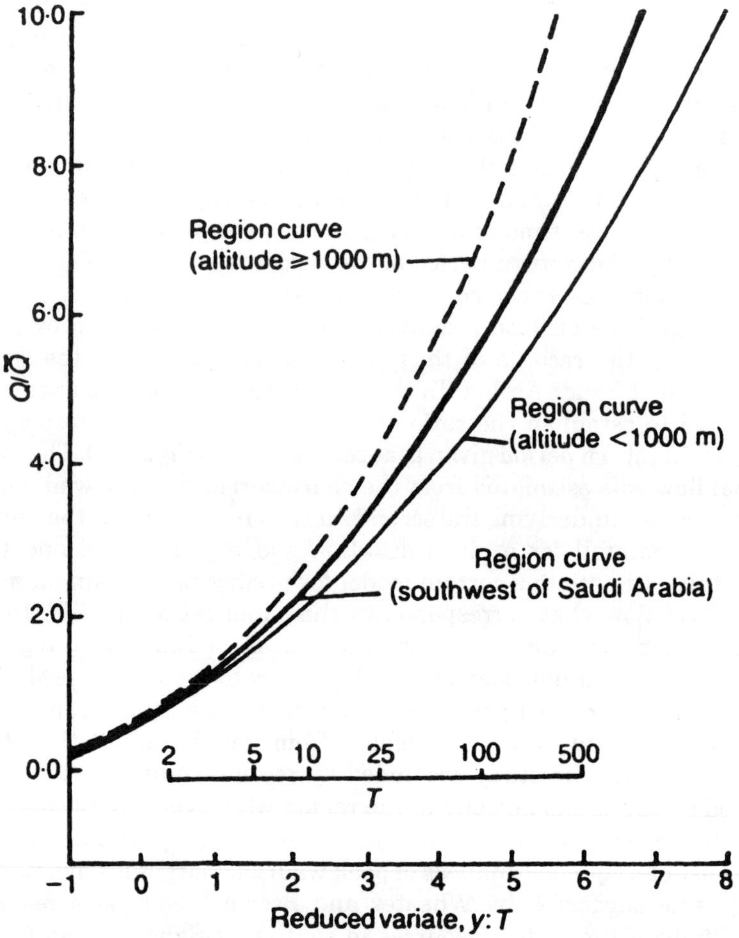

Figure 7. Regional curves for southwest Saudi Arabia (Nouh, 1988).

the enveloping curve of C = 80 narrowly exceeds the probable maximum floods estimated from the relationship:

$$\log Q = \log 1.3\ C + (0.935\ A^{-0.048} \log 0.358\ A) \qquad (10)$$

where A is the catchment area in km^2. So, for a catchment area of 1000 km^2 and C = 25, the flood peak should be about 390 $m^3\ s^{-1}$. The esti-

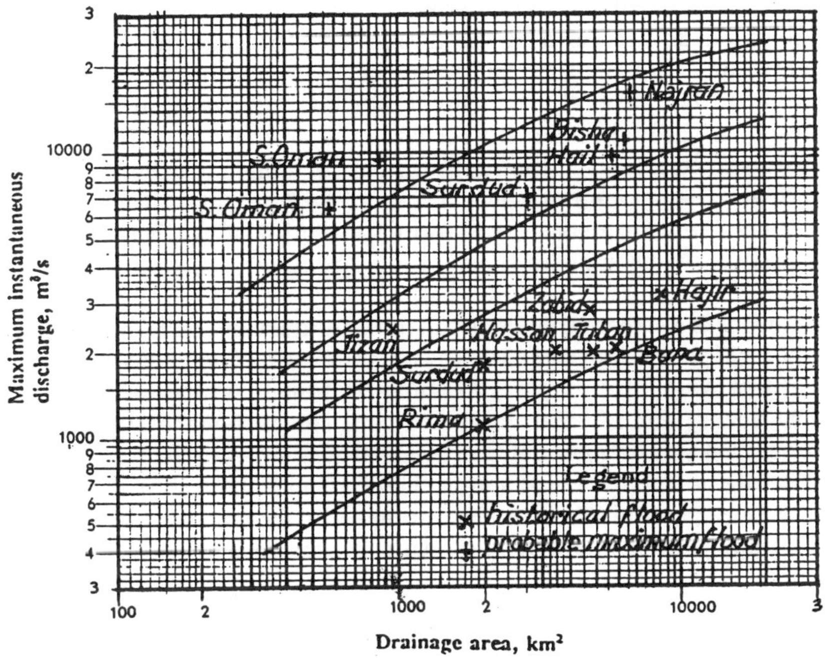

Figure 8. Historical and probable maximum floods of wadis in Saudi Arabia, Oman, and Yemen (ECOSOC ECWA, 1981; Saudi Arabia, 1986).

mated peak discharge increases to 2680 m^3 s^{-1} as the catchment area increases to 2500 km^2.

SURFACE WATER RESOURCES

Wadi Flows

In the section on annual rainfall it was made clear that with the exception of certain parts of Yemen and Oman the rain depth hardly reaches 100 mm y^{-1}, bringing the long-term average rainfall for the peninsula to just 72 mm. As the depth per rainy spell is limited, the water that reaches the ground disappears almost immediately by infiltration. This situation affects the process of surface runoff.

The runoff that is generated by the excess rainwater after infiltration, evaporation and other losses flows into a net-work of channels known as wadis. The density of the wadi network is not the same everywhere, but varies as described in the discussion of the physiography of the peninsula. High-density wadis can be found in the western part of Saudi Arabia parallel to the Red Sea coast increasing in a southerly direction. The highest density occurs in the southern part of the Tihama Plain and continues in the southern part of Yemen parallel to the Gulf of Aden. The network of wadis becomes less dense in the coastal plains of Oman and decreases further in the United Arab Emirates. The density of wadis appears to be strongly related to the annual rainfall depth.

A wadi is a stream that runs full of water during a certain rain storm and for a short time after the cessation of rain. The hydrograph (Figure 9) of Wadi Surdud at Faj Al-Hussein, Yemen, is an example of wadi flow during and after an intensive rain storm that took place 17–19 June 1984 (Ministry of Oil and Mineral Resources of Yemen and TNO Institute of Applied Geosciences, The Netherlands, 1995). A wadi seldom flows perennially. If this is the case, the flow is extremely variable from one year to another. The recorded annual flow in Wadi Dayqah

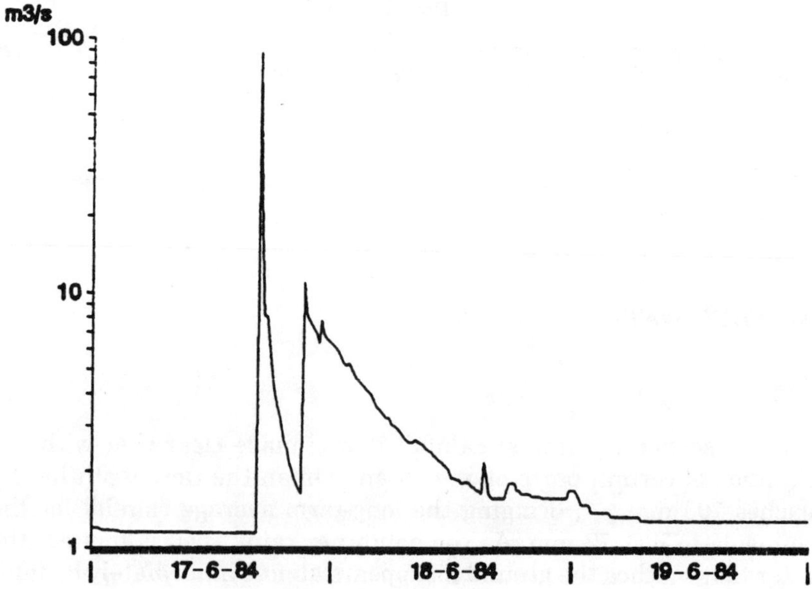

Figure 9. Hydrograph of Wadi Surdud at Faj Al-Hussein, Yemen (Ministry of Oil and Mineral Resources of Yemen and TNO Institute of Applied Geosciences, The Netherlands, 1995).

at Mazara, Oman, for example, varies from 3×106 m^3 to 192×10^6 m^3, with an annual average of 37×10^6 m^3, after excluding the exceptionally high flow of the year 1982 (Kaul, 1995).

Almost 60% of the volume of wadi flow in Saudi Arabia runs off the southwestern part, from the terrain between the Red Sea and the adjacent mountains, occupying 10% of the surface area of the country. Additionally, about 40% of the flow develops in the southern part of the Tihama Plain at the extreme south of the west coast, which represents no more than 2% of the total area of the kingdom. The total wadi flow in the Red Sea zone can be estimated at 39.8 m^3 s^{-1}, of which 27 m^3 s^{-1} (70%) occurs south of Jeddah, 8.2 m^3 s^{-1} (21%) north of Jeddah and the remaining 4.6 m^3 s^{-1} (9%) around Jeddah itself.

The wadi flow in the interior drainage system east of the Red Sea mountains is inferior in quantity compared to the runoff west of this mountain range. The largest wadi runoff in this system develops in the west and southwest where precipitation is relatively heavy. The largest volume of runoff in the interior drainage system occurs in the catchments of Wadi Ad-Dawasir and Wadi Nejran. The average flow of the former, estimated as 10.5 m^3 s^{-1}, depends essentially on the runoff of the tributary wadis such as Turabah, Ranya, Bishah and Tathlieth. The second largest yield, estimated at an average of 4.3 m^3 s^{-1}, is produced by the Asir-Najran group of subcatchments, of which the Najran group alone is responsible for 3.2 m^3 s^{-1}, or 75% of the total flow.

The wadi runoff for the different regions of Saudi Arabia is listed in the country report of Saudi Arabia (1986). Assuming that these estimates are fairly correct and complete, one can draw the conclusion that the annual total wadi surface runoff in Saudi Arabia is in the order of 2×10^9 m^3.

The available data on wadi flow in Oman suggest that the annual average runoff is 4.3 L km^{-2}. The corresponding figure for Saudi Arabia is barely 0.95 L, or just 0.22% the yield of the Omani wadis. This considerable difference between the two specific yields can be attributed in part to the difference in the annual rainfall in the two countries (see Table 1). More important in this respect, topography, land slope and other conditions needed for producing surface runoff are more favorable in Oman than in Saudi Arabia.

The country report of Oman (1986) gives the water balances of 145 wadis, which have been divided into nine groups: the Musandam Peninsula, 6 wadis; east and middle of the Batinah Plain, 11; north of the Batinah Plain, 22; the interior, 9; the eastern region, 13; Az'zahra (north interior), 29; Quryat and Sour, 4; Salalah and the southern region, 33; and Dhofar and the northern wadis, 18. From this report it is clear that the annual surface runoff of the Omani wadis is about 0.92×10^9 m^3.

Wadi catchments in Yemen are grouped into four major basins: Red Sea basin, 8 major and several minor wadis; Gulf of Aden basin, 7 major and some minor basins; Arabian Sea basin, 13 major and many smaller wadis; Ar-Rub Al-Khali basin, of which Wadi Najran is probably the most important.

According to Farquharson (1996), there are only three sites with more than ten years of record in northwest Yemen and two sites with twenty years of data from southern Yemen. At other sites the flow series are too short to be useful for planning purposes and the only method by which reasonable lengths of record can be obtained is by extending the measured flows using measured and generated rainfall data. This is different from the approach used by the water resources authority in the country.

As has already been mentioned, the flow of (ungauged) wadis is estimated as 0.055 times the areal average rainfall on the corresponding wadi catchments. The available data on wadi flows in Yemen, observed as well as estimated, suggest that the total wadi flow there is in the order of 2.15×10^9 m^3. This figure, when divided by the total surface area, 482,680 km^2, gives an average yield of 4.5 L km^{-2} y^{-1}. This is about the same as that produced by a unit surface area in Oman and about five times as much as that of Saudi Arabia. One should not forget, however, that the average areal rainfall on Yemen is about 140 mm y^{-1}, i.e., almost two times as much as the rainfall on Oman.

The surface flow in the United Arab Emirates mainly originates from flank spates in wadis, flow from springs, and the yield of the main Falajes (man-made conduits). The numerous wadi bed seepages are also included. These are generally subterranean flows forced to move up to the surface of a wadi bed by a constriction of the less permeable flanks or by a subsurface rock bar. The flow of these seepages may disappear again into the wadi bed gravels immediately downstream of the constriction.

Sometimes collection galleries are dug upstream of the spring into the wadi bed gravels to increase the magnitude of the seepages. These galleries are then roofed by stone slabs, covered with gravel. They are usually called Falajes or Aflaj in Oman and the U.A. Emirates. In Saudi Arabia and Iran these galleries are called Kanats.

It is thought that there are 150 Falajes in the Emirates. A report prepared by the Ministry of Agriculture and Fisheries of the United Arab Emirates (1982) mentioned that only 48 Falajes were located in the period 1977–1982 and these were used permanently for supplying irrigation water to the agriculture there. The flows in the existing Falajes are monitored and analyzed. These Falajes can be subdivided between the various drainage areas as follows:

1. The eastern drainage (southern area) is made up of two wadi catchments: the Wadi Ghor and Wadi Hatta/Masfut and their tributaries. These wadis incorporate 13 Falajes.
2. The western drainage (northern area) includes 5 Falajes.
3. The western drainage (central area) includes 9 Falajes.
4. The eastern drainage (eastern coast) includes 9 Falajes, of which 4 are draining Wadi Farfar and 1 is draining Wadi Al Basseirah.
5. The southern drainage (Al-Ain area) includes 10 Falajes.

The same report (1982) includes an average figure of 21.6×10^6 $m^3\ y^{-1}$ for the total flow of the Falajes in the U.A. Emirates for the period 1977–1982.

In another report (ECOSOC/ECWA, 1981) one finds old esimates of the peak flows of six major wadis and the corresponding probable flood flows dating back to 1965–1969. The sum of these volumes is about $25.3 \times 10^6\ m^3$. The total wadi flow in the Emirates was then estimated at $110 \times 10^6\ m^3$. In 1978 15 wadis out of a total of 27 wadis were monitored, 9 wadis draining their water to the west and the remaining 6 to the east. The total flow carried by those 15 wadis was estimated to fall between $102 \times 10^6\ m^3\ y^{-1}$ as an upper limit and $47 \times 10^6\ m^3\ y^{-1}$ as a lower limit. In a later study of the design of flood-retention dams (U.A. Emirates, 1986) the total annual flow of the wadis was estimated as $150 \times 10^6\ m^3$.

The above-mentioned figures show clearly that the volume of water carried each year by the wadis in the Arabian Peninsula is around 5×10^9 to $6 \times 10^9\ m^3$. This figure corresponds to a regional runoff coefficient of 2.5% to 3.0%, which seems reasonable given the harsh rainfall and physiographic conditions of the Peninsula.

Desalinated Water

The demand on water for various purposes, which has been estimated as $8.26 \times 10^9\ m^3$ for 1985 and $9.7 \times 10^9\ m^3$ for 1990, is expected to continue its rise at an average rate of 1.035 per year. This will bring the total demand in 2025 and 2030 to about 30×10^9 m^3 and 37×10^9 m^3, respectively (Shahin, 1989, 1996). These figures lead us to the conclusion that the useable rainwater, wadi flows and groundwater, as seen in the next section, are not and will not be sufficient to cover the demand on water for any of the countries in the peninsula. Other water resources are already used and will be used more in future.

Desalinated water, which is regarded as a surface water resource, is by all measures the primary freshwater source in Kuwait. Although desalinated water is so far a secondary source of water in Saudi Arabia, the United Arab Emirates, Bahrain and Qatar, the annual volume of desalinated water used in each of these countries is steadily increasing.

From the figures presented by Bushnak (1995), the total capacity of the desalination plants sold to the countries of the Arabian Peninsula in the period 1963–1993 augmented to 9.72×10^6 m^3 d^{-1}. Of this daily production, 51.6% represents the share of Saudi Arabia alone, 21.4% of the U.A. Emirates, and 15.7% of Kuwait, 6% of Qatar. The remaining 5.3% represents the combined share of Bahrain, Oman and Yemen.

As long as raw water is available and so is the source of energy, which in this case is gas and/or oil, the problems left are of economic and environmental natures. The cost-benefit analysis of water desalination is essentially influenced by the market price of the fuel, feasibility and economy of water importation and efficiency, life age and operation and maintenance costs of desalination plants. The environmental problem lies basically in the feasibility of dumping the residues left from the desalination process. The problem becomes more acute when a desalination plant is installed inland far from the sea coast.

Three major distillation processes are used commercially. These are the multiple-effect distillation (MED), multistage flash distillation (MSF) and distillation by reverse osmosis (RO). MED was the first process to be developed for large-scale operation. MED plants recover between 40 and 65% of the incoming water as desalinated product water. Their contribution to the world market in 1958 was about 60%, currently it is just 6% due to the heavy operation and maintenance problems associated with them. With the development of the membrane processes and the improved ability of membranes to deal with seawater, the plants working by reverse osmosis have taken over from the multistage flash plants since about 1980. However, RO plants are essentially used for providing small to medium supplies while the MSF plants are the giants of the desalination industry. The typical daily supply varies from 25 to 5000 m^3. This upper limit keeps rising.

Before a detailed review of water desalination in each of the countries sharing the Arabian Peninsula, reference will be made to the data given by Bushnak (1995). These data comprise the installed capacity of the desalination units countrywise and processwise as surveyed for 1995. The population per country for the same year and the share per capita of desalinated water are also provided. The data show clearly that the production capacity of desalination plants in the Arabian Peninsula has reached in 1995 some 2.83×10^9 m^3. This volume presents about 50% of the wadi flow discussed earlier.

The world's largest water desalination center is located in Al-Jubail in the eastern province in Saudi Arabia. Here 40 MSF units are producing altogether close to 1×10^6 m^3 d^{-1}. From this center, desalinated water runs into two transmission pipelines (1.5 m in diameter). The first has a capacity of about 0.83×10^6 m^3 d^{-1}, and carries water as far as Riyadh (465 km from Jubail). The second line has a capacity of 0.38×10^6 m^3 d^{-1}, and delivers its water to Qassim at a distance of about 390 km. Saudi Arabia has water transmission pipelines with a total length of over 3000 km and capacity of 1.8×10^6 m^3 d^{-1}. According to Bushnak (1995), an RO plant of about 130,000 m^3 d^{-1} capacity is under construction at Yanbu.

Water distillation in Kuwait began in the early 1950s by transforming saline water into vapor and then condensing it to a liquid state using the multi-effect submerged tube evaporation technique. This method helped to provide the cities of Kuwait and Al-Shuwaikh with distilled water. However, in view of its high consumption of heat energy, low efficiency and frequent maintenance, and need for a large number of skilled workers, this technique was soon abandoned and replaced by the flash-type evaporation. In this process the water pressure is reduced abruptly when its temperature is close to the vaporization temperature. Further development has led to the more efficient technique known as multistage flash distillation, which was introduced in the mid-1960s. These successive improvements have brought the daily production of desalinated water in Kuwait in 1985 close to 1×10^6 m^3.

The statistics of Kuwait show that the production capacity in the period 1957–1985 doubled every five years. Furthermore, a comparison between the unit cost of distilled water in 1983–84 and 1965–66 shows that when the price of fuel is excluded and the inflation of currency is taken into consideration, the cost price remains about the same. This means that the actual costs of operation and maintenance of the desalination plants are getting less with time, which in turn reflects the tremendous increase in the efficiency of the more recent techniques.

The difficulties associated with further extraction of fresh groundwater in Kuwait have made the installation of more desalination plants unavoidable. Experimental work on water distillation by means of reverse osmosis using different membranes began in 1984. The satisfactory results obtained from those experiments have encouraged the Kuwaiti authorities to add a few units to the existing plants thereby increasing the daily production to almost 1.5×10^6 m^3.

The United Arab Emirates is spending an increasing effort to satisfy a considerable part of the demand on water for domestic and industrial purposes in a serious attempt to reduce further extraction of groundwater. Desalinated water, especially after mixing it with fresh groundwater, is used for sprinkling public gardens and parks. By the mid-1980s there were eight installations with a total production capacity of about 232×10^6 m^3 y^{-1}; five (46.7%) in Abu-Dhabi, two (43.9%) in Dubbai and one (9.4%) in Al-Sharja (U.A. Emirates, 1986). Bushnak (1995) notes that the total capacity of the desalination plants sold to the United Arab Emirates in 1963–1993 was about 2.08×10^6 m^3 d^{-1}. The figure for the first half of the 1990s is roughly 1.65×10^6 m^3 d^{-1}, which is about four times as much as the wadi flows there. The 1989 capital cost of the seawater reverse osmosis desalination plant at Fujaira, U.A. Emirates, working with a capacity of 9090 m^3 d^{-1}, was US$ 1650 per m^3.

The annual volume of desalinated water per capita in Oman is the second lowest in the Arabian Peninsula. The daily capacity of the desalination plants is in the order of 159,000 m^3. The introduction of desalination technology into remote rural communities seems to solve a pressing water supply problem.

Al-Sajwany and Lawrence (1995) emphasized strongly the adverse environmental impact of desalination on the long run. The first environmental problem is associated with all desalination plants: disposal of the waste brine. Coastal plants dispose of their waste into the sea and measures have to be taken to ensure its mixing with the recipient seawater to avoid damage to the local ecosystem. Options of disposal of brine from inland plants are limited to evaporation, percolation or controlled deep injection. The second impact is coupled to the extraction level of groundwater, basically for crop irrigation. Whether this linkage will eventually lead to aquifer preservation or deterioration depends strongly on irrigation management practices.

Fifteen small units are producing desalinated water in northern Oman (Muscat/Musandam/Sharquia/Wusta). The largest at El-Sur has a capacity of 4550 m^3 d^{-1}. The price of desalinated water is U.S.$ 2.28 per m^3. The smallest plants such as A'zzahar with a capacity of just

50 m^3 produce water at an average cost of U.S.$11 per m^3. These figures show the inverse relationship between the unit cost price and the capacity of the production plant.

Water desalination in the state of Qatar began as early as 1953 by operating a small unit (680 $m^3\ d^{-1}$) to supply desalinated seawater to the capital city Al-Doha. Since then the number of operating units and their capacities have been increasing to bring the production of desalinated sea water in 1986 to $96 \times 10^6\ m^3$. The statistics of the state of Qatar show that desalinated water in 1984 covered about 93% of all drinking water supplies (Qatar, 1986).

The demand for potable water supply has risen from $4 \times 10^6\ m^3$ in 1964 to $48 \times 10^6\ m^3$ in 1980. The total production capacity has been expected to reach some 500,000 $m^3\ d^{-1}$ by the year 1995 so as to satisfy considerable proportions of the water requirements of the agricultural sector and the increasing domestic, commercial and industrial demands.

Hall and Hill (1983) reported that a detailed analysis of the unit costs of water from each potential source showed that groundwater was the cheapest. Desalinated water of high quality could be made available in almost unlimited quantities. But even the use of waste heat from other processes cannot mitigate the high cost of production, which amounts to 21 24 times that of groundwater.

The daily production of desalinated water in the state of Bahrain, which was limited to about 19,000 m^3 in 1976, has undergone a rapid growth since 1984. The use of desalinated water is enforced in an attempt to reduce further decline of the quality of groundwater there. As a result of the increase in the production of desalinated water to about 94,000 in 1985 and further to 170,000 m^3 in 1990, the salinity of water supplied for domestic and other purposes improved from 2800 ppm to 800 ppm in a few years. Further increase in the capacity of desalination plants has brought the total production in 1995 to about 300,000 $m^3\ d^{-1}$ or 60% of the total supply of water for all purposes.

Of all countries in the Arabian Peninsula, Yemen has the smallest total desalination plant capacity. The figure for 1993 was about 37,000 $m^3\ d^{-1}$. This makes the share per capita in desalinated water the least anywhere in the peninsula.

Treated Wastewater

Reuse of wastewater is gaining ground in many countries including those of the Arabian Peninsula. Upon reviewing country reports on water resources and their uses, one can easily conclude that recycled wastewater is regarded as a potential source for a number of purposes such as agriculture, industry and aquifer recharge.

The concept of wastewater reuse was given impetus by a legal ruling issued from a committee of eminent Muslim scholars from Saudi Arabia in 1979 stating that treated wastewater can be used for all religious rituals concerning cleanliness. This judgment was later contained in a *fatwa,* a legal decision based on Islamic rules and customs, having the power of Islamic law in the kingdom of Saudi Arabia and other Islamic countries. After this legal ruling, several projects for wastewater reuse have been envisaged.

The available information suggests that in 1990 the volume of wastewater reuse in the Arabian Peninsula reached 435 $\times$ 10^6 m^3 of which 50% was reused in Saudi Arabia alone, 19% in Kuwait, 14% in U.A. Emirates, 12.7% in Qatar and Bahrain, and 4.3% in Oman and Yemen.

The raw water that has to be treated or recycled comes from agricultural and/or urban drainage. The extent of treatment or reclamation depends essentially on the quality of the raw water, the purpose for which water is treated and the level of risk of environmental problems. The reclamation criteria used for irrigation and recreational impoundments in California, U.S., reported by Asano (1994), can be viewed as an example.

The country report of Saudi Arabia (1986) mentions that about 140 $\times$ 10^3 m^3 d^{-1} of the sewage water of Al-Riyadh, after being treated to an advanced stage, is used for irrigating about 4000 ha of agricultural land. These figures are expected to grow to 800 $\times$ 10^3 m^3 d^{-1} by the year 2000. It is also expected that by the year 2000 the sewage plants at Medinah Al-Munawwarah and Qussaim will be able to supply each day 140 $\times$ 10^3 and 98 $\times$ 10^3 m^3 of treated wastewater, respectively. These quantities will be made available for irrigating 3100 and 823 ha, respectively.

The same report expected that the total volumes of treated wastewater in Saudi Arabia will increase to 400 $\times$ 10^6 m^3 and 700 $\times$ 10^6 m^3 by 1990 and 2000, respectively. The reported volume of actually reused water in 1990 was only 217 $\times$ 10^6 m^3, or just 54% of the planned reuse (ESCWA/CEHA, 1993).

From cost analysis of treated water in Saudi Arabia it appears that the cost of reclaimed water depends on the size of the treatment plant, the composition of the raw water and the required quality of the treated water. Using the prices of 1987, and assuming that the treatment plant has a capacity of 37,850 m^3 d^1, the unit cost was found as SR/m^3 0.60

using activated sludge (secondary), 0.71 using secondary treatment plus filtration, 0.94 using secondary treatment plus activated carbon, and 2.81 using advanced treatment plus reverse osmosis. Futhermore, assuming 20 years of plant life, 5% interest rate, 60% plant capacity and 90% plant availability, the unit cost of treated effluent was presented by Ukayli and Hussein (1988) for selected plants. This was found to vary from a minimum of SR/m^3 0.44 (secondary treatment, Riyadh expansion I and II) to a maximum of SR/m^3 1.63 (tertiary with reverse osmosis, Jeddah).

As much as 70% of the total water supplied for domestic purposes terminates as wastewater. Some of this water, depending on the level of treatment, is used for irrigating certain crops. The 1990 data show that the share per capita of reused treated sewage water reached about 39 m^3 y^{-1}.

Research on irrigating certain crops like alfalfa, barley and corn in Kuwait by relatively untreated effluents from an oil refinery has been reported by Al-Nakshabandy (1979). Some of the results were as follows:

Substance	mg L^{-1}
TDS	648–2424
Phenol	3.2–10.7
COD	793–2816
Ammonia	223–1066
Total oil	28–324
Cyanide	2.3–21.3
Chromium	0.002–1.0

These figures and many more have served as guidelines for preparing irrigation water standards for agriculture in other countries in the peninsula.

The pressing demand for water in Qatar has urged the authorities in the early 1970s to lay down plans for utilizing treated urban drainage water. The pumped treated sewage water, which amounted to more than 60 × 10^6 m^3 d^{-1} in 1985 and was expected to exceed 120 × 10^6 m^3 d^{-1} before 1990, has been used ever since for municipal and agricultural activities (Figure 10). The figure for 1990, being about 44% of the then extracted groundwater, is strong evidence of the importance of treated sewage water for Qatar as a source of water.

The first sewage water treatment plant in Bahrain began operation in 1982 with a design capacity of 50,000 m^3 d^{-1} in the first phase. As

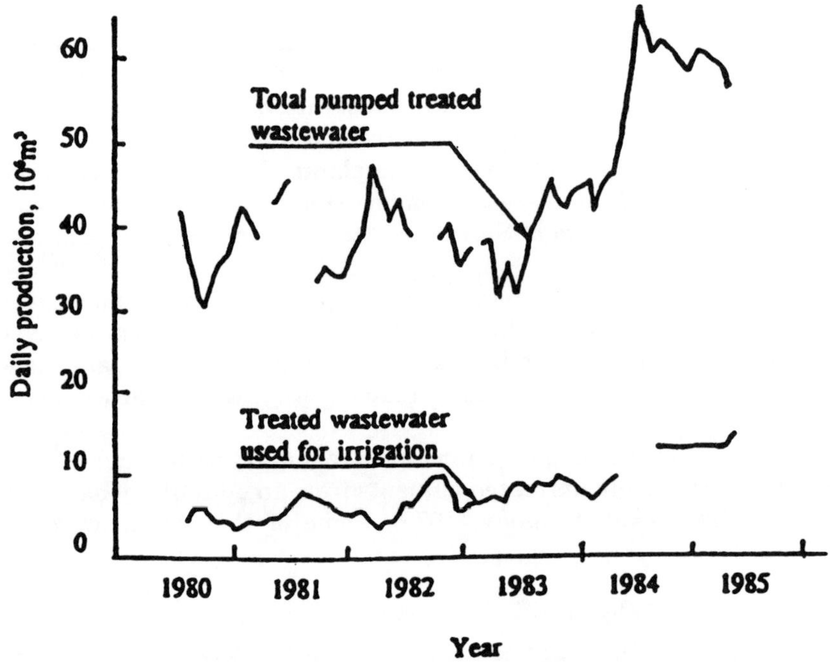

Figure 10. Development of treated wastewater reuse in Qatar (Qatar, 1986).

the organic content of the raw water is too low, it has been possible to operate the plant at a much higher capacity to produce 74,000 m^3 daily. Since then, the second phase of the project has been implemented and smaller plants have been added, bringing the total production in 1990 to 32×10^6 m^3. With this volume, the average share of the inhabitants of Bahrain in the treated wastewater becomes about 64 m^3 y^{-1}, the highest share per capita anywhere in the peninsula.

It has been reported by the responsible authorities that the quality of treated water is improving steadily, especially the water supplied for domestic purposes. Laboratory analysis of reclaimed water has shown reduction in the salinity of water from 4500 ppm in 1982 to 3000 ppm in 1985, and further to 2000 ppm by the end of the 1980s. The latter salinity figure is far better than the salinity of groundwater in several locations in Bahrain (Bahrain, 1986).

Since 1985 four sewage water treatment plants in the United Arab Emirates have been producing together 62×10^6 m^3 y^{-1} (U.A. Emirates, 1986). The same figure appears in the report of ESCWA-CEHA (1993), indicating that the increase in the volume of treated wastewater has come to a halt, at least during the last few years.

In addition to the treatment of some of its wastewater, recently estimated at 10×10^6 m^3 y^{-1}, investigations are carried out in Oman with the objective of reusing oil-contaminated water.

Reusing domestic wastewater can be critical depending on the level of pathogonic microorganisms and parasites, hence effective disinfection is a major concern. Such pathogons are unlikely to be present in water produced in conjunction with production of oil, but other substances are of importance when considering the potential for reuse. Organic substances comprise the total organic carbon, acetate, propionate, dissolved and colloidal aliphatic hydrocarbons, and other organic acid salts. Inorganic substances comprise sodium, calcium, and magnesium. The concentration of each of these substances varies from one location to another. For example, analysis of samples of produced water from Al-Nimr area oil fields, Oman, has shown a concentration of sodium from 1390 to 4400 mg L^{-1}, total dissolved solids from 5400 to 7200, and salinity as sodium chloride from 4100 to 6700 mg L^{-1} (Khan and Fuggle, 1995). These investigations have led to certain standards regarding irrigation of vegetables and fruits. Also, recommendations for upper limits of trace heavy metals and other constituents in reclaimed water for irrigation have been made as well.

In principle, the treatment procedure begins by separating oil from water. The water that is left is likely to contain dissolved hydrocarbons and nonhydrocarbon organic substances. Air stripping is introduced to help remove the volatile organic compounds. Biological treatment is applied to reduce the concentration of many of the dissolved organic compounds to a very low level. Some additional technologies such as activated carbon adsorption have to be used in order to treat those organic compounds that are not biodegradable. When a small portion, say one quarter, of the produced water is desalinated after the total flow receives preliminary, biological and activated carbon treatment, the cost per m^3 of treated water suitable for irrigation is between U.S.$1.0 and U.S.$2.0 (based on 1990–1995 prices).

Wastewater in Yemen that may be available for irrigation amounted to 23×10^3 m^3 d^{-1} in 1990 and is expected to reach 108×10^3 m^3 d^{-1} by the year 2005. In view of its quality, the use of the treated water in Sana'a, Yemen, so far is restricted to cooked vegetables and crops. As the effluents are still of unsatisfactory microbiological quality, the use of this water cannot be extended to raw eaten crops or vegetables. The problem lies essentially in the poor performance of the treatment ponds, which results in effluent with high concentration of BOD, and probably with a high degree of fecal coliforms (Al-Nozaily, 1992).

GROUNDWATER RESOURCES

Brief Hydrogeological Description

The geological history of the Arabian Peninsula dates back to the pre-Cambrian (more than 500×10^6 years BP[1]). The crystalline rock is exposed along the coast of the Red Sea forming the *Arabian Shield.* Thick layers of sand and sandstone suitable for groundwater storage and extraction were deposited on the base rock some hundred of million years ago.

From the hydrogeological point of view these sedimentary rocks can be grouped in the following three major categories:

1. Paleozoic continental sandy rocks
2. Mesozoic continental sandy rocks
3. Mesozoic and Tertiary carbonatic rocks

Additionally, there are two minor groups:

1. Tertiary and Quaternary volcanic rocks
2. Tertiary and Quaternary alluvium rocks

The water-bearing formations in the Arabian Peninsula are either of sandy or calcareous facies.

The *Saq* formation, which belongs to the Ordovician-Cambrian age rocks of the Primary era, can be regarded as one of the most important water-bearing formations in the peninsula. This is especially true for Saudi Arabia, where groundwater can be under high piezometric head causing water in some areas to flow under artesian pressure. In addition, the quality of water is favorable as the salinity is about 1000 ppm.

The areal extent of this formation can be as large as 160,000 km^2 and the exposed surface is 65,000 km^2, extending from Jordan in the north for a distance of 1200 km to the east and south in Saudi Arabia. It consists mainly of homogeneous sandstone. The average well yield

[1]BP = Before present.

in this formation is from 15 to 22 L s^{-1}, though in northern Saudi Arabia it can be as much as 100 L s^{-1}.

The formation can be found at a depth of 2400 m and is about 600 m in thickness. The transmissivity varies from 4×10^{-4} m^2 s^{-1} to 27×10^2 m^2 s^{-1} and the storativity from 13×10^{-4} for the semi-confined part of the aquifer to 2.5×10^{-5} for the more confined part. The volume of groundwater in storage above the the 300 m depth is estimated at 475×10^9 m^3 (Khoury, 1986).

The *Manjour* formation (Upper Teriasic) is a sandy formation that belongs to the Mesozoic era. The thickness of this formation in Saudi Arabia is between 350 and 400 m. In the Al-Riyadh area and environs it can be found at a depth of 3 km below the surface, a condition that renders waterflow under a high artesian pressure. In view of its high quality (400–1200 ppm), the groundwater has been subject to excessive extraction causing a decline in the water level of 70 m in less than thirty years.

The well yield of this formation is 15 to 40 L s^{-1}, with a maximum of 80 L s^{-1}. The aquifer's transmissivity is estimated as $(5–160) \times 10^{-4}$ m^2 s^{-1} and the storativity as 1.3×10^{-4}. The volume of water in storage above the 300 m level is 480×10^9 m^3 and the annual recharge to this aquifer is estimated at 80×10^6 m^3 (Saudi Arabia, 1986).

The *Al-Bayadh* (Barremian-Aptian) and *Al-Wasie* (Cenomanian-Turanian) formations belong to the Cretaceous age. These two formations together form one water resource penetrating with unequal proportions inside Saudi Arabia, southern Yemen, Qatar and Kuwait. Some of the wells dug in the Al-Wasie formation are characterized by their high yields in comparison to those dug in Al-Bayadh. The yield of the former can reach up to 130–170 L s^{-1} and can flow freely as in the Al-Hofuf area. The average yield of the two formations combined as one resource is between 40 and 60 L s^{-1}.

The Al-Bayadh formation varies in thickness from 30 to 600 m. Its transmissivity is about 1×10^{-2} m^2 s^{-1}, storativity about 0.12 and the salinity of water is between 550 and 900 ppm increasing up to 6000 ppm or more. The same characteristics for the Al-Wasie formation are 200–230 m thickness increasing up to 600 m in Ar-Rub Al-Khali, $(1.3–97) \times 10^{-3}$ m^2 s^{-1} transmissivity, 2×10^{-4} storativity and 1000–3000 ppm salinity. The total annual recharge to the two formations combined is estimated as 420×10^6 m^3.

The *Al-Wajid* is a sandy-facies formation belonging to the Paleozoic Lower Permian age. This formation can be found in south and east Saudi Arabia. The maximum width of the outcropping is 100 km and the thickness is from 200 to 420 m. The well yield varies from 15 to 40 L s^{-1}, with a maximum reaching 80 L s^{-1}.

The transmissivity of the aquifer is estimated as (6–210) $\times 10^{-4}$ m^2 s^{-1}. The volume of groundwater in storage above the 300 m level is estimated at 430 $\times 10^9$ m^3 and the annual natural recharge at 115 $\times$ 10^6 m^3.

The *Al-Tabuk* formation (Ordovician-Silurian) has been formed in the Palaeozoic era after the Al-Saq formation was formed. This formation can be described as a productive sandstone aquifer with flowing and nonflowing artesian conditions (ECOSOC/ECWA, 1981). The two formations (i.e., Al-Saq and Al-Tabuk) seem to enjoy more or less the same characteristics. The areal extent of the outcropping of the Tabuk is estimated as 77,000 km^2 and the thickness is widely variable, from 20 to 1050 m. The maximum well yield in this formation is 127 L s^{-1}, and the transmissivity is in the range of (1.5–15) $\times 10^{-3}$ m^2 s^{-1}.

The carbonatic formations belonging to the Mesozoic and Tertiary eras are, as a source of water, equally important to the formations of the sandy facies. The two major carbonatic formations are the *Umm Radhoma* and the *Dammam* formation.

The *Umm Radhoma* (Thanetian-Montian) formation was formed in the Paleocene age. Its vast outcropping, which extends over more than 1200 km, can be found in Bahrain, Saudi Arabia, Dhofar in Oman, Hadramout and southeast of Yemen. It disappears in the United Arab Emirates and Qatar. The thickness of this formation in Saudi Arabia is 300 m. The depth to water table varies from 60 to 100 m and the maximum depth can reach 700 m. The values of the hydrogeological parameters of this formation are widely variable. The coefficients of transmissivity and storativity are estimated at 4×10^{-5}–6×10^{-1} m^2 s^{-1} and 5×10^{-5}–1×10^{-3}, respectively. Accordingly, the well yield is between 10 and 70 L s^{-1}. The quality of water is fairly good (up to 1000 ppm) and poor (much more than 1000 ppm). Typical salinity of this formation in Bahrain is in the order of 12000 ppm.

The Umm Radhoma formation receives an annual recharge of about 2.4 $\times 10^9$ m^3; 190 $\times 10^6$ m^3 from direct rainfall, 220 $\times 10^6$ m^3 torrential flow and the rest from the surrounding formations.

The second carbonatic formation is the *Dammam* formation. Its outcropping occupies a surface area of about 20,000 km^2. This can be found in Kuwait, Qatar, Bahrain, Oman, and Al-Rub Al-Khali. The Dammam formation comprises five layers, of which Al-Alat and Al-Khobar are the principal water-bearing layers. The former has an average thickness of 83 m and can be found at a depth between 140 and 230 m. The Khobar, which is situated below Al-Alat, can be found at a depth of 200 to 300 m. It has an average thickness of 57 m. Despite its limited dimensions, the formation is a good water-bearing one. The well yield varies between 4 and 22 L s^{-1}. The coefficients of transmissivity and storativity are $6 \times 10^{-6} - 9 \times 10^{-2}$ m^2 s^{-1} and (3-83) $\times 10^{-4}$, respectively. The quality of water is relatively good (1000 ppm) in the western part. It has deteriorated in the south and east. It reaches 6 g L^{-1} or more in Al-Rub Al-Khali and Bahrain.

Last, the volcanic and alluvium formations of the hydrogeological groups belonging to the Tertiary and Quaternary ages are generally regarded as less important water resources for the Arabian Peninsula. However, these formations can be of some local importance. For example, the wells dug in the neogene aquifer in the Al-Hofuf area, Saudi Arabia, can yield up to 50 L s^{-1}. Some of these wells flow even freely by artesian pressure.

The water-bearing formations in the Arabian Peninsula are distributed among a number of groundwater basins of different dimensions, suitability and productivity. Ar-Rub Al-Khali in the south, Dabdaba and Al-Riyadh in the center and Al-Hammad in the north are examples of these basins. In the next subsections we will stress the hydrogeological and other relevant characteristics of the aquifers supplying each of the countries in the peninsula with groundwater. Additionally, some of the environmental impacts of large-scale abstraction projects are presented and discussed.

Groundwater in the Countries of the Arabian Peninsula

The aquifers currently under exploitation in Kuwait are members of the so-called Al-Kuwait and Al-Hasa groups. The Kuwait group (Eocene-Holocene) extends all over the country and varies in thickness from a few meters to more than 200 m. The values of the geohydrologic parameters of these aquifers vary widely from one location to another and the depth to which the aquifer is found can vary from a few meters close to the coast to 100 m or more in the west. The total dissolved salts are usually less than 1000 ppm but reaches up to 50,000 ppm in the northwest.

The Kuwait group consists of the recent deposits, the Dibdiba formation, the Fars and the Ghar formation. The Hasa group (Paleocene-Eocene age) comprises mainly the carbonaceous rock units of marine origin. The Dammam formation is the most important member of this group as it provides Kuwait with most of the groundwater it consumes.

The other two formations in descending order are the Ras and Umm-Radhuma. The Dammam formation is provisionally subdivided into three units depending on the limestone it contains and the relevant properties. Unit 1 contains shelly limestone, unit 2 contains chalky limestone and unit 3 contains dolomitic limestone.

Production wells generally tap units 2 and 3 of the Dammam aquifer, although some also penetrate unit 1. The historical development of the annual production of groundwater is summarized in Figure 11 (Sayed and Al-Ruwaih, 1995). The same reference gives the mean value of the transmissivity as 8.5×10^{-3} m^2 s^{-1}, with a standard deviation of 14.9×10^{-3} m^2 s^{-1} and the mean specific capacity of the wells as 8.45 L s^{-1} with a standard deviation of 13.65 L s^{-1}. The specific capacity in m^3 m^{-1} d^{-1}, q, of the production wells is related to the transmissivity in m^2 d^{-1}, T, by the power expression

$$q = 1.96\ T^{.887} \tag{11}$$

with a coefficient of correlation of 0.919 and a standard error of estimate of 0.235. Other regression relations have been developed between the well losses and the specific discharge.

The above-mentioned study (Sayed and Al-Ruwaih, 1995) terminated by asserting which of the well fields can be regarded as the most productive and which are the least productive.

The groundwater resources in Kuwait are subdivided into three classes depending on their quality. High-quality water is water with less than 1000 ppm total dissolved salts and is used for domestic and

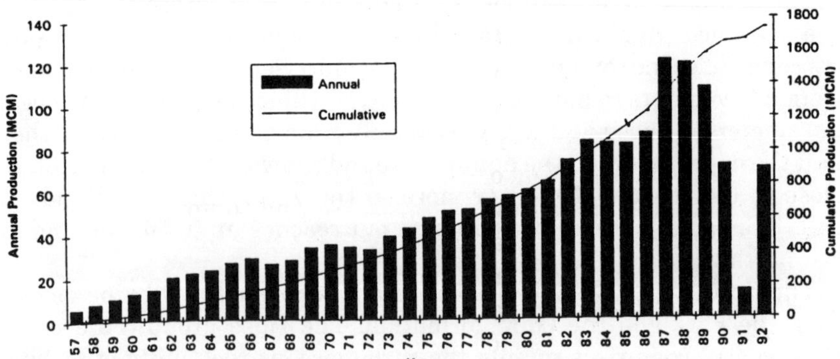

Figure 11. Historical development of groundwater in Kuwait (Sayed and Al-Ruwaih, 1995).

drinking purposes. Moderate-quality water has a salinity between 1000 and 10,000 ppm total dissolved salts and is used for agriculture and sheep breeding. It is a common practice to mix some of this water with desalinated water to get the acceptable quality. Poor-quality water, with more than 10,000 ppm, is not used currently except for very special reasons. The average annual production of moderate and good quality groundwater in the period 1957–1990 was 67×10^6 m^3. The rate in the two years 1988–1989 and 1989–1990, just before the outbreak of the Gulf War, reached 120×10^6 m^3 y^{-1}. In view of the limited recharge to the aquifers, there is an observed steady deterioration of the quality of water in some of the production wells. This situation is met with by limiting the production of some of the old wells, not renewing wells that run dry, instead, searching for new aquifers, where new wells can be drilled, and depending more and more on desalinated as well as treated wastewater.

Until 1975 the only source of water for domestic as well as agricultural purposes was groundwater. The total consumption of water in 1984 was estimated at 175×10^6 m^3, 90% of which (i.e., 153×10^6 m^3) was groundwater. The principal aquifer in Bahrain is the Al-Khobar water-bearing stratum of the Al-Dammam formation described in the previous section.

Historically, groundwater in Bahrain used to flow through natural springs. The flow of these springs declined from 57 in 1924 to 42 in 1953, then to 28 in 1966 and further to 8.1 in 1979–1980, all flows are in 10^6 m^3. These springs have been supported by pumped wells so as to be able to cover the increasing demand for water.

Development activities in Bahrain have led to increase in the volume pumped annually from 112×10^6 m^3 in the late 1960s to about 180×10^6 m^3 in the early 1990s. The aquifer recharge from the rain water and the areal flow from the upgradient Saudi aquifers is estimated at just 100×10^6 m^3 y^{-1}. The imbalance between the discharge and recharge has led to a decline of the aquifers' potentiometric levels by 0.5 m and 4.0 m depending on the location. The salinity level of water too is increasing due to seawater intrusion and to upward leakage of the brackish and/or saline water from the deep aquifers to the overlying ones. The extent of pollution of the Dammam aquifer and its salinization by the said causative factors have been investigated recently by Zubari et al. (1994). This study has shown that the total dissolved salts in about one third of the well water in Bahrain in the period 1986–1992 exceeded 4 gm L^{-1}. The deterioration of the quality of water in the

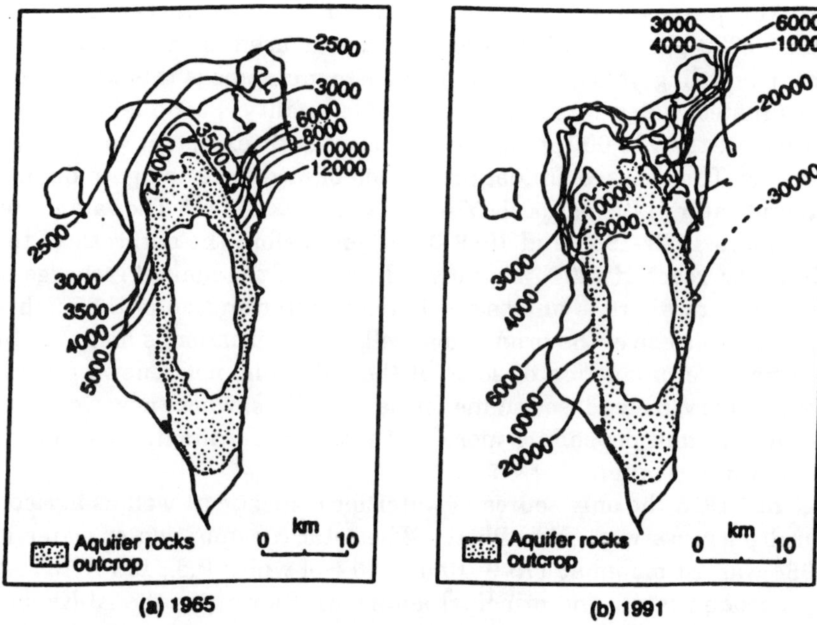

Figure 12. Salinity levels of groundwater in the Dammam aquifer in Bahrain in mg L^{-1} (Zubari et al., 1993).

Dammam aquifer in the mentioned period can be seen from the two salinity maps shown in Figure 12 (Zubari et al., 1993).

Similar to the situation in Kuwait, the problem can be mitigated by reducing groundwater abstraction in Bahrain in general, and in the polluted areas in particular, and depending more on alternative sources such as desalinated water in urban areas and treated wastewater in agricultural areas.

Groundwater in Qatar can be found in four different zones. The northern zone, which underlies the northern half of Qatar, contains the better quality or potable groundwater in the country. The groundwater in this zone can be found in the form of freshwater lenses within the limestone-dolomite succession of the Dammam-Rus formation overlying the saline water stored in the older rock units. In the southern zone the Rus becomes very thick and there is no extensive body of fresh groundwater. The limited number of wells draw their supplies from small, irregular unconfined or perched water bodies of varying quality. The third zone is in southwestern Qatar. Elements of the Dammam and Hofuf formations are present in this zone, creating an artesian aquifer whose recharge source is in Saudi Arabia. The Alat member of the Dammam formation is capable of yielding some 2×10^6 m^3 y^{-1}, but with a total dissolved solids of 3000–5000 mg L^{-1}. The underlying Umm-Radhuma formation is also artesian in this zone and could yield another 10×10^6 m^3 y^{-1}, but with a total dissolved solids of 17,000 mg L^{-1}. Despite their salinities, experiments have shown that a combination of both waters can be used successfully for raising a variety of crops to supply the zone of Al-Doha and its surroundings. The infiltration of water from the various sources is creating a freshwater lens above saline water intruding from the Arabian Gulf. It has been estimated that by the year 2000 some $(10\text{–}15) \times 10^6$ m^3 y^{-1} would be suitable for amenity watering. However, exploitation of this potential source would require careful management and is dependent upon bacteriological pollution being kept to acceptable levels (Hall and Hill, 1983).

The recharge to the aquifer in northern Qatar has been estimated as 11% of the long-term rainfall. When years with annual rainfall less than 25 mm are not taken into account, the recharge coefficient falls to just 1% for those years with 30 mm or so, and rises to 40% for years with an annual rainfall of 200 mm or more, a situation that develops once every twenty-five years. This brings the annual recharge to say 30×10^6 m^3 y^{-1}. Gischler (1979) estimated the long-term recharge at 22×10^6 m^3 y^{-1}.

The annual abstraction, on the other hand, increased rapidly from 18×10^6 m^3 in 1965 to 52×10^6 in 1975 and reached 105×10^6 m^3 in 1985. The increase in groundwater abstraction has become possible as a consequence of the increase in the number of production wells from about 110 to 330 and further to more than 700 for the same years in their respective order. As a result of the imbalance between the aquifer recharge and abstraction, the areal extent of the freshwater lens in northern Qatar has become much smaller in the course of time. This situation has been presented graphically by Lloyd (1992) as shown in Figure 13.

The principal formations in Saudi Arabia are Al-Saq, Wajid, Tabuk, Manjur, Bayadh and Wasie, Umm-Radhuma, Dammam and the Neogene. The secondary sources can be found in the aquifers of Al-Jawf, Berwath, Khuff, Jilh, Dhruma, Sakaka and Aruma. Hydrogeological properties of interest of most of these formations have been presented in the previous section. Detailed explanation of the occurrence of groundwater in these formations can be found in several publications of which the country report on the water resources of Saudi Arabia

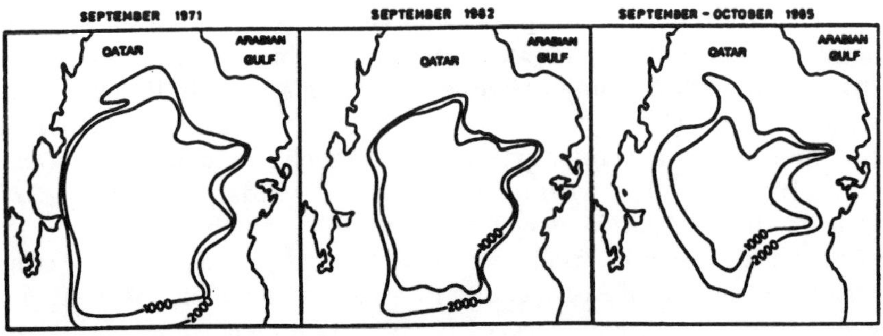

Figure 13. Decreasing size of the main fresh groundwater lens, Qatar; contour lines show total dissolved solids content, mg L^{-1} (Lloyd, 1992).

(1986) and the UNDP[2] document on groundwater in the East Mediterranean and West Asia (UNDP, 1982) are examples only.

Abstractions of groundwater in Saudi Arabia began in the 1950s. Since then the annual abstraction increased significantly, especially in the early 1980s following a governmental decree to subsidize wheat production. The report prepared by ACSAD[3] on the water resources in the Arab region (Khoury, 1986) estimates the total annual abstraction as 2.35×10^9 m^3, aquifer recharge as 2.338×10^9 m^3 y^{-1} and the volume of groundwater in storage as 354×10^9 m^3.

Next to the ever-growing demand for more freshwater for domestic and drinking purposes, there is a growing tendency to increase the acreage under irrigated agriculture, especially in certain areas in Saudi Arabia. The Paleozoic sandstone aquifer in northwestern Saudi Arabia and Al-Manjur in the Al-Riyadh area are examples of aquifers that have become overexploited. The decline of the potentiometric head in the Cambro-Ordovician sandstones as a result of groundwater mining in the northwestern part of the country can be seen clearly from the map shown in Figure 14. Additionally, as a consequence of the continuous abstraction of some 2.5 m^3 s^1 for supplying Al-Riyadh and environs with freshwater, the groundwater level has fallen by no less than 10 m during the last three decades. Adverse effects include decrease of water quality water drawn from the deep aquifer system.

In Al-Kharj oasis, southeast of Al-Riyadh, the recharge to the aquifer is small and water is extensively abstracted for agriculture. Surface

[2]UNDP = United Nations Development Program.
[3]ACSAD = Arab Center for Studies of Arid Zones and Dry Lands.

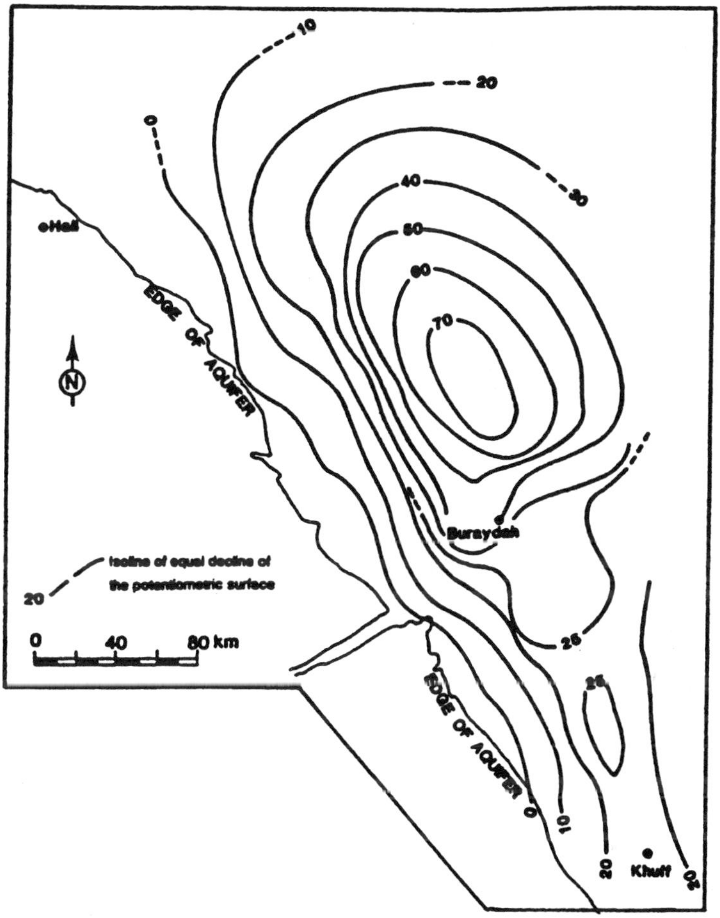

Figure 14. Decline of potentiometric surface in the Cambro-Ordovician sandstones as a result of groundwater mining, northwestern Saudi Arabia (BRGM, 1986).

rock strata have collapsed, and the water level has fallen by 4 to 5 m between 1970 and 1980. The water table in the areas of Wadi Fatima and Wadi Khulys (around Jeddah) has dropped rapidly in the past few years with the urban growth of the township of Jeddah. As a result, deeper wells and more powerful pumps are badly needed to support agriculture in the wadis. It has also been reported that withdrawing excessive quantities of water from the lower Wadi Fatima in the western region reduced the water table rapidly, causing all springs to dry up. Of 360 springs producing at one time, only four remain (Mohorjy, 1988).

The ongoing process of intensive exploitation of the aquifers has increased the annually abstracted volume of groundwater in Saudi Arabi from 2.35×10^9 m^3 in 1986 to 3.0×10^9 m^3 in 1995(UNESCO/ ACSAD, 1993). Measures for increasing aquifer recharge, especially the endangered aquifers, have to be enforced so as to restore the balance between abstraction and recharge.

The groundwater in the United Arab Emirates occurs in three aquifer systems: the shallow alluvial aquifer underlying the central gravel plain and desert foreland, the Batinah coastal plain aquifer and the deep carbonate aquifer. The gravel-conglomerate aquifer is in continuity with the underlying limestone-marl aquifer and is considered to be the main source of water. The major gravel aquifer is shallow and relatively thin. It is unconfined, with a storativity coefficient in the order of 1×10^{-2}. The transmissivity coefficient ranges from 2.9×10^{-3} to 2.9×10^{-2} m^2 s^{-1} in the northern part of the country, but has smaller values in the southern gravel plains. Yields from boreholes are variable, reflecting differences in hydraulic conductivity and consistently high well losses due to incomplete development after drilling (Naqash and Marrei, 1983). The Batinah coastal aquifer consists of coarse sand, gravel and boulders, ending up with fine-grained sabkha deposits in some localities along the coast of the Gulf of Oman.

The exploitation of the groundwater aquifers in many parts of the U.A. Emirates is beyond the safe yield, causing the water level to continue its decline. As an example, the water table in the Emirate of Al-Sharja dropped by 1.5 m y^{-1} in the period 1975–1979. The drop in 1980 exceeded 2 m and a number of wells since then have become dry. The country report on the water resources and their uses in the U.A. Emirates (1986) considers the central gravel plain as the principal aquifer and source of groundwater. The report mentions that the surface covered by this plain within the Emirates is 4000 km^2. The thickness of the aquifer varies between 50 and 80 m and the depth to water table between 18 and 25 m. Despite the incomplete survey of the production wells, the country report estimated the annual abstraction at about 0.42×10^9 m^3 y^{-1}. Other sources estimated the annual abstraction differently: 0.28×10^9 m^3 (UNESCO/ACSAD, 1993) and 0.39×10^9 m^3 (Shahin, 1996). The lowest of these figures is still far more than the annual recharge estimated at 0.1×10^9 m^3. Finally, analysis of water samples has shown that the salinity varies between 300 and 2000 ppm.

Hydrogeologically, rock units in Oman may be divided into three main groups: the alluvial plains including the coastal plains, the limestone mass with its substantial storage volume of groundwater, and the impermeable crystalline basement complex at the mountainous regions yielding most of the runoff.

Traditionally Oman is subdivided into four groundwater regions. The aquifer in the northern coastal region, of which the Batinah coastal plains are the most important, consists of water-bearing alluvial deposits. The average thickness of these deposits is 30–40 m, reaching a maximum of 100 m at some places. The permeability is about 6×10^{-4} m s^{-1} and the hydraulic gradient of the aquifer varies between 1×10^{-2} and 4×10^{-4}. The quality of water improves from more than 1000 ppm up to 1 km from the coast to less than 300 ppm 2–7 km farther from the coast, though at larger depths (ECOSOC/ECWA, 1981). Figure 15 shows the extent of drop of the groundwater level in the Batinah coastal aquifer (Poulter, 1995).

The interior region, due to its geographic location, receives summer rain, which provides the aquifer with some recharge. Groundwater in this region can be found in alluvial layers at a depth of 10 to 40 m and

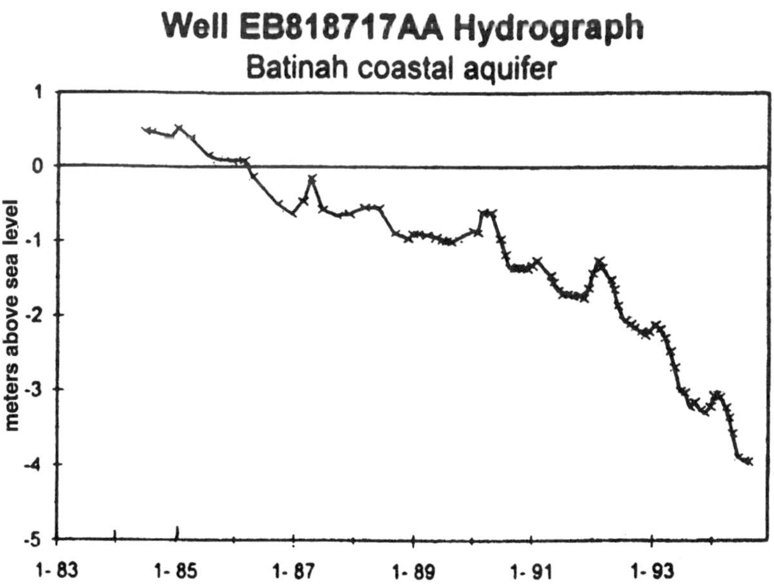

Figure 15. Decline of potentiometric level in the Batinah coastal aquifer (Poulter, 1995).

of quality varying from 300 ppm at the foothills to over 2000 ppm of total dissolved salts. Deep groundwater in the underlying aquifers in the region tends to be more saline. Groundwater abstraction in this region is so far accomplished through the *Falaj* systems and not by pumped wells. The total flow of the Falajes in the southern part is in the order of 3000 L s^{-1}. An important wadi in this area is Wadi Dayqah whose average discharge is about 360 L s^{-1}. Other areas of importance are the Buraimy Oasis where groundwater is encountered in relatively thin deposits of coarse clastic materials overlying impermeable marl. There is a marked decline of groundwater levels and deterioration of the water quality as a consequence of overdrafting the aquifer. Figure 16 shows the drop of the water level in the aquifer in the interior of the country (Poulter, 1995).

The mountainous region is characterized by groundwater resources that have been well developed in the narrow bottoms of deeply incised valleys. These valleys form natural line sinks for groundwater discharge from the hills. Groundwater discharges naturally through the

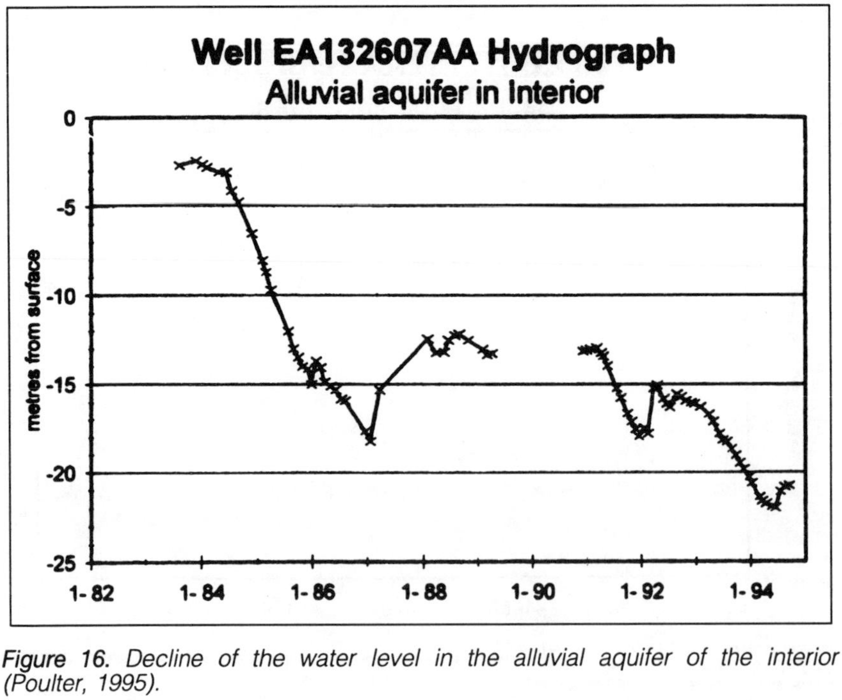

Figure 16. Decline of the water level in the alluvial aquifer of the interior (Poulter, 1995).

existing systems of Falajs or springs. This region is considered as the active recharging zone from which the groundwaters flow towards the Batinah coastal plain and the interior plains for the benefit of the users on the lower reaches.

The Dhofar region is reputed by the Salalah Plain, the most developed part of the region. This plain extends 40 km along the Arabian Sea and inland for about 12 km at its widest point and covers an area of 240 km^2. The plain is underlain by limestone with a thin covering of wadi alluvium and calcarnite. The thickness of the alluvium reaches 40 m under the major wadi channels.

Groundwater is contained extensively in the alluvium and limestones. The hydrogeological characteristics of the aquifer are complex and widely variable. The annual recharge to the plains (ECOSOC/ECWA, 1981) was estimated at 60×10^6 m^3 and the quantity actually used 15×10^6 m^3. Groundwater is generally poor except in the central part of the Salalah Plain. The properties of the aquifer and groundwater in the interior of Dhofar are not well known. The total annual abstraction and aquifer recharge in Oman have been estimated as 0.41×10^9 and 0.56×10^9 m^3, respectively (Oman, 1986).

In order to maximize the limited opportunities to augment its water resources, the government of Oman has built and is planning to build more recharge and flood retention dams. Although the impact of these dams on the sultanate's overall freshwater potential is small, they will enable flood flows, otherwise lost into the sea or into the desert, to be harnessed for the benefit of the users. They can also provide protection against floods. By 1985 it appeared that there were 56 suitable sites for recharge dams. Since that time, 15 dams, with a total storage volume of 57.1×10^6 m^3 have been completed. The recharge dam on the Wadi Al-Khawd near the Gulf of Oman, with a storage capacity of 12.5×10^6 m^3, is by far the largest (Ministry of Water Resources of Oman, 1995).

Based on the available hydrogeologic investigations, the principal aquifers in the northern part of Yemen can be subdivided into five aquifers: the Tihama Plain alluvial aquifer, the Kholan series, the Tawilah and Mejd-Zer series, the Yemeni volcanics and the Quaternary alluvium and Quaternary volcanics.

The Tihama Plain alluvial aquifer consists of conglomerate of silt, sand and gravel. It is mostly unconfined, with a thickness varying from a few meters to 600 m. The average regional transmissivity of this aquifer was estimated as between 1700 and 2500 $m^2\ d^{-1}$ and the storativity between 0.01 and 0.10.

Groundwater used to be of fair to good quality. The available data of the late 1980s show that in view of the increasing demand for water, essentially for agriculture, the number of pumped wells and consequently the annual abstaction is increasing constantly. As a result, the annual deficit between the aquifer recharge and discharge kept increasing till it reached some 150×10^6 m^3 in 1990. Van der Gun et al. (1992) reported that the upconing of saline or brackish water in some of the wells and the need to deepen those wells that have become dry are adding to the increase in the cost of pumping caused by the decline of the water levels. These are among the adverse effects of intensive aquifer exploitation.

The Kholan series represents an important aquifer in the interior areas of Yemen and it is expected to possess good aquifer potential. The aquifer is composed of a series of white and violet cross-bedded sandstone with several thin layers of conglomerates. The aquifer thickness varies between 100 and more than 500 m. Aquifer parameters are not well defined and water quality is uncertain.

The Tawilah and Mejd-Zir aquifer is composed of coarse-grained sandstones with conglomeratic layers totaling about 400 m in thickness. The depth to water table from the ground surface varies between 25 and 100 m. The areal extent of the aquifer is bounded from the west by the Red Sea escarpment and the escarpment facing the eastern desert from the east.

The aquifer's potential is good, particularly the lowest 300 m of its thickness. Despite the limited information known about its hydrogeological parameters, the aquifer is known to be characterized by a well-developed fracture system that results in good permeability. Several low-yielding springs of the water table type are fed by the aquifer. Large springs are reported to be fed by this aquifer along the Red Sea escarpment.

The Quaternary volcanic aquifer outcropping in the northeastern corner of Yemen is just a small part of the volcanic aquifer formed in the Tertiary and Quaternary periods. It has an areal extent of 1000 km^2 and consists predominantly of lava materials reaching a thickness of several hundred meters. The available information shows that the groundwater potential of this aquifer is too limited. However, those thin water-bearing layers constitute an important source for recharging the sandstone layers of the Tawilah and Mejd-Zir aquifer feeding the Sana'a basin.

In view of its importance as the source of water to the country's capital city, groundwater investigation of the Sana'a basin is receiving the utmost attention. This basin is located in the central plateau of the country at an elevation of about 2200 m amsl. It covers about 35,000 km^2 and comprises several wadis, which help recharge the basin with rainwater. In 1971 almost one third of the basin area was investigated geophysically, geologically and hydrologically. One of the results of that study was that the annual amount of water needed for Sana'a city has been estimated as 0.22×10^6 m^3 for 1975, 6.9×10^6 m^3 for 1985 and 21.5×10^6 m^3 for the year 2000. The first phase of the study, which lasted from 1971 to 1977, was followed by a second phase that continued up to 1982. The development of the water resources of Sana'a during those two phases comprised the drilling of 35 production wells and 26 observation wells. Additionally, private well drilling, basically for the purpose of agriculture, went on at an accelerating speed without much control from the state. The intensive exploitation of the aquifer resulted in such a rapid decline of the water level that 15 wells soon ran dry and the 1986 production reached 35% of the original design production (Republic of Yemen, 1986).

The drop in the water level during the period 1972–1985 varied between 16 and 41 m. The studies that followed included a survey of nearly 3000 wells of which 866 are drilled in sandstone. The volume that could be exploited annually was estimated then at 63×10^6 m^3; 26×10^6 for agriculture, 18×10^6 for water supply to Sana'a, 4×10^6 water supply to the country side and 15×10^6 m^3 as a reserve.

Later studies have shown that the transmissivity of the aquifer varies from 10 to 500 m^2 d^{-1} and the storativity from 1×10^{-4} in the confined part of the aquifer to 1×10^{-2} in the unconfined part. The aquifer is highly anisotropic and the permeability probably decreases with depth. The annual recharge has been estimated once as 50×10^6 m^3 and at another time as 28×10^6 m^3. The corresponding annual abstraction has been estimated as 30×10^6 m^3.

During the 1980s the groundwater levels have been declining generally at a rate of 0.2–0.5 m y^{-1}, and at 6 m near the well fields. This leads to the conclusion that despite recharge to the basin, overexploitation is causing widespread and substantial drawdown. The response of the aquifer in the Sana'a basin to alternative policies for its future management has been investigated by means of a mathematical model. The result is shown graphically in Figure 17 (UNDP, 1992).

The southern part of Yemen (formerly known as the People's Democratic Republic of Yemen) structurally lies in the southern part of the Arabian megablock of the Afro-Arabian platform. There, groundwater can be found in water-bearing formations consisting of sands, sand-

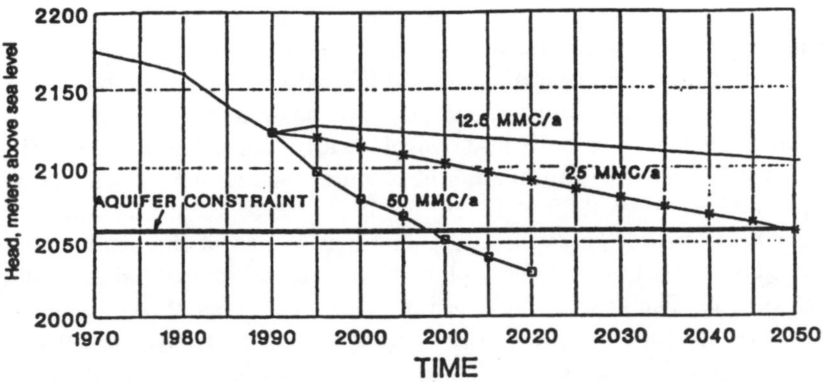

Figure 17. Observed drawdown, 1970–1990, and predicted drawdown for different model abstraction options, Sana'a basin, Yemen (UNDP, 1992).

stones, pebbles and some fractured basement rocks that are widely scattered in the valleys of the large wadis like Wadi Hadramaut, Tuban, Wadi Bana and Hassan, and so on.

Wadi Hadramaut and the northern areas occupy about 40% of the surface area of south Yemen. The principal aquifers are: the valley fill deposits (Quaternary), the Rus-Jeza deposits (Eocene), the Umm Radhuma deposits (Paleocene) and the Mukalla sandstone (Cretaceous). The valley fill deposits comprise the main water-bearing formation in that area. The thickness of the water-bearing layers has been estimated at about 30 m and the transmissivity is widely variable. The quality of groundwater in the flood plain of the wadi is fair, with salinity of 1000–3000 ppm. The central part of the wadi has a salinity of 3000–5000 ppm, and becomes more saline, 5000–14,000 ppm, with distance from the center. Contrarily, the groundwater in Al-Mukalla sandstone is of good quality, 400–1000 ppm. In the northern areas the groundwater tapped within the Rus-Jeza deposits is of moderate quality, with salinity of 1400–3700 ppm total dissolved solids. In 1981 it was reported that the amount of groundwater extracted for irrigation constituted 50% of the annual recharge to the aquifer in the area, roughly estimated as 195×10^6 m^3 y^{-1} (ECOSOC/ECWA, 1981).

The alluvial deposits of the Abien Delta are the most important water-bearing formation. The total thickness, which reaches 120 m, is divided almost equally between two water-bearing strata separated by a clay layer that is not continuous throughout its length. The annual groundwater extraction was estimated as 82×10^6 m^3 out of a recharge of 121×10^6 m^3.

The deserts with interior drainage towards Ar-Rub Al-Khali Desert and Sabatein wasteland are characterized by their deep groundwater, 200–300 m below ground surface. Several deep wells have been dug to reach Al-Mukalla sandy aquifer. Because of the salt domes formed by the Jurassic deposits, the water these wells yield is saline.

The recent volcanic rocks series are present in the provinces of Aden and Lahaj and along the coastal land strip. These rocks can be found at several locations in the tail reaches of the wadis like Tuban, Bana and Mefa'a, and are characterized by their fractures. The rock units in Wadi Tuban, for example, range from Pre-Cambrian complex to Quaternary volcanics and sedimentary formations. The depth to water below the surface varies from 20 m in the upper delta to 5 m in the lower delta of the wadi.

The principal aquifer can be found in the alluvial deposits, which consist of alternating series of highly transmissive coarse clastics and less permeable silt and clay. The aquifer thickness varies from approximately 20 m in the upper reach to 200–400 m in the central and lower reaches. The transmissivity has been estimated as $(1.1–2.4) \times 10^{-2}$ m^2 s^{-1} and the storativity as 0.07–0.12.

The recharge to the aquifers in Wadi Tuban catchment is about 80×10^6 m^3 y^{-1}. As the annually abstracted groundwater exceeds the recharge, a steady decline in the water level at a rate of 1.0–1.7 m has been reported (Republic of Yemen, 1986). The decline in groundwater level is associated with the deterioration of the quality of water and the seawater intrusion inland.

Summing up, the annual total groundwater abstraction from the aquifers in Yemen has been estimated as 1.4×10^9 m^3 by Shahin (1989), 1.2×10^9 m^3 by Al-Alawi and Abdul-Razzaq (1993), and as 1.25×10^9 m^3 (UNESCO/ACSAD, 1993). On the other hand, the aquifer recharge has been estimated as 1.55×10^9 m^3 (Al-Alawi and Abdul-Razzaq, 1993) and 1.4×10^9 m^3 (UNESCO/ACSAD, 1993). The narrow differences between all these estimates lead us to the conclusion that the present use of groundwater in Yemen is about the same as the aquifer recharge. This means that any increase in the currently extracted amounts will not be without adverse effects.

Before we bring this section to an end, it might be of interest to compare three sets of estimates of the annual groundwater use and recharge in the Arabian Peninsula as a region:

Item 10^9 m^3	Shahin (1989)	Al-Alawi and Abdul-Razzaq (1993)	UNESCO/ACSAD (1993)
Abstraction	5.00	(17.559)	4.63
Recharge	—	6.220	5.63

This author is inclined to believe that the total abstraction given by Al-Alawi and Abdul-Razzaq is not correct due to an error in the estimate, 14.43×10^9 m^3, of the use in Saudi Arabia. None of the available reports supports this figure, instead they all refer to some 2×10^9 m^3 as an annual abstraction. Therefore, we are going to consider that the annual abstraction of groundwater from the aquifers in the Arabian Peninsula amounts to about 5×10^9 m^3 and that the annual recharge to the same aquifers is about 6×10^9 m^3.

CONCLUDING REMARKS

After a brief description of the geography, physiography, landscape, demography, climate and hydrogeology of the Arabian Peninsula a detailed inventory of the water resources of the peninsula has been discussed.

The synthesis on aridity and hydrology of the water resources in the peninsula as given in this chapter is founded on data made accessible to investigators as they are mostly available in published material. However, there is evidences of the availability of much more data. The use of such classified data is often restricted to the respective government administrations, international organizations and consulting engineering offices. In a number of situations data are not available simply because there are no data, or if there are any, they are of poor quality and/or quantity, which renders their use unrecommendable. By data is meant here: meteorological data such as temperature, precipitation, humidity, radiation; hydrological data such as water levels and discharges of wadis, runoff coefficients; hydrogeological data such as aquifer thickness, type and hydraulic properties, groundwater levels and piezometric heads; and water quality data. For a number of reasons the quantity and quality of data in the Arabian Peninsula are far from adequate. It is necessary to improve and reinforce the existing measuring networks. Funds should be made available, and special care should be devoted to network design, monitoring, data collection, analysis, processing and publication.

Precipitation is the source of all waters appearing and flowing on the earth's surface or beneath it. The long-term average rainfall has been estimated as 219×10^9 m^3 y^{-1} corresponding to a uniform rain depth of 72 mm y^{-1} distributed over the surface area of the peninsula. A substantial part of this amount is lost by the prevailing high evaporation, and the rest flows along the surface and/or disappears in the ground.

The Arabian Peninsula is probably the largest surface on earth without any perennial river. Instead, in several locations the surface of the peninsula is intersected by ephemeral streams or wadis that run full during and shortly after heavy rains. The annual total surface runoff carried by the wadis can be estimated at $(5–6) \times 10^9$ m^3. A more or less similar figure can be considered to represent the abstraction from the aquifers mainly by pumped wells and the rest through natural springs.

The estimated volume of desalinated water in 1995 was about 2.85×10^9 m^3 and reuse of treated wastewater for 1990 was 0.435×10^9 m^3. The sum of all these figures suggests that the annual total volume of water used in the Arabian Peninsula in the first half of the 1990s is in the range of $(13.5–15) \times 10^9$ m^3. This amount used to serve the interests of some 34×10^6 inhabitants, at an average rate of about 420 m^3 y^{-1} per capita. This rate is about 14% larger than the 370 m^3 y^{-1} per capita one can obtain for 1990 using the figures reported by Engelman and Le Roy (1994). The reasonable agreement between these two independently obtained figures leads us to conclude that the water used per capita in the Arabian Peninsula in the first half of the 1990s is in the order of 400 m^3. This rate is far below the upper limit for water scarcity, which has been set at 1000 m^3 per capita.

All available sources of information point clearly to the ever-increasing demand per capita for freshwater. This situation is leading to excessive exploitation of some of the groundwater resources due to the imbalance between the recharge to and the yield of some of the aquifers. The decline of water levels in those aquifers is causing many of the naturally flowing springs and some of the pumped wells to dry up. Deeper wells have to be drilled and more powerful pumps have to be installed. The unit cost of water is getting higher with time. Aquifer contamination caused by seawater intrusion and upconing of brackish water is a serious threat to the environment.

Desalinated sea and brackish waters constitute an unbounded resource as long as energy is available. Most of the countries sharing the Arabian Peninsula are privileged by having adequate sources of energy. This explains the continuous rise of the volume annually desalinated. Uninterrupted research is unavoidable to improve the efficiency and therefore the production capacity of the desalination plants. The objective is to try to bring the unit cost of desalinated water down to a reasonable price. The issue of the unit cost price is associated with the transport distance of desalinated water to the centers of consumption and the removal of the salt left from the desalination process.

The present synthesis does not include any mention of rain augmentation or rain harvesting. Both are too new to the peninsula, and are as yet in an experimental phase. We have also not dwelt upon the issue of water importation from outside the peninsula. This is essentially a question of water resources development policy, which is based on a number of factors some of which are beyond the scope of this chapter.

REFERENCES

Al-Alawi, J. and M. Abdul-Razzaq. 1993. Water in the Arabian Peninsula: Problems and perspectives. *Proceedings of the Conference on Water in the Arab World, Perspectives and Prognoses,* Harvard, October 1–3:171–202.

Allen, R. G. 1994. An update for the calculation of reference evapotranspiration. *ICID Bul.* 43(2):35–92.

Al-Nakshabandy, G. A. 1979. Preliminary investigation into the effects of refinery wastewater and the single effects of its major pollutants on commercially interesting crops. Paper presented at the International Association on Water Pollution Research Workshop on Treatment of Domestic and Industrial Wastewater in Large Plants, Vienna, 3–7 September.

Al-Nozaily, F. A. 1992. Reuse of wastewater in Sana'a. *Proceedings of the National Seminar on Water Reuse,* Sana'a, 9–11 May: 103–109.

Al-Sajwany, T. A. and R. J. Laurence. 1995. Small Desalination Plants in Rural Communities in Oman. *Proceedings of the International Conference on Water Resources Management in Arid Countries,* Oman, 12-16 March, 2:617–626.

Asano, T. 1994. Reusing urban wastewater—An alternative and a reliable water resource. *Wat. Inter.* 19(1):36–42.

Bagnouls, F. and H. Gaussen. 1953. Saison Sèche et Régime Xèro-thermique. Documents pour les Cartes des Productions Végetales, Art 8, Toulouse 3:1–49.

Bahrain. 1986. Water resources and their uses in Bahrain. Country Report (in Arabic) submitted to the Symposium on Water Resources and Their Uses in the Arab Region, Kuwait, ACSAD/AFESD/KFAED, 17–20 February: 23 plus annexes.

Beaumont, P., G. Blake, and J. M. Wagstaff. 1976. *The Middle East: A Geographical Study.* John Wiley & Sons, New York, pp. 49–88.

Budyko, M.I. 1956. The heat balance of the Earth's surface (translated from Russian). United States Weather Bureau, Washington, D.C.

Bureau de Rechèrches Géologiques et Minières (BRGM). 1986. Water agriculture and soil studies of the Saq and overlying aquifers. Consultants' Report to the Ministry of Agriculture and Water, Riyadh, Saudi Arabia.

Bushnak, A. A. 1995. Putting a price on water. *Proceedings of the Middle East Consultants Conference on Desalination Technology and Economics,* Bahrain, Dar Al-Taqniya, Jedda, 19 January: 50.

Doorenbos, J. and W. O. Pruitt. 1992. Guidelines for predicting crop water requirements. FAO Irrigation and Drainage Paper 24, Food and Agriculture Organization of the United Nations, Rome.

ECOSOC/ECWA. 1981. Assessment of water resources in the ECWA region; United Nations Commission for Western Asia, Natural Resources, Sci. and Tech. Div., Water Programme, Document E/ECWA/NR/L/1, New York, p. 290.

Engelman, R. and P. Le Roy. 1994. Sustaining water: An update to population and the future of renewable water supplies (updated version of the 1993 report). Population and Environment Programme/Population Action International, Washington, D.C., pp. 2–6.

ESCWA/CEHA. 1993. Water desalination: The experience of GCC countries. *Proceedings of the Regional Seminar on Water Use and Conservation,* Amman, 28 November–2 December: 63.

FAO. 1981. Arid zone hydrology for agricultural development. FAO Irrigation and Drainage Paper 30, Food and Agriculture Organization of the United Nations, Rome.

Farquharson, F.A.K., D. S. Plinston, and J. V. Sutcliffe. 1996. Rainfall and runoff in Yemen. *J. Hydrological Sciences* 41(5):797–811.

Gischler, C. E. 1979. *Water Resources in the Arab Middle East and North Africa.* Cambridge, Middle East and North African Studies Press Ltd., pp.73–91.

Hall, M. J. and N. A. Hill. 1983. A master water resources and agricultural plan for the State of Qatar. Part I: Physical setting and resources. *J. Water Resources Development* 1(1):15–30.

Hargreaves, G. H. and Z. A. Semani. 1982. Estimating Potential Evapotranspiration. *J. Irr, & Dr. Div. ASCE* 108(3):225–230.

Hargreaves, G. H. And Z. A. Semani. 1985. Reference crop evapotranspiration from temperature. *Applied Engineering in Agriculture,* Trans. ASAE, 1(2):96–99.

Jensen, M. E. and H. R. Haise. 1963. Estimating evapotranspiration from solar radiation. *J. Irr. & Dr. Div. ASCE* 89(4):15–41.

Jensen, M. E. 1966. Empirical methods of estimating or predicting evapotranspiration using radiation. *Proceedings of the Conference on Evapotranspiration and Its Role in Water Resources Management* Chicago, ASAE, pp.49–53.

Kaul, F. J. 1995. Investigation of potential for storage dams in Wadi Dayqah, Oman. *Proceedings of the International Conference on Water Resources Management in Arid Countries,* Oman, March 12-16 (1):166–175.

Khan, M. O. and R. W. Fuggle. 1995. Beneficial re-use of oil contaminated water in Oman. *Proceedings of the International Conference on Water Resources Management in Arid Countries,* Oman, March 12–16, 2:666–675.

Khoury, J. 1986. Water resources in the Arab world and their future perspectives. Paper submitted to the Symposium on Water Resources and Their Use in the Arab Region, Kuwait, ACSAD/AFESD/KFAED, February 17–20, p.110.

Köppen, W. 1936. Das Geographische System der Klimate. In Handbuche der Klimatologie W. Köppen and R. Geiger (Eds.), Berlin, Gebrüder Bornträger, 1(C).

Kuwait. 1986. Water resources and their uses in the State of Kuwait. Country Report submitted to the Symposium on Water Resources and Their Uses in the Arab Region, Kuwait, ACSAD/AFESD/KFAED, 17–20 February: 33 plus annexes.

Lloyd, J. W. 1992. Groundwater management problems in the developing world. *J. Applied Hydrogeology* 4:35–48.

Lockwood, J. C. 1974. *World Climatology: An Environmental Approach.* Edward Arnold Publishers, London, UK, pp. 44–77.

Martyn, D. 1992. *Climates of the World.* Elsevier Publ.Co. and PWN Polish Sc.Publ., Amsterdam-Warsaw, pp. 178–185.

Ministry of Oil and Mineral Resources of Yemen and TNO Institute of Applied Geosciences, The Netherlands. 1995. The Water Resources of Yemen, Report Wray-35, Sana'a and Delft.

Ministry of Water Resources of Oman. 1995. Water Resources of the Sultanate of Oman, An Introductory Guide, RUWI, Oman, pp. 15–50.

Mohorjy, A. M. 1988. Water resources management in Saudi Arabia and water use. *Water International* 13(3):161–171.

Naqash, A. B. and S. H. Marrei. 1983. The occurrence of groundwater resources in arid zones, with emphasis on United Arab Emirates. *Proceedings of the International Symposium on Groundwater in Water Resources Planning,* UNESCO/IAH/IAHS, Kolenz, 28 August–3 September: 159–170.

Nouh, M. A. 1988. On the prediction of the flood frequency in Saudi Arabia. *Proc. Inst. Civ. Engrs,* Part 2, 85:121–144.

Oman, 1986. Water resources and their uses in the Sultanate of Oman. Country Report submitted to the Symposium on Water Resources and Their Uses in the Arab Region, Kuwait, ACSAD/AFESD/KFAED, 17–20 February: 43.

Penman, H. L. 1948. Natural evaporation from open water, bare soil and grass. *Proceedings Royal Meteorological Society,* London, 193:120–145.

People's Democratic Republic of Yemen. 1986. Water resources and their uses in (The People's Democratic Republic of Yemen). Country Report submitted to the Symposium on Water Resources and Their Uses in the Arab Region, Kuwait, ACSAD/AFESD/KFAED, February 17–20: 29.

Petrov, M. P. 1976. *Deserts of the World* (translated from Russian by IPST Staff). John Wiley & Sons, New York, pp. 47–49.

Poulter, S. 1995. Regional Monitoring on Unconfined Aquifers. *Proceedings of the International Conference on Water Resources Management in Arid Countries,* Oman, 12–16 March (2):457–465.

Republic of Yemen. 1986. Water resources and their uses in the (Arab) Republic of Yemen. Country Report submitted to the Symposium on Water Resources and Their Uses in the Arab Region, Kuwait, ACSAD/AFESD/KFAED, 17–20 February: 16.

Qatar. 1986. Water resources and their uses in Qatar. Country Report submitted to the Symposium on Water Resources and Their Uses in the Arab Region, Kuwait, ACSAD/AFESD/KFAED, 17–20 February: 55 plus annexes.

Salih, A. M. and U. Sendil. 1984. Evapotranspiration under extremely arid climates. *J. Irr. & Dr. Div. ASCE* 110(3):289–307.

Saudi Arabia. 1986. Water resources and their uses in the Kingdom of Saudi Arabia. Country Report submitted to the Symposium on Water Resources and Their Uses in the Arab Region, Kuwait, ACSAD/AFESD/KFAED, February 17–20: 67 plus annexes.

Sayed, S. A. and F. M. Al-Ruwaih. 1995. Relationships among hydraulic characteristics of the Dammam aquifer and wells in Kuwait. *J. Hydrogeology* 3(1):57–70.

Shahin, M. 1989. Review and assessment of water resources in the Arab Region. *Water International* 14(4):206–219.

Shahin, M. 1996. *Hydrology and Scarcity of Water Resources in the Arab Region.* A. A. Balkema, Rotterdam-Brookfield, p. 137.

Shahin, M. 1998. Estimation of reference evapotranspiration and crop water use in the Arabian Peninsula. *Arabian Journal of Science and Engineering,* Part II, June, p. 27.

Sha'lan, S. A. and A. M. Salih. 1987. Evapotranspiration estimates in extremely arid areas. *J. Irr. & Dr. Div. ASCE* 113(4):564–574.

Thornthwaite, C. W. 1948. An approach toward rational classification of climate. *Geogr. Rev.* 38(1):55–94.

Ukayli, M. A. and T. Hussein. 1988. Comparative evaluation of surface water availability, wastewater reuse and desalination in Saudi Arabia. *Water International* 13(4):218–225.

UNDP. 1982. Groundwater in the Eastern Mediterranean and Western Asia. Study Series on Water (9), ST/ESA/112 (in Arabic), New York, p. 560.

UNDP. 1992. Groundwater Resources in the Yemen: Report to Technical Secretariat of the High Water Council, UNDP/DTCD Project Report Yem/88/001, New York-Sana'a.

UNESCO. 1977. Map of the World Distribution of Arid Regions. MAB Tech. Note 7, United Nations Educational, Scientific and Cultural Organisation, UNESCO, Paris.

UNESCO/ACSAD. 1993. Groundwater protection in the Arab world. Paper prepared as a contribution to UNESCO's IHP-IV Programme, Paris/Cairo/Damascus, p. 158.

United Arab Emirates. 1982. Report on traditional water irrigation systems in the United Arab Emirates UAE. Progress Report by the Ministry of Agriculture and Fisheries, Water and Soil Department, Dubbai, p. 19 plus annexes.

United Arab Emirates. 1986. Water resources and their uses in the United Arab Emirates. Country Report (in Arabic) submitted to the Symposium on Water Resources and Their Uses in the Arab Region, Kuwait, ACSAD/AFESD/KFAED, February 17–20: 78 plus annexes.

Van der Gun, J. A., W. I. Elderhorst, and G. P. Kruseman. 1992. Predicting long-term impacts of groundwater abstraction from an intensively exploited coastal aquifer. *Proceedings of 23rd International Congress of the International Association for Hydrogeologists,* Tenerife, I. Simmers, F. Villarroya, and L. F. Rebollo, (Eds.) 3:303–318.

Wallén, C. C. 1968. Agroclimatological Studies in the Levant. *Proceedings of the Reading Symposium,* Agricultural Methods, Reading, pp. 225–233.

Wan, P. 1976. Point rainfall characteristics of Saudi Arabia. *Proc. Inst. Civ. Engrs,* Part 2, 61:179–187.

Wheater, H. S. and N. C. Bell. 1983. Northern Oman flood study. *Proc. Inst. Civ. Engrs,* Part 2, 87, Paper 8659:453–473.

Wheater, H. S. and R. P. C. Brown. 1989. Limitations of design hydrographs in arid areas—An illustration from South West Saudi Arabia. *Proc. 2nd Nat. Hydroi. Symp.,* Sheffield 4–6, British Hydrological Society, London: 3.49–3.56.

Wilson, E. M. 1969. *Engineering Hydrology.* The MacMillan Press Ltd., London, pp. 91–117.

Zubari, W. K., A. M. Mubarak, and I. M. Madany. 1993. Development impacts on groundwater resources in Bahrain. *Water Resources Development* 9(3):263–279.

Zubari, W.K., I. M. Madany, S. S. Al-Junaid, and S. Al-Manaly. 1994. Trends in the quality of groundwater in Bahrain with respect to salinity, 1941–1992. *Environment International* 20(6):739–746.

Appendix 1. *Monthly mean climatic data for four stations in the Arabian Peninsula.*

Station	Month	T,°C	RH, %	S, h d^{-1}	R, g cal $cm^{-2}d^{-1}$	U, m s^{-1}	P, mm mo^{-1}	Eo, mm d^{-1}
	Feb.	17.3	70	7.0	413	1.3	21.5	3.1
	Mar.	20.6	64	6.2	454	1.4	24.2	4.7
	Apr.	25.4	55	7.3	517	1.4	11.2	7.1
	May	30.6	48	8.3	573	1.6	0.4	10.1
	Jun.	32.9	47	9.4	599	2.1	0	12.2
R. Al-Faras,	Jul.	34.3	50	8.1	557	1.6	0	11.3
Qatar	Aug.	34.0	57	8.6	544	1.3	0	9.8
	Sep.	31.9	62	8.8	507	1.0	0	7.6
	Oct.	28.1	64	8.5	456	0.9	4.3	5.5
	Nov.	23.0	69	7.6	389	1.0	5.0	3.8
	Dec.	18.5	73	6.3	330	1.1	17.1	2.6
	Year	26.1	61	7.8	473	1.3	100.1	6.7
	Jan.	12.8	42	10.7	331	1.0	3	4.5^{++}
	Feb.	15.9	56	9.5	364	1.3	7	4.8
	Mar.	18.6	55	8.3	396	1.3	36	5.3
	Apr.	17.3	66	6.6	359	1.2	53	5.0
	May	21.3	42	9.8	426	1.5	28	6.7
Sanáa	Jun.	22.8	42	8.4	387	1.9	4	7.1
Airport,	Jul.	22.5	48	7.4	378	1.6	25	5.9
Yemen	Aug.	21.3	49	8.4	384	1.9	38	6.2
	Sep.	20.1	42	10.0	442	0.8	2	6.2
	Oct.	16.6	54	10.7	409	1.7	13	6.1
	Nov.	15.9	40	10.5	318	1.1	3	5.0
	Dec.	14.4	50	9.2	436	1.1	1	4.4
	Year	18.3	49	9.[illegible]	386	1.4	213	5.6
	Jan.	14.1	53	7.5	323	1.33	14.1	4.0†
	Feb.	16.2	45	8.3	394	1.53	7.1	5.7

continued

Appendix 1. *(continued).*

Station	Month	T,°C	RH, %	S, h d^{-1}	R, g cal $cm^{-2}d^{-1}$	U, m s^{-1}	P, mm mo^{-1}	Eo, mm d^{-1}
	Mar.	20.5	41	7.9	451	1.64	22.6	7.8
	Apr.	25.3	38	8.2	486	1.58	33.5	9.7
Ar-Riyadh,	May	30.2	32	8.7	525	1.53	10.0	12.1
Saudi	Jun.	33.0	21	10.7	580	1.72	0.1	14.6
Arabia	Jul.	34.2	20	10.4	570	1.75	0.5	14.8
	Aug.	33.8	20	10.3	544	1.50	0.9	13.9
	Sep.	30.9	24	9.6	513	1.19	0	11.7
	Oct.	25.5	31	8.9	461	1.00	0.3	8.9
	Nov.	19.8	44	7.9	385	1.06	7.9	5.3
	Dec.	15.4	52	7.1	320	1.17	11.0	3.9
	Year	24.9	35	8.8	463	1.42	114.2	9.4
	Jan.	21.1	64	8.2	325	2.41	12.3	6.1*
	Feb.	21.9	66	8.3	384	2.74	25.5	6.2
	Mar.	24.8	58	8.3	451	2.58	15.2	8.3
	Apr.	29.4	47	9.6	526	2.41	19.9	12.5
	May	33.8	43	11.2	583	2.49	0	15.4
Al-Seeb,	Jun.	35.7	46	11.1	574	2.66	0	14.2
Oman	Jul.	34.6	58	9.9	524	2.92	0	11.4
	Aug.	31.2	72	9.1	506	2.49	0.2	6.0
	Sep.	31.2	64	10.0	485	2.32	0	8.1
	Oct.	29.4	58	10.3	438	2.33	2.2	9.2
	Nov.	25.9	61	9.6	365	1.97	0.2	7.3
	Dec.	23.0	66	8.9	315	2.23	16.4	6.1
	Year	28.5	59	7.6	457	2.46	91.9	9.2
	Jan.	16.2	73	6.6	341	1.2	16.4	2.4+

+Class A pan evaporation
++Penman evaporation
*Pich evaporation

Appendix 2. Monthly maximum rainfall at selected stations in Kuwait, Saudi Arabia and Oman.

Country and Station	Period of Record	Maximum Rainfall, in mm, During the Period of Record for											
		Jan.	Feb.	Mar.	Apr.	May	Jun.	Jul.	Aug.	Sep.	Oct.	Nov.	Dec.
Kuwait													
Ahmadi	1961–1980	77.7	73.5	136.2	114.9	36.7	0	0	0	0	53.6	110.9	58.9
Kuwait Airport	—	73.2	95.9	50.0	67.0	19.0	0	0	0	0	56.6	107.6	57.9
Shuwaikh	—	97.3	60.4	45.9	74.5	21.4	0	0	0	0	48.0	110.7	52.5
Saudi Arabia													
Dhahran	1961–1980	58.7	66.7	77.4	75.6	22.8	0	0	0	0	18.0	27.4	94.0
Jeddah	—	124.7	98.9	3.0	93.0	20.0	0	2.8	0	1.2	4.0	83.0	55.5
Riyadh	—	111.1	39.8	99.6	124.3	69.0	2.2	8.8	17.4	0	6.0	44.0	70.0
Oman													
Masierah	1971–1980	29.6	12.5	29.5	43.8	0	430.6	33.0	0	4.2	0	14.6	99.8
Salalah	1971–1980	1.4	10.8	4.4	123.0	59.6	124.0	31.7	30.2	11.0	34.0	2.1	6.0
Muscat	1893–1959	143.0	98.6	70.4	38.3	8.9	64.0	37.1	14.7	0.0	44.5	77.2	171.2

Appendix 3(a). Estimates of grass reference evapotranspiration, ET_o, at some locations in the Arabian Peninsula.

Station No.	Method of Estimation	Grass Reference Evapotranspiration, ET_o, (mm.mo^{-1}) Jan	Feb	Mar	Apr	May	Jun	Jul	Aug	Sep	Oct	Nov	Dec	Year (mm)	Years of Record	Source
1	FAO B-C[a]	80	106	144	178	259	297	306	311	250	200	125	81	2337	1965–89	
	Hg, 1982[b]	73	93	139	167	237	264	280	280	220	166	97	66	2082		
	Hg, (1985)[c]	69	87	137	182	237	267	282	263	210	153	95	76	2058		
	FAO P[e]	88	115	148	179	237	254	274	271	225	183	133	98	2205		
	FAO P-M[f]	79	107	132	164	218	231	244	247	208	161	119	88	1998		
2	FAO B-C	83	93	122	179	252	296	291	263	225	180	136	86	2207	1989–93	
	Hg, 1982	75	82	112	148	193	221	234	222	184	151	108	82	1802		
	Hg, (1985)	77	84	123	163	207	222	233	212	170	138	101	79	1809		
	Hg, 1985[d]	56	65	96	127	165	165	181	169	141	115	77	62	1419		
	FAO-P	89	94	148	205	283	295	332	307	247	180	144	86	2410		
	FAO P-M	74	81	120	172	247	263	305	278	231	177	122	76	2146		
3	FAO B-C	79	85	116	145	217	259	237	224	197	157	117	84	1917	1972–92	
	Hg, 1982	71	83	106	155	200	219	212	203	174	143	99	74	1739		
	Hg, 1985	79	91	134	177	231	234	247	232	205	162	109	82	1983		
	FAO P	80	92	131	178	231	243	234	224	188	161	101	76	1939		
	FAO P-M	78	91	125	150	210	223	227	205	178	154	108	88	1837		
4	FAO B-C	82	90	120	154	217	230	226	207	189	153	109	95	1872	1976–92	
	Hg, 1982	67	73	101	128	167	176	174	183	158	135	93	71	1526		
	Hg, 1985	76	81	123	164	194	192	206	196	177	147	99	83	1738		
	FAO-P	97	110	141	194	235	239	240	236	202	162	103	83	2042		
	FAO P-M	96	110	139	165	209	221	222	213	185	165	120	91	1936		
5	FAO B-C	—	—	—	—	—	—	—	—	—	—	—	—	2372	1969–76	Anonymous
	FAO P	—	—	—	—	—	—	—	—	—	—	—	—	2396		
	FAO P-M	—	—	—	—	—	—	—	—	—	—	—	—	2223		

Appendix 3(a). *(continued)*.

Station No.	Method of Estimation	Jan	Feb	Mar	Apr	May	Jun	Jul	Aug	Sep	Oct	Nov	Dec	Year (mm)	Years of Record	Source
		Grass Reference Evapotranspiration, ET_o, ($mm.mo^{-1}$)														
6	FAO B-C	89	112	147	191	248	298	341	293	247	176	126	88	2356	1970–81	Salih and Sendil, 1984; Sha'lon and Salih, 1987
	Hg, 1982	93	112	154	182	220	248	257	243	210	172	120	90	2101		
	Hg, 1985	88	103	145	182	232	250	263	258	216	169	114	91	2111		
	FAO P	95	123	167	194	231	263	278	243	210	158	102	81	2145		
	FAO P-M	115	134	202	237	285	342	384	322	264	205	156	118	2764		Anonymous
7	Hg, 1982	98	116	166	198	237	258	268	259	210	166	120	92	2188	1973–81	Salih and Sendil, 1984; Sha'lan and Salih, 1987
	FAO P	90	112	158	188	186	237	248	222	191	147	98	82	1959		
8	Hg, 1982	102	124	178	209	253	270	278	271	228	185	131	105	2334	1977–81	Salih and Sendil, 1984; Sha'lan and Salih, 1987
	FAO P	112	133	183	219	254	276	285	260	216	155	117	93	2303		
9	Penman (E_o)	$E'_o = -5.092 + 0.145\,T + 0.01142 + 0.012\,RC - 0.012\,RH$ where: E'_o = predicted Penman evaporation, (E_o), RC = incoming radiation (cal. $cm^{-2}\,d^{-1}$), and coefficient of determination, $r^2 = 0.968$.													1970–72	FAO, 1981
10	E_o	148	170	223	237	279	288	298	282	258	211	165	143	2702	6 years	FAO, 1981
	$ET_o = \alpha\,E_o$	104	128	178	201	251	274	283	254	232	179	132	100	2316		
	α	0.7	0.75	0.8	0.85	0.9	0.95	0.95	0.9	0.9	0.85	0.8	0.7	$\alpha = 0.857$		
11	FAO B-C	112	131	167	200	258	252	258	228	218	169	147	121	2261	1981–82	
	Hg, 1982	101	115	157	189	233	211	215	196	175	160	117	99	1968		
	Hg, (1985)	95	107	145	175	212	189	187	171	156	139	109	89	1774		
	FAO P	118	143	171	213	272	260	269	247	203	180	138	119	2333		
	FAO P-M	115	128	166	185	248	250	262	235	211	162	139	112	2213		

continued

Appendix 3(a). (continued).

Station No.	Method of Estimation	Jan	Feb	Mar	Apr	May	Jun	Jul	Aug	Sep	Oct	Nov	Dec	Year (mm)	Years of Record	Source
		Grass Reference Evapotranspiration, ET_o, ($mm.mo^{-1}$)														
12	FAO B-C	148	142	164	159	217	205	210	194	185	177	158	143	2102	1982–83	
	Hg, 1982	121	122	143	141	177	163	156	149	142	155	133	114	1716		
	Hg, (1985)	115	116	140	146	163	163	175	164	149	136	119	113	1699		
	Hg, 1985	87	95	131	132	151	157	153	143	141	135	109	97	1531		
	FAO P	127	134	159	154	193	183	184	174	160	153	150	127	1898		
	FAO P	121	127	155	167	186	177	176	182	167	154	133	120	1865	unknown	Farguharson et al., 1996
	FAO P-M	19	120	140	144	151	157	163	150	141	138	128	116	1667	1982–83	
13	FAO B-C	143	166	191	207	238	205	223	213	211	222	193	170	2382	1983–90	
	Hg, 1982	108	125	149	163	174	152	151	165	159	169	142	124	1781		
	Hg, (1985)	113	116	142	154	171	173	182	180	170	152	124	110	1787	1983–90	
	Hg, 1985	104	105	130	141	163	161	173	166	152	132	105	96	1628		
	FAO P	146	135	165	175	192	174	196	194	182	183	152	142	2036		
	FAO P-M	109	132	147	163	179	162	164	169	160	162	152	136	1835		
14	FAO B-C	135	115	149	157 (128)	209 (171)	238 (195)	230 (189)	227 (186)	211 (173)*	166	152	126	2115 (1885)	1982–86	
	Hg, 1982	121	114	138	142	148	166	149	160	166	143	132	109	1688		
	Hg, 1985	129	130	154	162	190	188	190	170	165	154	131	122	1885		
	FAO P	124	139	161	141	200	212	188	190	180	181	150	125	1991		
	FAO P	140	137	165	151	209	212	183	192	186	190	149	136	2050	unknown	Farquharson et al., 1996
	FAO P-M	115	118	137	141	171	198	189	181	153	156	125	105	1789		

[a]FAO-Blaney-Criddle
[b]Hargreaves and Semani (1982)
[c]Hargreaves and Semani (1985 combined with 1982)
[d]Hargreaves and Semani (1985)
[e]FAO-Penman
[f]Penman-Monteith
*Figures between parentheses are estimates of ET_o without elevation correction

Appendix 3(b). Measurement and estimates of Alfalfa reference evapotranspiration at some locations in the Arabian Peninsula.

Station No.	Measurements or Estimation Method	Alfalfa Reference Evapotranspiration, ET_o, ($mm.mo^{-1}$)													Years of Record	Source
		Jan	Feb	Mar	Apr	May	Jun	Jul	Aug	Sep	Oct	Nov	Dec	Year (mm)		
1	J-H, 1963[a]	73	97	158	207	309	356	382	382	293	209	112	68	2646	1965–89	
	J, 1966[b]	77	101	159	200	292	333	355	356	275	201	111	71	2531		
2	J-H, 1963	77	87	126	179	249	292	312	297	242	191	132	92	2276	1989–93	
	J, 1966	72	81	115	159	217	253	272	256	210	168	117	84	2004		
3	J-H, 1963	75	90	122	191	230	291	284	272	229	182	118	82	2196	1972–92	
	J, 1966	87	104	136	205	272	300	292	279	238	192	129	93	2327		
4	J-H, 1963	71	79	117	158	215	230	232	243	207	170	113	78	1913	1976–92	
	J, 1966	76	83	126	158	213	226	227	238	204	170	115	82	1918		
5	Lysimeters[c]	115	136	209	238	319	369	389	346	212	158	110	79	2680	1970–71	Salih and Sendil, 1984; Sha'lan and Salih, 1987
6	J-H, 1963	106	134	189	239	284	332	344	322	270	217	148	106	2691	1970–81	
	J, 1966	109	127	172	202	240	266	276	264	225	180	132	108	2301		
														2200 ÷ 0.87 = 2528	unknown	Anonymous

continued

Appendix 3(b). (continued).

Station No.	Measurements or Estimation Method	Alfalfa Reference Evapotranspiration, ET_o, (mm.mo^{-1}) Jan	Feb	Mar	Apr	May	Jun	Jul	Aug	Sep	Oct	Nov	Dec	Year (mm)	Years of Record	Source
7	J-H, 1963	124	147	214	270	310	351	366	347	282	217	156	118	2902	1973–81	Salih and Sendil, 1984; Sha'lan and Salih, 1987
8	J-H, 1963	133	155	220	270	326	354	366	350	293	234	161	124	2986	1977–81	Salih and Sendil, 1984; Sha'lan and Salih, 1987
11	J-H, 1963	119	140	196	245	309	280	286	260	230	192	146	118	2521	1981–92	
	J, 1966	105	122	170	211	264	239	244	222	197	166	127	104	2171		
12	J-H, 1963	149	152	179	172	229	214	205	194	186	195	166	141	2182	1982–83	
	J, 1966	116	118	139	133	176	164	157	148	143	150	134	110	1688		
13	J-H, 1963	133	155	188	210	228	202	200	218	209	219	179	153	2294	1983–90	
	J, 1966	116	136	164	182	196	173	172	187	159	159	156	134	1934		
14	J-H, 1963	117	120	154	122	173	198	177	187	190	153	139	111	1841	1982–86	
	J, 1966	126	131	167	165	189	216	194	204	207	165	150	119	2033		

[a]Jensen and Haise (1963)
[b]Jensen (1966)
[c]measured ET_r using lysimeters

10

Methods of Estimating Sediment Inflow to Reservoirs

TARIQ J. CHEEMA
Department of Earth Sciences
Sultan Qaboos University
P.O. Box 36, Al-Khod 123
Sultanate of Oman

INTRODUCTION

GLOBAL ECONOMIC AND energy crises have forced us to reevaluate the water resources potential of our rivers. Many river management projects are being implemented and many more are under consideration. Unpredictable climatic conditions and the complex geologic environment of drainage basins result in varying degrees of sediment inflow to reservoirs. The problem confronting the project planner is how to estimate the sediment inflow in order to determine the useful life of the reservoir.

In a country like Oman, the problem of estimating sediment inflow is further compounded by the fact that sufficient water is never available and that many areas have started experiencing a deficit in groundwater recharge (Sultan, 1995). Therefore, almost all of the dams in Oman impound water for recharging the groundwater resources. An under-estimation of the sediment inflow in these circumstances can significantly reduce the useful life of the dam and may endanger the design and purpose of the project, causing huge economic losses.

Sediment is the end product of erosion, and the total amount of erosion in a watershed is the gross erosion. However, all the eroded material does not enter the stream system. Some of it is deposited at natural or man-made barriers within the watershed, and some is deposited within the channels and their flood plains. The portion of the eroded material that does travel through the drainage network to a downstream measuring or control point is referred to as the sediment yield.

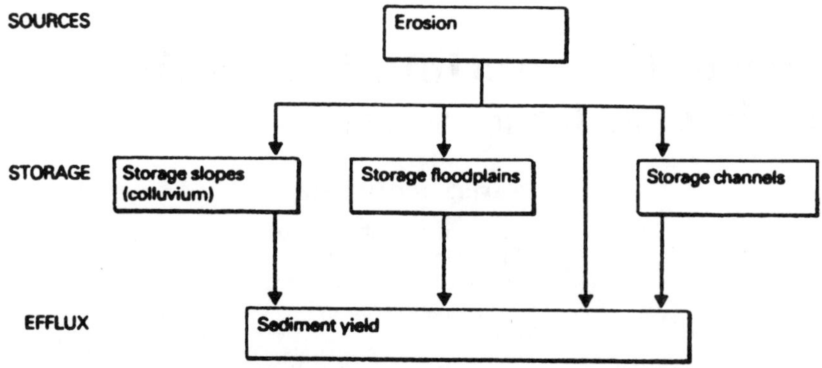

Figure 1. General model of sediment movement in a drainage basin [after Cooke and Doornkamp, Geomorphology in Environmental Management (1990) by permission of Oxford University Press].

The degree of accuracy in estimating sediment yield depends entirely on how well the sediment movement in a drainage basin can be modeled. Figure 1 shows a general model of sediment movement through a drainage basin. This model looks simple but within any one basin, the details of sediment movement are extremely complex. Figure 2 illustrates some of the complexities in modeling reservoirs and pathways of Rock Creek basin, in Oregon, U.S. As a result of such a complexity, the precise routes of sediment through a drainage basin are unknown. Thus, estimation and prediction of sediment yield is commonly based on general but useful concepts and techniques. Based on these concepts, all methods of sediment yield estimation can broadly be classified into the following three categories:

- predictive equations
- gross erosion and sediment delivery ratio computations
- direct measurements

PREDICTIVE EQUATIONS

The sediment yield of a drainage basin can be estimated by using one or more predictive equations. Predictive equations are statistical equations developed from easily measured watershed parameters. Some of these parameters include rainfall amount and intensity, soil type and geologic formation, ground cover, land use, runoff, and the geometry of the drainage basin. The number of predictive equations

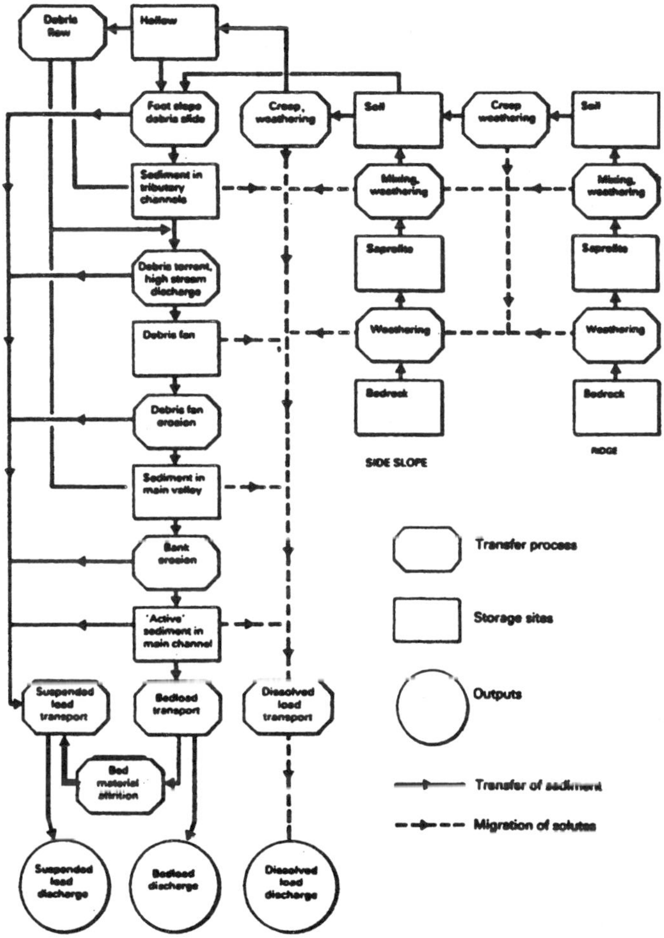

Figure 2. Sediment budget model for Rock Creek basin, Oregon (after Dietrich et al., 1982).

based on these variables is large. Following are some of the predictive equations and graphs that can be used in estimating the sediment yield of a drainage basin.

Brune's Nomogram

Empirical relations between sediment yield, drainage area, percentage of cultivated land, and runoff were summarized by Brune (1951)

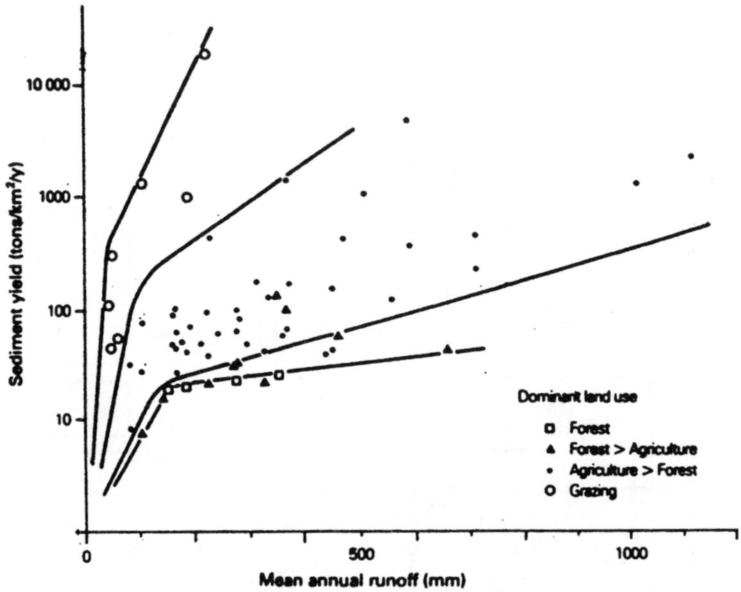

Figure 3. Relationship between sediment yield and annual runoff (after Douglas, 1967).

for the upper Mississippi region of the U.S. (Figure 3). The Brune's nomogram shown in Fgure 3 explains how to calculate sediment yield for a 10-square mile basin, 30% of which is composed of cultivated and idle land. The total annual runoff in this example is 400 acre feet per square mile. The nomogram can also be used for other basins showing variable amount of total runoff and idle land.

Fournier's Method

Fournier's method (1960) has been used at regional and local scale to predict specific values of the sediment yield. This method requires a classification of the drainage basin according to its relief and climate into the following four categories:

- Ia low relief, temperate climate
- Ib low relief, tropical, subtropical, semi-arid climate
- II high relief, humid climate
- III high relief, semi-arid climate

Once this classification is done, the sediment yield is calculated for each class by using the following regression equations:

Ia $Y = 6.14\,X - 49.78$
Ib $Y = 27.12\,X - 475.4$
II $Y = 52.49\,X - 513.21$
III $Y = 91.78\,X - 737.62$
Y = sediment yield (in metric units)
$X = p^2/P$
p = rainfall (mm) in the month of greatest precipitation
P = mean annual precipitation (mm).

Douglas Equation

The amount of effective precipitation is the only input data needed for this equation to predict the sediment yield for a drainage area. The equation developed by Douglas (1967) for major Asian rivers is as follows:

$$E = \frac{1.631\,(.03937\,P)^{2.3}}{1+0.0007\,(0.03937\,P)^{3.3}}$$

E = suspended sediment yield in m^3/km^2/year
P = effective precipitation (mm).

Similar equations can be derived for other areas. Figure 4 shows the relationships between sediment yield and annual runoff for major Asian, Australian and U.S. rivers. The peak of the curves declines both towards areas of more and less effective precipitation. Langbein and Schumm (1958) explained this variation by the operation of two variables. First the erosive influence of precipitation increases with its amount. Opposing this influence is the protective effect of vegetation, which also increases with precipitation. This model is not universally applicable because it ignores the influence of drainage basin and land use on the sediment yield (Cooke and Doornkamp, 1990).

Jansen and Painter's Equation

This predictive equation, developed in 1974, incorporates five easily measurable watershed parameters (Jansen and Painter, 1974).

$$\text{Log sediment yield} = 12.133 - 0.340 \log Q + 1.590 \log H + 3.704 \log P + 0.936 \log T - 3.495 \log C$$

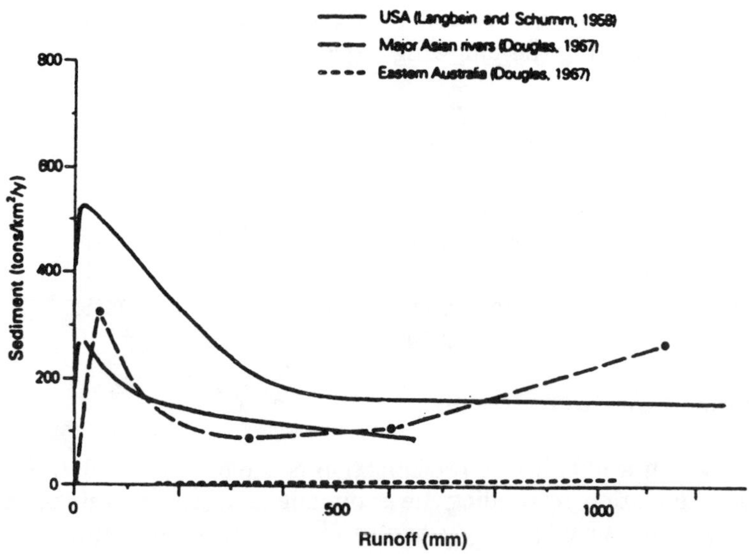

Figure 4. The effect of vegetation cover on sediment yield (after Dunne, 1979).

Where:

- Q = annual discharge in $10^3 m^3/km^2$
- H = altitude (m, above mean sea level)
- P = annual rainfall (mm)
- T = average annual temperature (°C)
- C = natural vegetation index

Dunne's Predictive Equation

The influence of different land uses upon sediment yield for a given runoff was studied by Dunne (1979). The findings of his studies are summarizd in Figure 5. The lines shown in this figure separate basins with different vegetation cover. As can be seen in this figure, for a single land-use type, there is a general increase of sediment yield with runoff. As the average density of cover decreases from forest to range lands, the sediment loss becomes more sensitive to variations in runoff.

GROSS EROSION AND SEDIMENT DELIVERY RATIO COMPUTATIONS

The gross erosion and sediment delivery ratio technique of estimating sediment yield is a two-step procedure. At first, the gross erosion in a

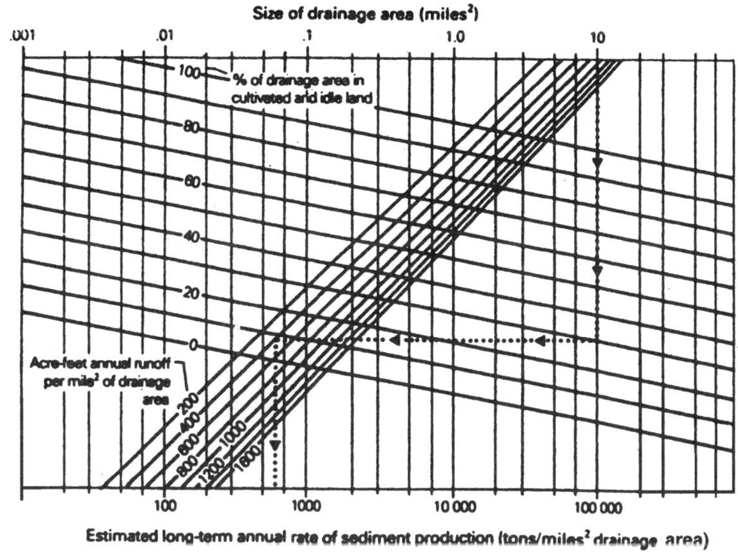

Figure 5. Brune's nomogram (after Brune, 1951).

drainage area is computed from the Universal Soil Loss Equation (USLE); then, a sediment delivery ratio is determined. The sediment delivery ratio is a fraction of gross erosion, that which is expected to be delivered at a specific point of the drainage area under consideration. Sediment yield is a product of gross erosion and the delivery ratio.

In 1974, the Soil Conservation Service (SCS) of the U.S. designed a form (SCS-ENG-309) for predicting erosion losses occurring in a drainage area. This form is reproduced in Figure 6. The SCS (1971) has also developed a general sediment delivery ratio versus drainage area relationship from the data of earlier studies (Figure 7). This relationship shows that the sediment delivery ratio varies approximately inversely as the 0.2 power of the drainage area. Fine-textured soils (clays and silts) show high delivery ratios compared to coarse-textured soils (sand). If no soil textural data are available, the median curve can be used to estimate the delivery ratio.

Modified USLE for Estimating Sediment Yield

Wischmeier and Smith (1965) proposed USLE for predicting total soil loss. Since then, many modifications of USLE have been proposed

SEDIMENT YIELD BY SOURCES (AVERAGE ANNUAL)

		PRESENT CONDITIONS			FUTURE (AFTER CONS. TREATMENT)		
		ACRES	SOIL LOSS (TONS/AC)	TOTAL (TONS)	ACRES	SOIL LOSS (TONS/AC)	TOTAL (TONS)
SHEET EROSION	CULTIVATED LAND 1/	540	18.5	2/ 9,990	500	12.9	2/ 6,450
	IDLE LAND	150	8.1	1,215			
	PASTURE - RANGE	880	4.1	3,610	950	2.9	2,755
	WOODLAND	630	2.3	1,450	750	1.5	1,125

	DELIVERY RATIO (%)		TONS DELIVERED	DELIVERY RATIO (%)		TONS DELIVERED
SHEET EROSION - TOTAL	20	16,265	2/ 3,255	20	10,330	2/ 2,065
GULLY EROSION	80	6,600	5,280	80	3,160	2,530
STREAMBANK EROSION	90	1,200	1,080	90	1,200	1,080
FLOODPLAIN SCOUR						
CONSTRUCTION	50	2,000	1,000	50	800	400
		TOTAL	3/ 10,615		TOTAL	4/ 6,075

[1]Includes row crops, small grains, meadow.
[2]Product of proceding two columns rounded to nearest 5 tons.
[3]Enter under "Sediment Delivered to Site" in "Present" space of "Deposition" section.
[4]Enter under "Sediment Delivered to Site" in "Future" space of "Deposition" section.

Figure 6. SCS-ENG-309 form for sediment yield data (after SCS, 1974).

for estimating the sediment yield of a reservoir. Some of these modifications are as follows:

- $A = (0.224)\,(RKLSCP)Ec$ (Renard et al., 1974)
  Ec Channel erosion factor
  RKLSCP USLE factors
- $Y = 11{,}800\,(Qq_p)^{0.56}\,KCPLS$ (William and Berndt, 1976)
  Y Sediment yield from an individual storm (in kg)
  Q Storm runoff volume (cubic m/sec)
  q_p Peak runoff rate (cubic m/sec)
  KCPLS USLE factors.
- $A = (0.224)\,WKCPLS$ (Onstad et al., 1976)
  W Hydrologic factor
  $W = aRST\,(1\text{-}a)(0.4Qq_p)^{0.333}$
  a Coefficient of rainfall energy
  Rst Rainfall storm factor
  Q Runoff volume
  qp Peak runoff rate.

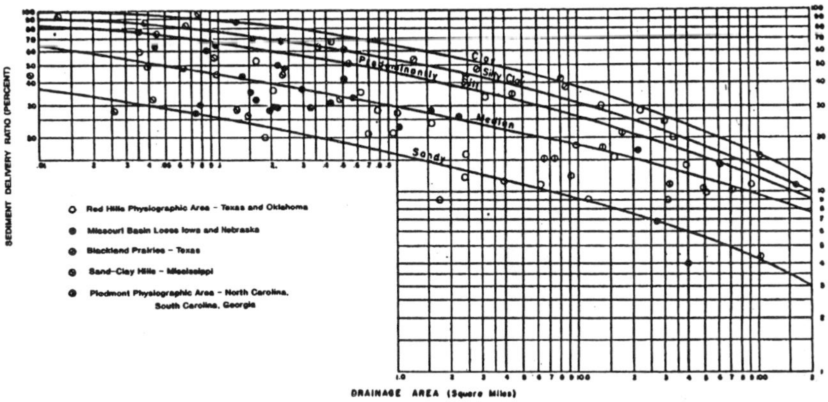

Figure 7. Sediment delivery ratio versus drainage area curves (after SCS, 1971).

The applicability of the above equations is limited to similar drainage basins and hydrologic conditions for which they were derived.

Generalized Sediment Yield Maps

A convenient way of obtaining sediment yield information is by analyzng all available sediment data and extending them over a national or international scale (Annandale, 1987). Equal erosion potential areas are delineated without having to put yield figures to the areas. A map of sediment yield may then be prepared and may be calibrated by all the available sediment yield data. Figure 8 is an example of a sediment yield map for South Africa (Rooseboom, 1975). Once such a map is available, the area under consideration can be plotted on the map and sediment yield predicted.

DIRECT MEASUREMENTS

Reservoir Sediment-Deposition Method

The reservoir sedimentation-deposition method (Cooke and Doornkamp, 1990) or reservoir resurvey method (Strand and Pemberton, 1982) involves determining the volume of sediment deposited in a reservoir at known intervals. The volume may be derived in several ways, including surveying, the use of electronic echo sounding, satellite re-

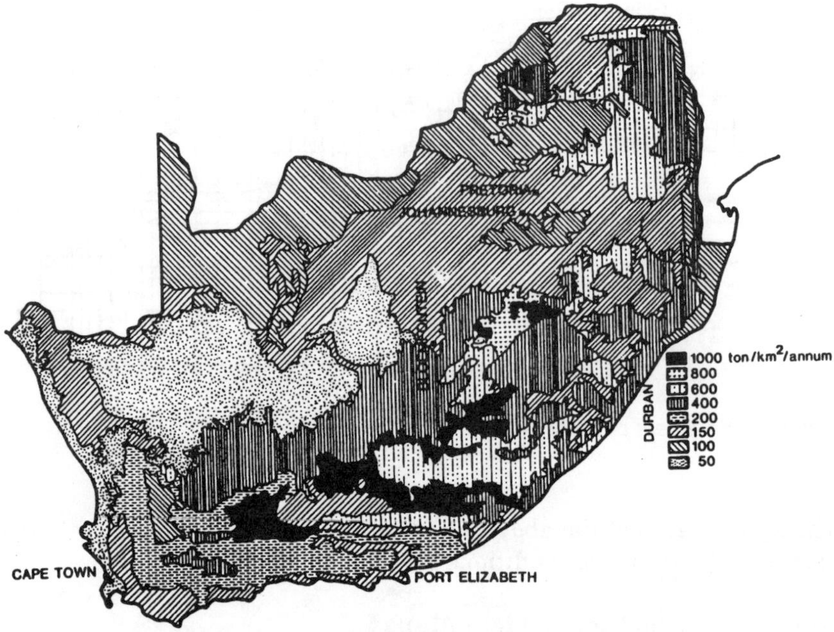

Figure 8. Sediment yield map of South Africa (after Rooseboom, 1975).

mote sensing mapping techniques, and some empirical methods. Surveying the reservoir is the most common and considered the best method for estimating sediment yield (Strand and Pemberton, 1982).

The main purpose of a reservoir survey is to determine the storage capacity at the time of survey which, when compared to an earlier survey (usually the original survey), gives the sediment accumulation. Area and capacity curves are then plotted for the original and new surveys. An example of such a plot is shown in Figure 9. It is suggested that the method selected to compute capacities from contour areas be the same for both the surveys. This helps eliminate any differences in technique having undue influence on the sediment volume computations. Annandale (1987) recommended investigating older reservoirs in order to make better inference.

Once a calculation of the sediment volume is done, an estimation of the average density of the deposited sediment is required to convert the measured volume to a mass yield. Due to the fact that in-situ

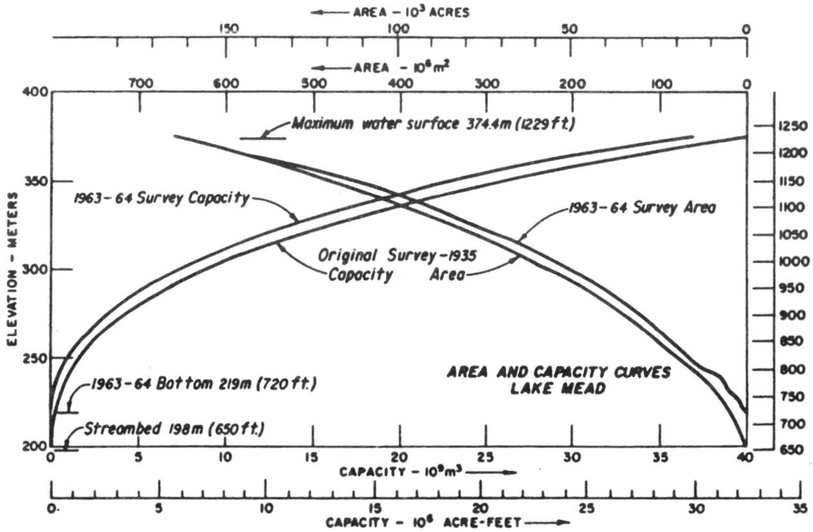

Figure 9. Area and capacity curves for Lake Mead (after Strand and Pemberton, 1982: Furnished by the U.S. Department of the Interior, Bureau of Reclamation).

tests are very expensive and difficult to perform (Sabtan, 1988), the use of empirical data is recommended to obtain density values. Depending upon the size distribution of the sediments (% of sand, silt and clay), an average density of the sediment may vary from 1100 kg/m^3 to 1350 kg/m^3 for 50-year-old deposits. The sediment yield (tons/km^2/year) is determined by multiplying the density by the volume of the deposited sediment and dividing the answer by the size of the catchment and the age of the reservoir. This information can be transferred to the catchment area under consideration. The reservoir sediment deposition method requires that hydrological factors such as rainfall, geology, soil type, vegetation, catchment size, runoff, and so on, be similar for the catchment under consideration. Strand and Pemberton (1982) suggesed the use of Figure 10 for estimating sediment yields for drainage areas that differ from the actual measured drainage area. This model was developed from the data of selected reservoirs located in arid climate of the southwestern United States. However, it is best to make a calibration with a known sediment yield of a specific location and evaluate the effects of different hydrologic factors. Parallel lines similar to those shown in Figure 10 can be

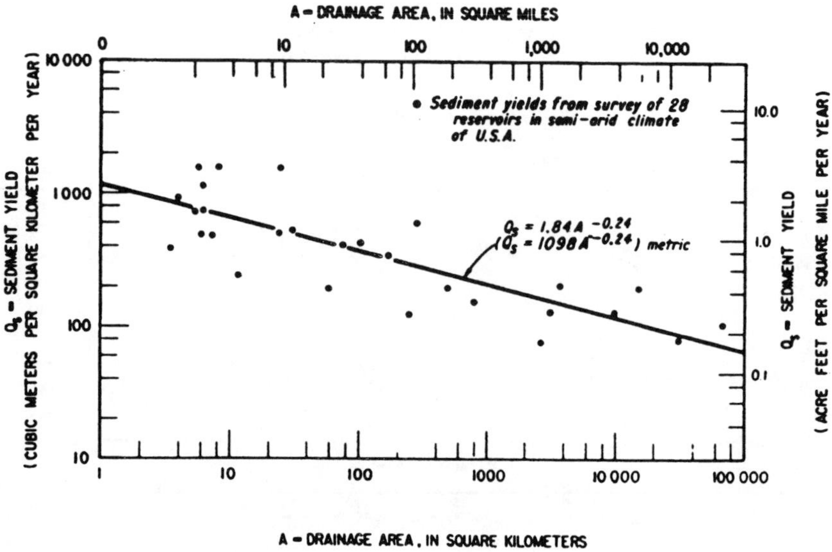

Figure 10. Average annual sediment rate versus drainkage area (after Strand and Pemberton, 1982: Furnished by the U.S. Department of the Interior, Bureau of Reclamation).

drawn through any measured data point to obtain the sediment yield of the drainage area in question.

Sediment Sampling Method

Sampling is the surest method of obtaining an accurate determination of the suspended sediment load being carried by a stream at a particular location. Suspended sediment sampling in combination with total load computations is the preferred method for planning studies to determine the sediment inflow to a proposed reservoir (Strand and Pemberton, 1982) and the most useful means of predicting sediment yield (Cooke and Doornkamp, 1990).

The total sediment transported in a stream consists of bedload and the suspended sediment load. The objective of a sediment sampling program on a river is to collect sufficient samples of sediment both as suspended load and as bedload to define the total sediment being transported.

Suspended sediment load is determined by measuring the suspended sediment concentration, C, in mg/L from water samples. The water discharge (Qw) is measured in m³/s. Suspended sediment load (Qs) can be calculated in tons/day by the equation:

$$Qs = 0.0864\ C\ Qw$$

A suspended sediment sampler of any shape and size can be used; however, the D-43 sampler is normally used for a large stream and U.S. DH-48 sampler for a small stream (Linsley et al., 1949). It is important that samples represent an integration with width across the channel as well as depth from the water surface to the river bed.

A suspended sediment rating curve is established as a first step toward computing the sediment yield of a drainage basin. This rating curve is normally a plot on logarithmic paper of water discharge (Qw) versus sediment load (Qs). Figure 11 is a plot of suspended sediment rating curve for Rio Toa Vaca near Villaba, Puerto Rico (Strand and

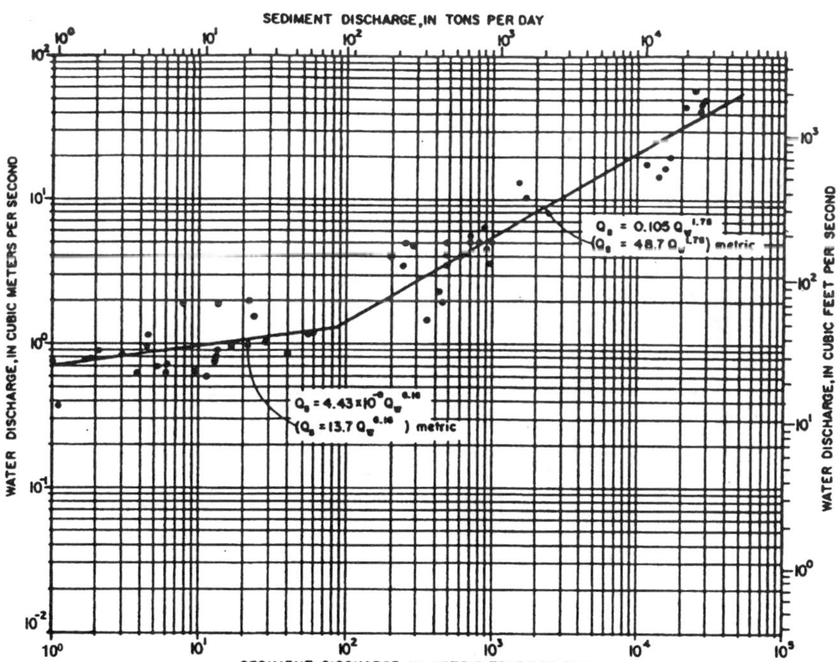

Figure 11. Sediment suspended rating curve for Rio Toa Vaca, Puerto Rico (after Strand and Pemberton, 1982: Furnished by the U.S. Department of the Interior, Bureau of Reclamation).

Pemberton, 1982). The equation for any segment of this plot is in the form:

$$Qs = a\,Q^{b}w$$

Qs = suspended transport ton/day

Qw = water discharge, ft^3/s

a = coefficient

b = exponent

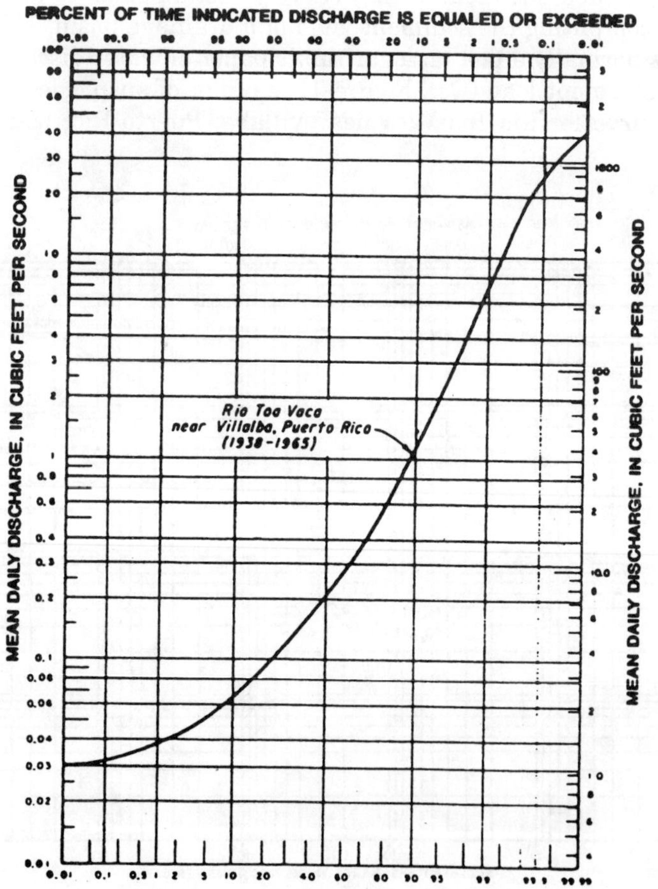

Figure 12. Flow duration curve for Rio Toa Vaca, Puerto Rico (after Strand and Pemberton, 1982: Furnished by the U.S. Department of the Interior, Bureau of Reclamation).

Table 1. *Sediment load computation for Rio Toa Vaca, Puerto Rico (after Strand and Pemberton, 1982: Furnished by the U.S. Department of the Interior, Bureau of Reclamation).*

Project: Puerto Rico
Stream: Rio Toa Vaca
Period of record: Streamflow 1938–1965
Computed by:

Reservoir: Toa Vaca
Section:
Sediment: 1963–1967
Checked by:

Date: May 1968

(1) Limits, (%)	(2) Interval	(3) Middle ordinate	(4) Q_w		(5) Q_s		(6) $Q_w = \frac{\text{col.(2)} \times \text{col.(4)}}{100}$		(7) $Q_s = \frac{\text{ccl.(2)} \times \text{col.(5)}}{100}$	
			ft³/s	m³/s	ton/d	t/d	ton/d	t/d	ton/d	t/d
0.00–0.02	0.02	0.01	1,412.0	40.0	34,151	30,984	0.282	0.008	6.83	6.20
0.02–0.1	.08	.06	1,037.8	29.4	19,925	18,077	0.830	.024	15.94	14.46
0.1–0.5	.4	.3	617.8	17.5	8,038	7,292	2.47	.070	32.15	29.17
0.5–1.5	1.0	1.0	250.6	7.1	1,657	1,504	2.51	.071	16.57	15.04
1.5–5.0	3.5	3.25	115.4	3.27	426.6	387	4.039	.114	14.93	13.55
5–15	10	10	44.1	1.25	59.7	54.2	4.410	.125	5.97	5.42
15–25	10	20	20.8	0.59	0.583	0.531	2.080	.059	0.058	0.053
25–35	10	30	14.1	.40	.053	.049	1.410	.040	.005	.005
35–45	10	40	10.6	.30	.009	.008	1.060	.030	.001	.001
45–55	10	50	8.1	.23	.002	.002	0.810	.023	.0002	.0002

continued

Table 1. (continued).

Project: Puerto Rico
Stream: Rio Toa Vaca
Period of record: Streamflow 1938–1965
Computed by:

Reservoir: Toa Vaca
Section:
Sediment: 1963–1967
Checked by:

Date: May 1968

(1) Limits, (%)	(2) Interval	(3) Middle ordinate	(4) Q_w ft³/s	(4) Q_w m³/s	(5) Q_s ton/d	(5) Q_s t/d	(6) $Q_w = \frac{col.(2) \times col.(4)}{100}$ ton/d	(6) t/d	(7) $Q_s = \frac{col.(2) \times col.(5)}{100}$ ton/d	(7) t/d
55–65	10	60	6.4	.18	.0004	.0004	.640	.018		
65–75	10	70	4.6	.13			.460	.013		
75–85	10	80	3.5	.10			.350	.010		
85–95	10	90	2.4	.068			.240	.007		
95–96.5	3.5	96.75	1.7	.048			.060	.002		
98.5–99.5	1.0	99.0	1.3	.038			.013	.0004		
99.5–99.9	0.4	99.7	1.2	.034			.005	.0001		
99.9–99.98	.08	99.94	1.1	.030			.001			
99.98–100	.02	99.99	1.1	.030			.00			
						Total	21.67	0.615	92.45	83.90

Annual discharge = Total Q_w = 21.67 × 365 × 1.9835 = 15,700 acre-ft/yr = Total Q_w = 0.615 × 365 × 86.4 × 10³ = 19.4 10⁶ m³/yr
Annual sediment load = Total Q_s = 92.45 × 365 = 33,700 ton/yr = Total Q_s = 83.90 × 365 = 30,600 t/yr

$$\text{Average concentration, } C = \frac{Q_s}{Q_w \times 0.0027} = \frac{92.45}{21.67 \times 0.0027} = 1580 \text{ mg/L} = \left[\frac{Q_s}{Q_w \times 0.0864} = \frac{83.90}{0.615 \times 0.0864} = 1580 \text{ mg/L}\right]$$

Table 2. Bedload correction (after Strand and Pemberton, 1982: Furnished by the U.S. Department of the Interior, Bureau of Reclamation).

Condition	Suspended sediment concentration, mg/L	Streambed material	Texture of suspended material	Percent bedload in terms of suspended load
[1]1	<1000	Sand	20 to 50% sand	25 to 150
[1]2	1000 to 7500	Sand	20 to 50% sand	10 to 35
3	>7500	Sand	20 to 50% sand	5
[2]4	Any concentration	Compacted clay gravel, cobbles, or boulders	Small amount up to 25% sand	5 to 15
5	Any concentration	Clay and silt	No sand	<2

[1]Special sampling program for Modified Einstein computations required under these conditions.
[2]A bedload sampler such as the Helley-Smith bedload sampler may be used, or computations made by use of two or more of the bedload equations when bed material is gravel or cobble size.

The suspended sediment rating curve can be combined with available water discharge records to determine the long-term average sediment yield. The longer the period of the discharge records, the more reliable the results. An example of the flow-duration curve for the station used to develop the sediment rating curve (Figure 11) is illustrated in Figure 12. Table 1 shows the computation of suspended sediment load at the gauge based on combining the sediment rating curve with the flow-duration curve.

Measuring the bedload component of sediment transport is a difficult operation. Therefore, Gregory and Walling (1976) were of the opinion that no single apparatus or procedure, whether theoretical or empirical, has been widely accepted as completely adequate for the determination of bedload discharge. Correction of suspended sediment load for the bedload to give total load can be made with the help of Table 2. When the bedload is thought to be a significant portion of the total load, this correction is strongly recommended by Mitchell and Bubenzer (1980).

CONCLUSIONS

The choice of method to estimate the sediment yield of a reservoir depends on the nature of the problem and time and funds available for solving the problem. Predictive equations for calculating sediment yield make use of easily determined watershed parameters and provide only a rough estimate of sediment yield of a drainage basin. The application of the Universal Soil Loss Equation (USLE) holds good only for small drainage areas. Direct measurements provide the most accurate determination of sediment yield. The reservoir sediment-deposition method requires that the drainage area considered and the drainage area to be investigated have same hydrologic parameters. Although time-consuming sampling program and bedload correction makes the sediment sampling method very expensive, it is the most accurate available method for calculating the sediment inflow to a reservoir.

REFERENCES

Annandale, G. W. 1987, *Reservoir Sedimentation,* Elsevier, New York.

Brune, G. M. 1951. Sediment record in midwestern U.S. *Int. Assoc. Sci. Hydrology* 3:29–38.

Cooke, R. U. and J. C. Doornkamp. 1990. *Geomorphology in Environmental Management.* Clarendon Press, Oxford.

Dietrich, W. E., T. Dunne, N. F. Humphrey, and L. M. Reid. 1982. Construction of sediment budgets for drainage basins. USDA Forest Service, Tech. Report PNW-141.

Douglas, I. 1967. Man, vegetation and the sediment yield of rivers. *Nature* 215:925–928.

Dunne, T. 1979. Sediment yield and land use in tropical climates. *J. Hydrol.* 42:281–300.

Fournier, F. 1960. *Climat et erosion.* Presse Université Francaise, Paris, France.

Gregory, K. J. and D. E. Walling. 1976. Drainage Basin Form and Process. Edward Arnold, London.

Jansen, J. M. and R. B. Painter. 1974. Predicting sediment yield from climate and topography. *J. Hydrol.* 21:371–380.

Langbein, W. B. and S. A. Schumm. 1958. Yield of sediment in relation to mean annual precipitation. *Trans. Am. Geophys. Union* 39:1074–1084.

Linsley, R. K., M. A. Kohler, and J. L. Paulhus. 1949. *Applied Hydrology.* McGraw Hill Book Comp.

Mitchell, J. K. and G. D. Bubenzer. 1980. Soil Loss Estimation. In: Kirby and Morgan (Eds.). *Soil erosion,* Wiley Publishing, Chichester, PA, pp. 17–62.

Onstad, C. A., R. F. Piest, and K. E. Saxton. 1976. *Watershed Erosion Model.* Water Research Council, Washington.

Renard, K. G., J. R. Simanton, and H. B. Osborn. 1974. Soil erosion and sediment transport. *Proceedings of the Arizona Academy of Science,* Flagstaff, AZ.

Rooseboom, A. 1975. Sediment transport in rivers and reservoir basins. D.Sc. Thesis, University of Pretoria, South Africa.

Sabtan, A. 1988. Determination of longitudinal and lateral sediment distribution in reservoirs. Ph.D. Thesis, South Dakota School of Mines and Technology, South Dakota (U.S.).

Soil Conservation Service. 1971. Sediment sources, yields and delivery ratios. *SCS Nat. Eng. Handbook,* USDA, Washington, D.C.

Soil Conservation Service. 1974. Sediment storage requirements. Technical Release—309, USDA, Washington, D.C.

Strand, R. I. and E. L. Pemberton. 1982. *Reservoir Sedimentation.* USBR Publications, Denver, CO.

Sultan, K. A. 1995. Effective management plan needed. *Oman Daily Observer,* 11 March.

William , J. R. and H. D. Berndt. 1976. *Sediment Yield Prediction Based on Hydrology.* American Society of Agriculture Engineers, St. Joseph, Michigan.

Wischmeier, W. H. and D. D. Smith. 1965. Predicting rainfall erosion losses from cropland east of Rocky Mountains. *United States Department of Agriculture Handbook,* Little Rock, Arkansas.

11

Design Criteria and Effectiveness of Recharge Dams in Arid Climates

NASSER AL MAQBALI AND VINCENT KOTWICKI

Dams Department
Ministry of Water Resources
P.O. Box 2575
Ruwi, Sultanate of Oman

INTRODUCTION

In hot and arid climates, potential evaporation can exceed rainfall by several orders of magnitude. Surface storage in such a situation is often impractical due to large water losses to the atmosphere. Practically, there is no technology available at present for reducing evaporation from large free water surfaces. The few available methods, floating plastic covers, thin layers of cethyl alcohol over the surface, or sheltering of the reservoir by vegetation, cause many undesirable side effects like rising water temperature, pollution and increased biological activity. Further, these methods are usually not effective in high winds, and their maintenance is difficult. Avoiding evaporation by storing water underground is very effective.

The process of artificially recharging groundwater may be defined as follows: The planned activity of man whereby surface water is made to infiltrate the ground, commonly in rates and in quantities many times in excess of natural recharge. Artificial recharge is most commonly executed with the help of water spreading, pits, wells and recharge dams.

Todd (1980) identified the following purposes of recharging:

1. Maintain or augment the groundwater as an economic resource
2. Coordinate the operation of surface and groundwater reservoirs

3. Combat adverse conditions such as progressive lowering of groundwater levels, unfavorable salt balance, and saline water intrusion
4. Provide subsurface storage for local or imported surface waters
5. Reduce or stop significant land subsidence
6. Provide a localized subsurface distribution system for established wells
7. Provide treatment and storage for reclaimed wastewater for subsequent reuse
8. Conserve or extract energy in the form of hot or cold water

Although the concept of a recharge is relatively new, it is rapidly gaining wide acceptance throughout the world (Abdulrazzak, 1997). Many recharge programs are being implemented in European countries, Australia, U.S. and other parts of the world. Many recharge dams have also been constructed in Arab countries. There are already 200 dams in Saudi Arabia and seventy dams in the UAE which have recharge as the main or supplementary function. Many more schemes are planned.

Artificial recharge to underground storage yields several benefits. Groundwater storage capacity exceeds that of most surface structures; the method is relatively cheap; silting difficulties are largely avoided; water supplies are naturally purified for drinking purposes; and evaporation losses are minimized.

STUDY AREA

The Sultanate of Oman is situated in the southeastern corner of the Arabian Peninsula and characterized by arid climate, with hot and dry summers and mild winters, except the region in the south of the country, which is affected by a monsoon climate during the months of June to September each year.

The northern part of Oman is mostly mountainous, where Al Hajar Al Gharbi ranges, exceeding 3000 m, run parallel to the coast and separate the most fertile Batina region from the interior. The massifs of Al Hajar Al Gharbi and Al Hajar Al Sharqi are the generating source of runoff after rains, either toward the sea or the desert. Extensive Ar Rub Al Khali and Ash Sharqiya deserts and widespread gravel plains covering large areas of the country are extremely arid.

The Sultanate in general is confined to a 100 mm rainfall per year. Higher rainfall occurs only in the mountains: the Saiq Plateau receives some 350 mm per year whereas the capital receives some 75 mm and parts of the interior much less than that. In the Sultanate, with overall high temperatures and low humidity, conditions for evaporation are favorable. The potential evaporation is estimated between 3000 mm in the interior through 2100 mm on Al Batina Coast to some 1700 mm on the Salalah Plain.

Occasional flows to the sea occur only during exceptional floods, usually lasting several hours every few years. Runoff coefficients for particular gauging stations range from 5 to 10% in central reaches of wadis.

DESIGN ASPECTS OF RECHARGE DAMS IN OMAN

Groundwater

Since the groundwater assessment is based on only short available rainfall and groundwater level records of less than twenty years, to study long-term effects additional rainfall data have to be generated. Models are used to simulate rainfall, runoff and recharge. Different scenarios for rainfalls and abstractions are being studied. This allows for analyzing the long-term efficiency of dams and its effect on the groundwater regime. The results have to be carefully interpreted, taking into account the insufficient data input. Further, it is difficult to model the aquifer in a realistic way. The hydrogeology of Northern Oman is often characterized by complex limestone and ophiolite formations overlain by gravel layers. Groundwater flow is not restricted to the aquifer but is often strongly influenced by fracture flow in the limestone base. Its incorporation into the model often requires a good deal of assumptions.

Evaluation of the technical, economical, and social feasibility of the dam finally provides the basis for deciding the appropriate water management means. It also includes actual or potential water usage downstream of the target areas. Possible negative impacts on well abstractions and Aflaj (qanat) yields have to be also assessed and considered.

Topography

Most of the recharge projects are concentrated in the coastal Al Batina Plains, which include the largest cultivated areas in Oman. The relief of the Al Batina landscape is very uniform with typical widespreading wadis towards the coast. The recharge dams, which are supposed to be most effective if located 10 to 15 km from the coast, in general lie within these alluvial wadi fans. Here, active wadi channels have to be identified and considered in the arrangement of the outlets and the spillway. The outlets should discharge to active channels that provide good infiltration conditions. The spillway must cover the main wadi channels in order not to change the downstream flow regime. Floods must not be artificially diverted to other wadi channels.

Another consequence of the topography is the distinctive shape of almost all recharge reservoirs in Al Batina. Whereas the width of these reservoir surfaces (corresponding to the dam length) is up to several kilometers, their length might only be a few hundred meters. Wing embankments often have to be constructed to define the sides of the reservoirs. With this reservoir shape and a maximum spillway height of 5 to 8 m, the optimal ratio of reservoir to embankment volume can be achieved. In addition, this shape allows integration of the highest number of active wadi channels and thus maximizes the catchment area with the least construction costs. It is, therefore, the most economic solution.

The topography is more variable for recharge dams in the interior. Some sites are similar to the uniform coastal topography, whereas others are typical mountainous sites with steep rock abutments forming elongated reservoirs.

Geology

To be efficienct recharge dams require deep and wide alluvial layers. Since the recharge dams are in general close to the recharge area, the same geology applies to the dam site too. The embankments are founded on cemented or loose gravel layers; rock foundations are rare exceptions. However, gravel foundations also allow for the construction of concrete structures like weir-type spillways. Thus, combinations of concrete spillways and earthfill embankments are feasible.

With respect to the stability of the foundation, it is important to keep leakage losses at a safe level. The alluvial underground typically contains buried wadi channels consisting of coarse gravel. Locally, the permeability may reach 10^{-3} m/s or more. If located in upper layers, they are likely to provide seepage paths under the embankment which might initiate piping and erosion phenomena. Therefore, a key trench or cutoff is indispensable.

Hydrology

Rainfall patterns of northern Oman are extremely variable. Most of the precipitation comes from short but intense rainstorms, with rainfalls of more than 250 mm/day reported during the last 100 years. Together with the poor vegetation cover, these rainstorm events lead to extreme runoff peaks flows.

For illustration, rivers like the Amazon have a highest flood peak of the order of 0.04 $m^3/s/km^2$. Very similar values have been recorded for mighty ephemeral rivers of arid central Australia. This is because the catchments of such rivers are immense, and the unit discharge from a massive flood is relatively low. Time from the occurrence of rainfall to a flood is measured in days, weeks or even months. For average world conditions, and catchments of a small to medium size, a probable maximum flood of the order of 1 $m^3/s/km^2$ is probably a workable approximation. However, Omani wadis are capable of providing flood peaks of the order of 20 $m^3/s/km^2$, and possibly more during rare and extreme events. Muscat Flood Alleviation Dam has a design flood peak of over 100 $m^3/s/km^2$.

Time from the occurrence of rainfall to a flood is measured in minutes. Therefore, flood protection dams have to be durable, and their spillways able to withstand ferocious floods of very high magnitude. For these reasons, the spillway becomes a major cost factor for the recharge dams. Dictated by economic considerations, the reservoir storage needs to be designed for high return probabilities (10–20 years). For extreme flood events, the reservoirs have only a marginal retention effect and cannot provide significant protection for the downstream habitants. However, they prevent them from the more frequent small and medium floods that have already occurred at certain sites. In general, the spillway design flood is chosen to be a proportion of the Probable Maximum Flood (PMF), often a 0.5 PMF, depending on downstream safety considerations.

Dam Types

The recharge dams in Oman are located on wadi gravel deposits, with the objective of retaining floods and allowing increased infiltration into the gravel aquifer by slow releases into the downstream wadi channel. The period of water detention in the reservoir is usually less than ten days, which is dictated by health requirements. An "average" recharge dam in the sultanate is ten meters high and has a crest 3.5

km long. It has an "average" reservoir of four million cubic meters, which in an "average" year stores more than four million cubic meters of water.

Material

The types of dams chosen are governed principally by the construction materials available. The large and flat wadi fans with deep alluvial deposits provide sufficient fill material for the construction of sand, silt and gravel embankments. In general, the grading and permeability allow for processing filter and fill material within the reservoir area. Even with more distant borrow areas, the fill embankment is generally the most economic dam type. Rock material for protection of the embankment slopes is usually not available at site but has to be transported from distant quarries in the mountains.

Spillway Design

A spillway is a part of the dam designed to safely pass flood waters. The spillway is designed for overtopping, whereas the remainder of the dam is usually not. The spillways of the recharge dams in Oman need to be large because the dams are generally located in the lower reaches of the wadis with relatively large catchment areas of arid mountains. Flash floods produce extremely high flow peaks in these conditions.

Spillway-specific discharge varies widely, depending mainly on the type of material used. In Oman, gabion constructions have a low specific discharge of 1 to 5 $m^3/m/s$. Rock armor spillways have a specific discharge of 5 to 10 $m^3/m/s$ and concrete spillways of up to 25 $m^3/m/s$. For all recharge dams, the spillway is the dominating structural element. The spillway design flood reaches several thousands of cubic meters per second for the large Al Batina dams. The most economical solution is to integrate the spillway in the dam structure as an overtoppable embankment. Suitable means to protect the embankment in the overtoppable section are needed. Three types of protection have been applied: gabion lining, rock armor and concrete lining. The appropriate type is selected based on economic considerations. Since each type of protection offers its specific resistance against erosion, the maximum flow velocity and unit discharge vary widely for the different spillway types.

The topography in general imposes no restriction for the overflow length. Once the spillway type has been selected, its overflow length can be fixed mainly according to economic and safety aspects including future maintenance costs.

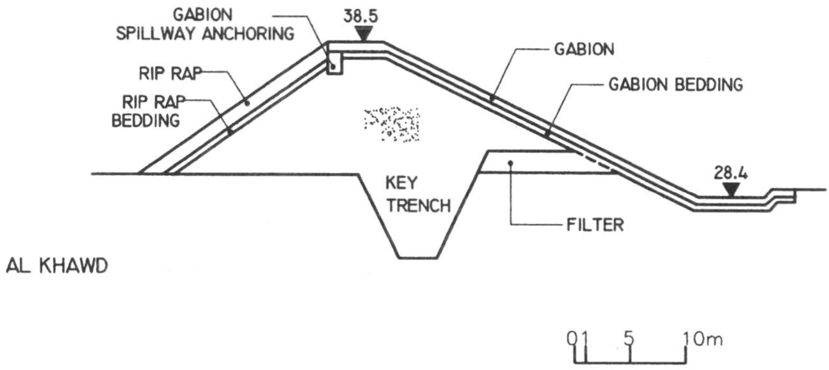

Figure 1. Gabion spillway.

GABION SPILLWAY

An example of a gabion spillway is presented in Figure 1. The overflow profile corresponds to a broad-crested weir type with low discharge coefficients. An important detail of the overflow crest is the concrete unit that is keyed into the core of the embankment. It defines the exact overflow level of the spillway and elongates the seepage path in the contact zone of filter and core material.

The downstream slope of the spillway is protected by gabion mattresses, arranged on filter layers. In the lower pan of the dam the gabion mattress is doubled to resist the accelerated flow. The main design criterion for the thickness of the gabion layer is the seepage flow velocity in the contact zone between gabion mattress and embankment. The seepage velocity needs to be controlled in order to prevent erosion of the fill material. The design criteria for the gabion boxes consist of the maximum unit discharge and the flow velocity. For specified gabion characteristics, critical and limit velocities have been derived from laboratory tests. Above the critical velocity, severe damages and deformations are to be expected. The design is adjusted to the critical velocity in order to provide a safety reserve against extreme flood events, which is essential if considering the hydrological uncertainties.

ROCK ARMOR SPILLWAY

The overflow profile for a rock-armour spillway corresponds to a broad-crest weir type. Figure 2 shows a typical cross-section. A U-shaped concrete unit is covering the entire crest. The side walls cut into the embankment to protect the core. The downstream embankment

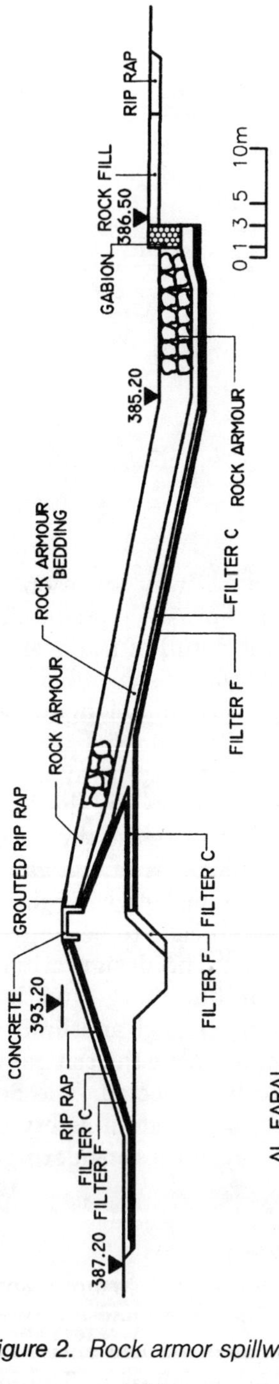

Figure 2. Rock armor spillway.

slope is covered by loose rocks, which are individually positioned to achieve a uniform consistency and a constant spillway slope.

Main design parameters for the rock armor include the unit discharge and the flow velocity for the design flood. The range of the maximum rock sizes can vary widely, which provides flexibility in adjusting the overflow length and height. However, experience has shown that in Oman rock sizes bigger than one meter are difficult to obtain in sufficient quantities and acceptable quality. The stilling basin floor is also protected by rock armor, whereas the end sill of the stilling basin is usually designed as a gabion wall. Problems can arise from locating and establishing suitable quarries for rocks and the difficulty to work with rocks of this size. Environmental aspects like disturbance of landscape, noise and dust pollution or long haulage distances might offset ostensible cost advantages.

CONCRETE SPILLWAY

This type of spillway, shown in Figure 3, allows the highest unit discharge. The spillway slope can be chosen to be very steep and adapted to stability requirements and construction methods. Thus, the spillway length becomes distinctively shorter than for the other types, which partly compensates for the higher material costs. If combined with an ogee weir type, the overflow length can be shortened by one third without reducing the spillway capacity. If further reduction of length is required, for example, due to topographic restrictions, the unit discharge is to be increased by heightening the nonoverflow embankment.

Depending on the flow velocity, the stilling basin might be of concrete or gabion-lined. An important aspect to be considered when selecting this spillway type is the sealing element of the embankment. In order to avoid uplift pressures under the concrete slab, only sealing elements

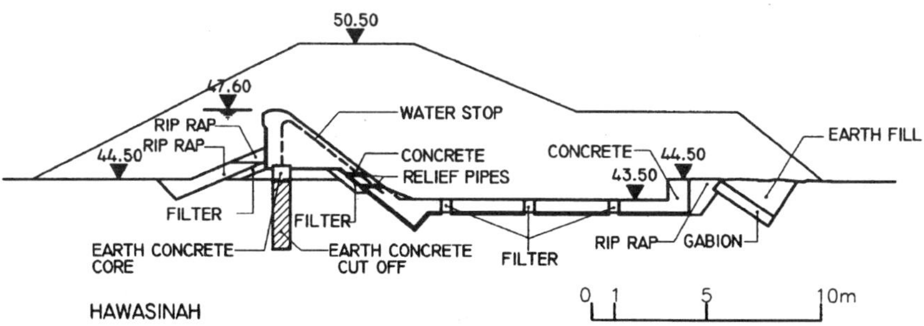

Figure 3. Concrete spillway.

with very low permeability ($< 10^{-6}$ m/s) are appropriate. If the unit discharges are very high ($> 5m^3/s/m$) and flow velocities exceeding 6.5 m/s, the concrete-type spillway is the only option left.

An important advantage of the concrete type is its high safety reserve in case of flood events exceeding the design flood. Whereas for rock armor or gabion spillways the failure of the system is likely to start at the spillway itself, the structural integrity of the concrete spillway will not be endangered by floods above the design flood. In this case, erosion of the wadi bed might occur downstream of the stilling basin due to insufficient energy dissipation. However, this will not immediately endanger the safety of the dam. The failure might therefore start only with the overtopping of the nonoverflow embankment. This guarantees the use of the maximum retention volume and thus the maximum safety level for the dam.

Embankment Design

The sealing element of the embankment is in most cases a natural silt and gravel core. Seepage losses can be tolerated as long as the safety of the dam is not affected. Therefore, the grading envelopes have to be determined and controlled, and adequate drainage and filter systems are essential. To keep under-seepage to a safe level, a key trench or a cutoff wall has to be provided. Alternative options that have been successfully used in Oman include artificial sealing elements like earth concrete cores (Wadi Hawasina Dam), asphaltic membranes (Wadi Ghul and Wadi Tanuf) or asphaltic core sealing (Muscat Flood Alleviation Dam).

Outlets

The number of outlets to be provided must be closely related to the number of active wadi channels to be served. For the typical Al Batina dam, five to eight controllable outlets are suitable, with a pipe diameter of 600 to 800 mm.

Recharge Process

Contrary to popular belief, recharge mainly occurs downstream of the dam, not in the reservoir itself. The reservoir bed becomes quickly sealed by silt, so the recharge through it becomes low. However, water released from the reservoir has little sediment, and easily infiltrates

Table 1. Recharge dams in the Sultanate of Oman.

	Recharge Scheme	Completion Year	Catchment area km^2	Reservoir Capacity 10^6 m^3	Mean Annual Rainfall mm	Crest Length m	Dam Height m	No of Outlets	Spillway Design Flood m^3/s	Spillway Length m	Dam Type	Spillway Type
1	Al Khoud (Seeb)	1985	1635	11.5	110	5100	1[illegible]	11	12500	3000	ZE	gabion
2	Hilti Salahi (Sohar)	1985	554	0.55	126	9063	4.5	31	2200	1120	ZE	gabion
3	Quriyat (Bahla)	1986	427	0.13	131	1620	5.3	2	1280	300	ZE	gabion weir
4	Al Jizzi (Sohar)	1989	812	5.4	128	1234	20.4	2	4700	184	ZE	main: concrete weir
									3100	278		aux: concrete weir
5	Tanuf (Nizwa)	1989	181	0.68	160	135	17	2	3000	118	GD	concrete lining
6	Ghul (Bahla)	1989	173	0.45	164	415	7.6	2	2800	330	GD	gabion cascade weir
7	Al Kabir (Ibri)	1991	757	0.5	148	2664	8.9	1	4200	600	ZE	rock armor
8	Ma'awil (Barka)	1991	555	10.0	120	7500	8.3	10	4000	4040	ZE	gabion
9	Fulayj (Sur)	1991	677	0.78	140	530	7.5	2	3300	430	ZE	concrete lining
10	Al Fara'i (Rustaq)	1992	201	0.6	140	638	12	2	3000	430	ZE	rock armor
11	Taww (Barka)	1992	259	5.1	120	4500	9.2	8	3100	2500	ZE	gabion
12	Fulayj (Halban)	1993	117	3.7	115	3000	7.7	2	2200	1150	ZE	gabion
13	Sahalnoot (Salala)	1993	258	6.4	150	3315	21.8	3	850	103	ZE	main: concrete weir
									2390	982		aux: concrete slabs
14	Ahin (Saham)	1994	829	6.8	120	5640	8.0	5	5600	4000	ZE	gabion
15	Hawasina (Kabura)	1994	869	3.7	120	5900	6.8	5	6000	1135	ECC	concrete weir

ZE: Zoned Embankment, GD: Gabion Dam, ECC: Earth Concrete Core

into gravel aquifers downstream of the dam. The dams are designed in such a way that the maximum discharge of outlets ensures that the total wetted contact area in the channel downstream is sufficient to infiltrate the total release volume. The rate of infiltration into the alluvium downstream, therefore, does not usually present a constraint in utilizing all the available water for recharge.

Usually there is more than sufficient storage capacity in the alluvium for a single recharge event. Preferred locations can accommodate several recharge events in a short period of time. For illustration, an aquifer with a porosity of 10% measuring ten by ten kilometers could store ten million cubic meters of water with a one-meter rise in the water table. Due to slow groundwater movement, most of this water may be in time withdrawn for use.

Reservoir

Due to topography, recharge dams have relatively small-capacity reservoirs. The current tendency is to design storages that can contain a flood that may occur, on average, once every five to ten years, and in certain conditions, twenty years.

Flood Protection

Each dam that has an ungated spillway reduces the incoming flood peak. How much the peak is reduced depends on the size of the reservoir, the magnitude of the flood and the geometry of the spillway. Typically, recharge dams in Oman are designed to fully contain a flood that is likely to occur every five to ten years. Larger dams, some of them built specifically for flood protection, are effective in reducing floods that statistically occur once in a century, and to a lesser extent reduce more rare floods.

Monitoring Network

In all studies carried out in past years, most of the design criteria have been subject to assumptions because of lack of long-term accumulated hydrological and hydrogeological knowledge, as well as the appropriate measurement network. Therefore, the Ministry of Water Resources has designed and implemented a very dense meteorological and hydrological network in the catchments where dams are constructed to enable an evaluation of the performance of the dams as well as the functioning of the recharge process downstream of the dams.

EFFECTIVENESS OF DAMS

Volumes of Water Stored

The total volume of water stored temporarily in a recharge dam over a period of time provides a measure of its effectiveness, since most of this water infiltrates the alluvium downstream of the dam. Just how much water additionally infiltrates the aquifer in various synoptic and hydrogeological conditions is a matter of considerable interest.

Several data collection and modeling studies are under way to determine the volumetric effectiveness of recharge. For the purpose of this evaluation it is assumed that the actual recharge is 75% of water stored in the dam, which is an order of magnitude suggested by preliminary evaluations.

A three-year study in Saudi Arabia by Al Muttair et al. (1989) showed that relatively high efficiencies can be obtained when dams are used for recharging groundwater. Al Muttair and Sendil (1994) gave a range of recharge efficiencies from 68% for small dams (reservoir of 250,000 cubic meters) to 92% for larger dams (reservoir of 1 million cubic meters). The efficiency of recharge in the reservoir was found to be 76% and the efficiency of recharge in the downstream channel 96%.

Cost Effectiveness

The parameter that is perhaps the most difficult to quantify is the percentage recoverability of the increased recharge that is induced by the dam structure. Most of the costs associated with operation of the recharge scheme is the capital expenditure for construction of the recharge scheme. Experience shows that maintenance costs on average amount to some 3% of annualized capital costs. Table 2 presents the economical effectiveness of recharge activities. The overall cost of 1 m^3 of water captured by recharge dams is U.S. $0.15. This compares favorably with the marginal unit cost of desalinated water, which is the only feasible alternative source of additional water in the Sultanate. These costs are U.S. $2.00 per cubic meter at a discount rate of 7%, with U.S. $1.20 due to direct operational costs and U.S. $0.80 due to capital costs. For comparison, the cost of treated water in the Sultanate is on average U.S. $0.23 per cubic meter.

Table 2. Effectiveness of recharge.

	Recharge Scheme	Completion Year	Cost Million U.S.$	Reservoir Capacity 10^6 m^3	Unit Storage Cost U.S.$/$m^3$	Design Annual Recharge1 10^6 m^3	Actual Recharge1 10^6 m^3 Per Year	Annual Costs Million U.S.$	Cost of One m^3 of Water U.S.$
1	Al Khoud (Seeb)	1985	15.600	11.5	$1.36	3.0	1.64	0.97	$0.59
2	Hilti Salahi (Sohar)	1985	1.248	0.55	$2.27	0.6	0.73	0.18	$0.25
3	Quriyat (Bahla)	1986	1.260	0.13	$9.69	0.3	0.26	0.10	$0.38
4	Al Jizzi (Sohar)	1989	7.280	5.4	$1.35	3.5	8.13	0.51	$0.06
5	Tanuf (Nizwa)	1989	3.380	0.68	$4.97	1.5	2.33	0.24	$0.10
6	Ghul (Bahla)	1989	3.200	0.45	$7.11	1.2	1.62	0.23	$0.14
7	Al Kabir (Ibri)	1991	3.250	0.5	$6.50	0.5	0.64	0.19	$0.30
8	Ma'awil (Barka)	1991	7.296	10.0	$0.73	2.2	3.75	0.42	$0.11
9	Fulayj (Sur)	1991	3.333	0.78	$4.27	2.5	1.92	0.23	$0.12
10	Al Fara'i (Rustaq)	1992	2.961	0.6	$4.94	0.9	1.09	0.19	$0.17
11	Taww (Barka)	1992	4.680	5.1	$0.92	1.4	1.42	0.34	$0.24
12	Fulayj (Halban)	1993	3.640	3.7	$0.98	0.5	1.16	0.23	$0.20
13	Sahalnoot (Salala)	1993	13.130	6.4	$2.05	0.6	0.46	0.96	$2.09
14	Ahin (Saham)	1994	8.320	6.8	$1.22	5.0	8.2	0.56	$0.07
15	Hawasina (Kabura)	1994	7.670	3.7	$2.07	3.0	5.8	0.46	$0.08

The capital cost is annualized over a 30-year period at a 7% discount rate. The annual costs include maintenance, which is on average 3% of annualized costs.

Volumes Lost to the Sea

It would be impractical, or even difficult, to construct recharge dams that could intercept every flood that may occur. For example, it is estimated, that in March 1997 some 120 million m^3 spilled over Wadi Ahin Dam and 80 million m^3 over Wadi Hawasina Dam. This water was lost to the sea. Existing topography of both catchments precludes the construction of reservoirs that could match such a flood volume.

During the period investigated, some 500 million m^3 was lost to the sea, compared to 370 million m^3 retained in recharge dams. Methods that could reduce loss to the sea, such as water spreading are under consideration.

CONCLUSIONS

Recharge dams play an important role in augmentation of water resources in arid countries. While desalination provides an unlimited water supply at a cost, recharge dams provide a limited water supply relatively cheaply. Both methods should therefore be optimized, desalination for lowering unit costs and recharge for increasing volumetric capacity.

Additional benefits of recharge dams such as flood protection and arresting saline intrusion in coastal areas make this technology even more attractive. In Oman, a very important benefit of storing floodwaters underground is the reduction of seawater intrusion in coastal areas. The latter has become a serious problem in many parts of the country, especially in the Al Batina Plain.

In an arid country, the amount of additional water available through the recharge process is limited by topography, geology, weather, and human activities (scale of recharge schemes, wastewater recharge, underground dams, to name but a few). The cost of recharge water is very low compared to other methods of increasing water supplies. Therefore, consideration should be given to making full use of this option if conditions are suitable.

REFERENCES

Abdulrazzak, M. J. 1997. Water supply augmentation through artificial groundwater recharge techniques. *Proceedings of the Third Gulf Water Conference, Muscat,* Vol. 1, pp. 241–281.

Al Muttair, F. F., S. S. Al Turbak, U. Sendil. 1989. Management of water stored behind recharge dams in central Saudi Arabia. Final report AR-6–18 prepared for King Abdulaziz City for Science and Technology (KACST), Riyad, Saudi Arabia.

Al Muttair, F. F. and U. Sendil. 1994. Management of recharge dams in Saudi Arabia. *J. of Water Resources Planning and Management* 120(6):749–763.

Todd, D. K. 1980. *Groundwater Hydrology,* Second Edition. John Wiley & Sons, New York.

12

Irrigation and Wastewater Reuse: A Case Study

ALAA ELDIN ZL-ZAWAHRY AND AMER A. AL-RAWAS
Department of Civil Engineering
Sultan Qaboos University
P.O. Box 33, Al-Khod 123
Sultanate of Oman

INTRODUCTION

WATER IS BECOMING an increasingly scarce resource in most arid and semi-arid areas. Any source of that which might be used effectively and economically to promote further development must be considered. At the same time, with high rates of population expansion, the need for increased food production is vital. The potential for irrigation to raise agricultural productivity has long been recognized. Wherever good quality water is scarce, water of less (marginal) quality has to be considered for use in agriculture. From the viewpoint of irrigation, use of marginal quality water requires more complex and more stringent monitoring than when good quality water is used. This study addresses the agricultural use of domestic sewage with or without a proportion of industrial effluents that are discharged to public sewers in arid areas with emphasis on the sultanate of Oman. The quantity of wastewater available in most areas accounts for only a small fraction of the required total irrigation water. The use of municipal wastewater in irrigation not only conserves valuable water resources but also takes advantage of the nutrients contained in sewage to grow crops. Moreover, the nitrogen and phosphorous content of sewage may reduce or eliminate the requirements for commercial fertilizers. However, sewage techniques for effluent discharge to surface waters are not always appropriate for agricultural use.

Some countries (for example Jordan and Saudi Arabia) have a national policy to reuse all treated wastewater effluents and have already made considerable progress. The sultanate of Oman is known to be one of those arid countries that have been environmentally conscious and given conservation and prevention their utmost attention. Most of the wastewater treated in Oman is used for irrigation purposes especially in Muscat, the capital area.

WASTEWATER CHARACTERISTICS

Domestic wastewater is comprised of water (about 99.9%) with relatively small amounts of suspended and dissolved solids (organic and inorganic). The levels of the major constituents for strong, medium, and weak domestic wastewaters are summarized in Table 1 (WHO, 1989).

In arid and semi-arid areas, water consumption is often fairly low, in the order of 100 L/capita/day, and sewage tends to be strong or even very strong. From the point of view of health, the presence of pathogenic micro- and macroorganisms is the key factor that may limit the use of wastewater in agriculture. Figure 1 shows an example of strong wastewater for Amman, Jordan, where the water consumption is low, 90 L/capita/day (Al-Salem, 1987). The raw municipal water may include pathogenic viruses, bacteria, protozoa, and helminths. Table 2 summarizes the possible levels of those pathogens in wastewater. The pathogens may survive in the environment for a long time as indicated in Figure 2. For example, the survival time of *Ascaris Lunbricoides* eggs

Table 1. Major constituents of typical domestic wastewater.

Constituent	Concentration mg/L		
	Strong	Medium	Weak
Total solids	1200	700	350
Dissolved solids (TDS)*	850	500	250
Suspended solids	350	200	100
Nitrogen (as N)	85	40	20
Phosphor (as P)	20	10	6
Chloride*	100	50	30
Alkalinity (as $CaCo_3$)	200	100	50
Grease	150	100	50
BOD_5**	300	200	100

*The amounts of TDS and chloride should be increased by the concentrations of these constituents in the carriage water.

**BOD_5 is the biochemical oxygen demand at 20°C over 5 days and is a measure of the biodegradable organic matter in the wastewater.

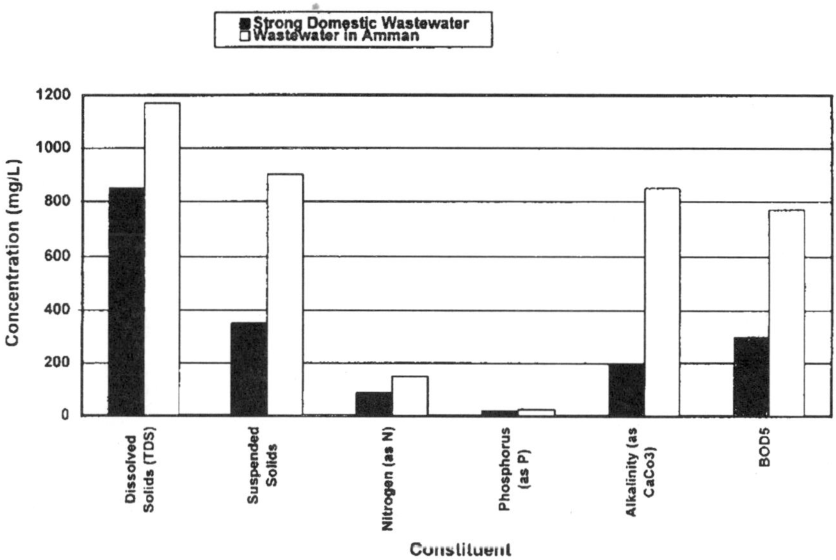

Figure 1. Strong domestic wastewater for Amman, Jordan.

is estimated by many months, for nightsoil, sewage, and soil (Feachem et al., 1983).

WASTEWATER QUALITY PARAMETERS IN AGRICULTURAL USE

The quality parameters of importance in agricultural use of wastewater may be classified into two main groups, which are based on their significance to health and agriculture. The principal health hazards associated with organic chemical constituents of wastewater arise from the contamination of crops or groundwater. There is particular concern attached to the cumulative poisons, heavy metals, and carcinogens, mainly of organic nature (Hillman, 1988). The guidelines for organic and toxic substance (WHO, 1989) may be adopted directly for groundwater. The possible accumulation of certain toxic elements (for example, cadmium and selenium) in crops irrigated with contaminated wastewater must be carefully assessed.

The greatest health concern in agricultural use of wastewater is related to the pathogenic organisms. In areas where raw untreated

Table 2. Possible levels of pathogens in wastewater (Feachem et al., 1983).

Type of Pathogen		Possible Concentration per liter in Municipal Wastewater[1]
Viruses	*Enteroviruses*[2]	5000
Bacteria	Pathogenic *E. coli*[3]	uncertain
	Salmonella spp.	7000
	Shigella spp.	7000
	Vibrio cholerae	1000
Protozoa	*Entamoeba histoytica*	4500
Helminths	*Ascaris Lunbricoides*	600
	Hookworms[4]	32
	Schistosoma mansoni	1
	Taenia saginata	10
	Trichuris trichiura	120

[1]Based on 100 L/capita/day of municipal sewage and 90% inactivation of excreted pathogens
[2]Includes polio, echo, and coxsackieviruses.
[3]Includes enteroxigenic, enteroinvasive, and enteropathogenic *E.coli.*
[4]Anglostoma duedenale and necator americanus.

sewage is used to irrigate salad crops and/or vegetables eaten uncooked, transmission of helminthic diseases is likely to occur through the consumption of such crops. The negative health effects have only been detected in association with the use of raw or poorly settled wastewater. However, inclusive evidence suggests that appropriate wastewater

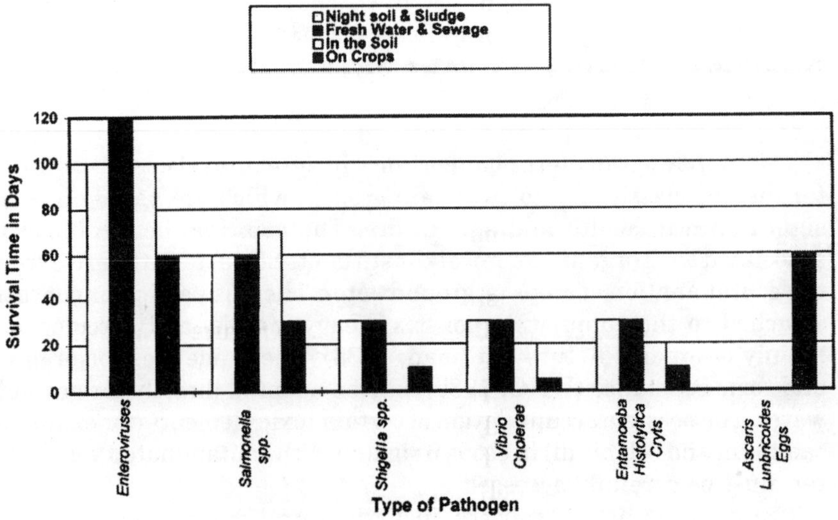

Figure 2. Pathogen survival time.

treatment may provide a high level of health protection (Shuval et al., 1986; Shuval et al., 1985). Important agricultural water quality parameters include several physical and chemical characteristics that are briefly discussed in the next section.

WASTEWATER QUALITY GUIDELINE FOR AGRICULTURAL USE

Wastewater treatment, crop restriction, control of wastewater application, and control of human exposure are the health protection measures that may be applied in agricultural use of wastewater. Microbiological quality guidelines for wastewater use in agriculture are shown in Table 3.

The adequacy of water for irrigation relies on many parameters such as climatic conditions; soil, physical and chemical properties; and grown crop salt tolerance. The irrigation water is classified into three groups based on salinity, sodicity, toxicity, and miscellaneous hazards as shown in Table 4 (Ayres and Westcot, 1976).

In arid and semi-arid areas, a successful irrigation system based on the use of treated wastewater requires basic conditions to be met. The required amount of water should be applied, the water must have acceptable quality, proper application schedule and irrigation method should be adopted, salt accumulation must be prevented by leaching, groundwater table must be controlled by drainage, and plant nutrients should be managed.

WASTEWATER REUSE IN OMAN

Irrigation Water Requirements

The sultanate of Oman, situated between latitudes 16°40′N and 26°20′N and longitudes 51°20′E and 59°40′E (Figure 3), has an arid climate with evaporation rates higher than the annual rainfall. Consequently, it has limited water resources represented mainly by rainwater. Therefore any additional source of water that might be utilized economically and effectively for further development must be considered. Water of low (marginal) quality has to be considered for use in agriculture. From the viewpoint of irrigation, use of low quality water requires more strict and complex standards and monitoring. Hence,

Table 3. Recommended microbiological quality guidelines for wastewater use in agriculture (FAO, 1985).[a]

Category	Reuse Condition	Exposed Group	Intestinal Nematodes[b] (arithmetic mean no. of eggs per liter[c])	Fecal Coliforms (geometric mean no. per 100 mL)	Wastewater Treatment Expected to Achieve the Required Microbiological Quality
A	Irrigation of crops likely to be eaten uncooked, sports fields, and public parks[d]	Workers, consumers, public	≤ 1	≤ 1000[d]	A series of stabilization ponds designed to achieve the microbiological quality indicated, or equivalent treatment
B	Irrigation of cereal crops, industrial crops, fodder crops, pasture, and trees[e]	Workers	≤ 1	No standard recommended	Retention in stabilization ponds for 8–10 days or equivalent helminth and fecal coliform removal
C	Localized irrigation of crops in category B if exposure of workers and the public does not occur	None	Not available	Not available	Pretreatment as required by the irrigation technology, but not less than primary sedimentation

[a]In specific cases, local epidemiological, sociocultural, and environmental factors should be taken into account, and the guidelines modified accordingly.
[b]*Ascaris* and *Trichuris* species and hookworms.
[c]During the irrigation period.
[d]A more stringent guideline ($\leq$ 200 fecal coliforms per 100 mL) is appropriate for public lawns, such as hotel lawns, with which the public may come into direct contact.
[e]In the case of fruit trees, irrigation should cease 2 weeks before fruit is picked, and no fruit should be picked off the ground. Sprinkler irrigation should not be used.

Table 4. Guidelines for interpretation of water quality for irrigation.

Potential Irrigation Problem	Units	Degree of Restriction on Use: None	Slight to moderate	Sever
Salinity				
EC_w[a]	ds/m	< 0.7	0.7–3.0	> 3.0
TDS	mg/L	< 450	450–2000	> 2000
Infiltration				
SAR[b] = 0–3		And EC_w => 0.7	0.2–0.7	< 0.2
3–6		=> 1.2	0.3–1.2	< 0.3
6–12		=> 1.9	0.5–1.9	< 0.5
12–20		=> 2.9	1.3–2.9	< 1.3
20–40		=> 5	2.9–5	< 2.9
Specific Ion Toxicity				
-Sodium (Na)				
Surface irrigation	me/L	< 3	3–9	> 9
Sprinkler irrigation	me/L	< 3	≥ 3	
-Chloride (Cl)				
Surface irrigation	me/L	< 4	4–10	> 10
Sprinkler irrigation	me/L	< 3	≥ 3	
-Boron (b)	mg/L	< 0.7	0.7–3.0	> 3.0
Miscellaneous Effects				
Nitrogen (NO_3-N)[c]	mg/L	< 5	5–30	> 30
Biocarbonate (HCO_3)	mg/L	< 1.5	1.5–8.5	> 8.5
pH		6.5–8.4	6.5–8.4	6.5–8.4

[a]EC_w: electrical conductivity in deciSiemens per meter at 25°C.
[b]SAR: sodium adsorption ratio.
[c]NO_3-N: nitrate nitrogen reported in terms of elemental nitrogen.

some strategies are adopted to supplement the water resources such as new groundwater exploration programs, use of desalinated seawater, and reuse of wastewater.

The total Omani area under production is given in Table 5 with a breakdown of crops and annual yield. Out of the 54,900 hectares (only about 6.9 % of the suitable agricultural area) irrigated in the sultanate, 5370 ha are uncultivated, 9750 ha under field crops, and 33,100 ha under fruit trees. Tomatoes and watermelon account for about 46% of the area under vegetable production, whereas 91% of the area under field crops produce alfalfa. Date palms account for about 75% of the area under fruits, with a percentage of 45.5% of the total area (MWR,

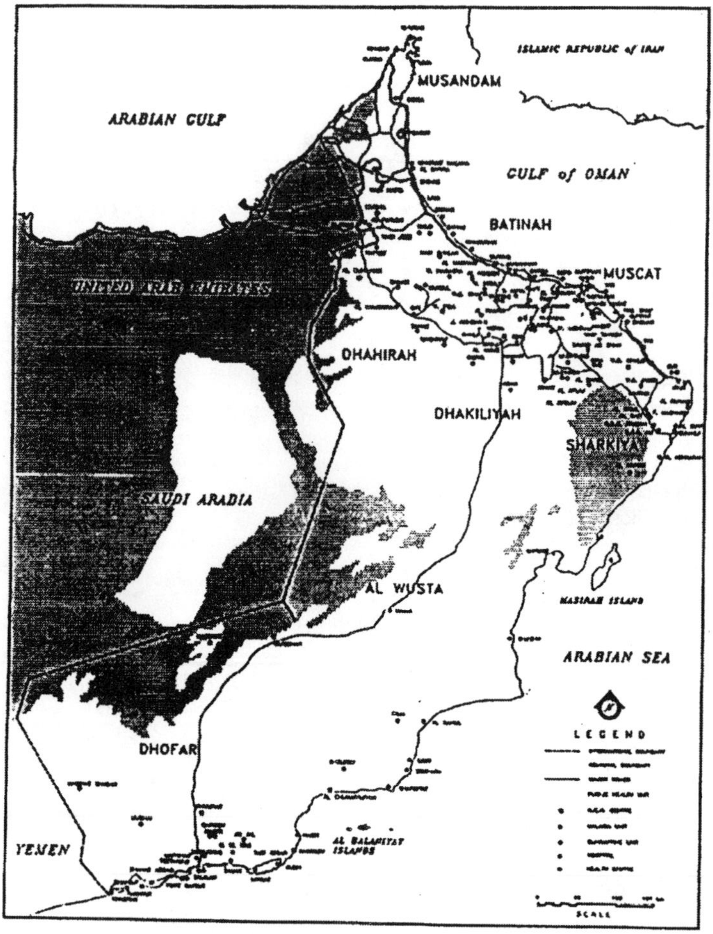

Figure 3. Map of the Sultanate of Oman.

1990). In any water conservation plan it is important to know the varieties and areas of crops grown.

Figure 4 gives the 1990 estimates for the crop water requirements in the different regions of Oman. The Batinah region required 531 million m^3/year while the Salalah Plain required 33 million m^3/year. The crop water requirements of the rest of the country were 561 million m^3/year. With these values, groundwater reservoirs will be depleted at rates higher than estimated (Abdel-Rahman and Abdel-Magid, 1993).

Table 5. Crop production by areas and yields for 1988 (MWR, 1990).

Crop	Area (ha)	Yield (Tons/ha)
Vegetables		
Tomato	1212	22.2
Pepper	610	9.0
Onion	560	13.8
Garlic	150	8.0
Okra	53	13.2
Water melon	1250	19.0
Musk melon	625	13.1
Cabbage	770	23.3
Potato	140	25.0
Subtotal	5370	
Field Crops		
Alfalfa	8870	38.0
Wheat	468	1.5
Tobacco	409	4.9
Subtotal	9747	
Fruits		
Dates	25,000	4.0
Lime	2400	10.8
Mango	3780	2.0
Banana	1625	13.6
Coconut	328	16.8
Subtotal	33,133	
Others	6651	
TOTAL	54,901	

Wastewater Reuse Regulations

In the case of wastewater and sludge treatment and disposal, the Omani regulations mainly have been divided into three broad groups. The first group indicates parameters to be considered for wastewater reuse, discharge, and sludge disposal (MRME, 1993). Table 6 outlines wastewater maximum quality limits set for concentrations of elements in wastewater and irrigation water. This table shows that more emphasis was placed on chemical substances. Radioactive materials are covered in a separate article in previous regulations MEWR (1986), which do not permit reuse of wastewater or sludge containing radioactive material. The discharge of wastewater or sludge containing radioactive material is also subject to regulations laid down by the International Atomic Energy Agency. Biological and bacteriological characteristics as

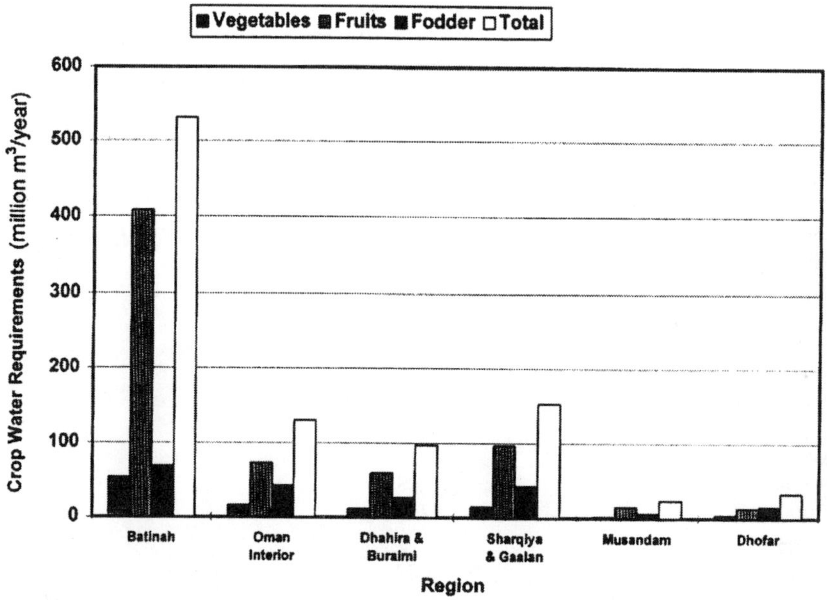

Figure 4. Crop water requirements for different regions of Oman (MWR, 1990).

laid down by the Omani wastewater standards (MEWR, 1986) stressed fecal coliform bacteria and viable nematodes.

In the case of sludge reuse in agriculture, conditions for application to land have been divided according to three parameters that relate to the maximum concentration based on dry solids, maximum annual application rate based on land area to be irrigated, and maximum permitted concentration in soil based on dry solids value. Table 7 gives the concentration of metals for each of the aforementioned categories.

After the spreading of sludge, there must be a minimum period of three weeks before grazing or harvesting of forage crops. Sludge use is prohibited on soils while fruit or vegetable crops, other than fruit trees, are growing or being harvested; for six months preceding the harvesting of fruits or vegetables that grow in contact with the soil and that are normally eaten raw; and on soils with a pH < 7.0.

Reuse and discharge regulations further classify wastewater use for crops and vegetables, grass and ornamental areas, aquifer recharge, and method of irrigation. Table 8 illustrates these areas of application of wastewater. Any other reuse application is subject to the approval of the Ministry of Regional Municipalities and Environment.

Table 6. Concentration of elements in wastewater and irrigation water (all units are in mg/L except if otherwise stated).

	Wastewater (MRME, 1993)		Irrigation Water (Ayres and Westcot, 1976)	
Parameter	Standard A[a]	Standard B[a]	Max-CT[b]	Max-20[c]
Physical				
Total dissolved solids	1500	2000	—	—
Suspended solids	15	30	—	—
Electrical conductivity (μs/cm)	2000	2700	—	—
Chemical				
Ag	0.01	0.01	—	—
Al	5	5	5	20
As	0.1	0.1	0.1	2
B	0.5	1	10.75	2
Ba	1	2	—	—
Be	0.1	0.3	0.1	0.5
BOD_5	15	10	—	—
Cd	0.01	0.01	0.01	0.05
Cl	650	650	—	—
COD	150	200	—	—
Co	0.05	0.05	0.05	5
Cn	0.05	0.1	0.2	5
Cr	0.05	0.05	0.1	1
Cu	0.5	1	—	—
F	1	2	1	15
Fe	1	5	5	20
Hg	0.001	0.001	—	—
Li	0.07	0.07	2.5	2.5
Mg	150	150	—	—
Mn	0.1	0.5	0.2	10
Mo	0.01	0.05	0.01	0.05
N Ammonia	5	10	—	—
N Kjeldahl	5	5	—	—
N as No_3	50	50	—	—
Na	200	300	—	—
Ni	0.1	0.1	0.2	2
Oil and grease	0.5	0.5	—	—
Pb	0.1	0.2	5	10
pH	6–9	6–9	6–8.5	6–8.5
Phenol	0.001	0.002	—	—
S-	0.1	0.1	—	—
Sodium Absorption Ratio (SAR)	10	10	—	—
Se	0.02	0.02	0.02	0.02
SO_4	400	400	—	—

continued

Table 6 (continued).

Parameter	Wastewater (MRME, 1993)		Irrigation Water (Ayres and Westcot, 1976)	
	Standard A[a]	Standard B[a]	Max-CT[b]	Max-20[c]
V	0.1	0.1	—	—
Zn	5	5	2	10
Bacteriological				
Fecal coliform bacteria (/100 mL)	200	1000	—	—
Viable nematode ova (per liter)	<1	<1	—	—

[a]Refer to Table 8.
[b]Waters used continuously on all soils.
[c]Waters used up to 20 years on fine textured soils of pH 6–8.5.

In general, standards, regulations, or guidelines for wastewater and sludge disposal are usually classified into categories that embody relevant areas. Suggested categories may include pretreatment waste disposal in natural bodies of water, industrial waste disposal in sewers, water reuse for agriculture, water use for swimming and recreation, and water use for specific intentions.

Table 7. Reuse of sludge in agriculture and conditions for application to land (MRME, 1993).

Metal	Maximum Concentration (mg/kg of dry soil)	Maximum Application Rate (kg/ha/yr)*	Maximum Permitted Concentration in Soil (mg/kg dry soil)
Cadmium	20	0.15	3
Chromium	1000	10.0	400
Copper	1000	10.0	150
Lead	1000	15.0	300
Mercury	10	0.1	1
Molybdenum	20	0.1	3
Nickel	300	3.0	75
Selenium	50	0.15	5
Zinc	3000	15.0	300

*Based on a 10-year average and a soil pH > 7.0.

Table 8. Areas of application of Standards A and B for wastewater reuse (MRME, 1993).

ITEM	Standard A	Standard B
Crops	Vegetable likely to be eaten raw. Fruit likely to be eaten raw within 2 weeks of any irrigation	Vegetables to be cooked or processed. Fruit if no irrigation within 2 weeks of cropping. Fodder, cereal and seed crops
Grass and ornamental areas	Public parks, hotel lawns, recreational areas. Areas with public access. Lakes with public contact (except places that may be used for hand washing before praying)	Pastures Areas with no public access
Aquifer recharge	All controlled aquifer recharge	
Method of irrigation	Spray or any other method of aerial irrigation not permitted in areas with public access unless with timing control	
Any other reuse applications	Subject to the approval of the Ministry	

Wastewater Treatment

SEWAGE TREATMENT PLANTS

The 1993 estimates of the inhabitants of Oman reached 2,071,591, of which 1,480,531 are Omanis and 537,060 are expatriates (Al-Sabahi, 1997). The highest population densities occur along the north coast and the capital, Muscat. The main contributors to the wastewater used in irrigation are the domestic and industrial sources. Figure 5 shows the present and future estimated domestic and industrial water demands in Oman for several regions (Al-Sabahi, 1997; McDonalds and Hawksley, 1993). The data are extrapolated up to year 2025 to give a total demand of 2,260,000 m^3/d.

The approximate total number of sewage treatment plants (STP) in Oman is about 262. The capacity of the STPs ranges from a private STP for a building to an STP with a capacity of 11,000 m^3/d in Darsait. Figure 6 shows the distribution of STPs by number across different regions of Oman (Al-Sabahi, 1997). It is worth noting that about 50% of the STPs are located in Muscat, while 20% are equally distributed between Dhofar and Al-Batinah.

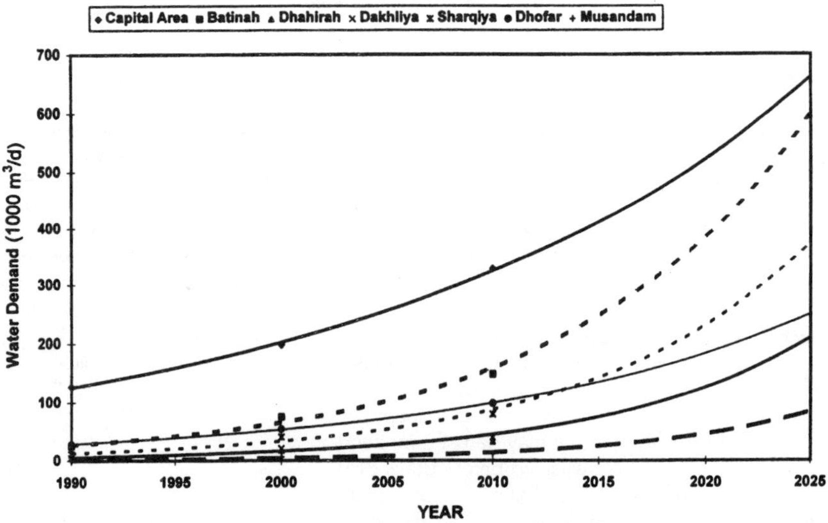

Figure 5. Domestic and industrial water demand for different regions of Oman.

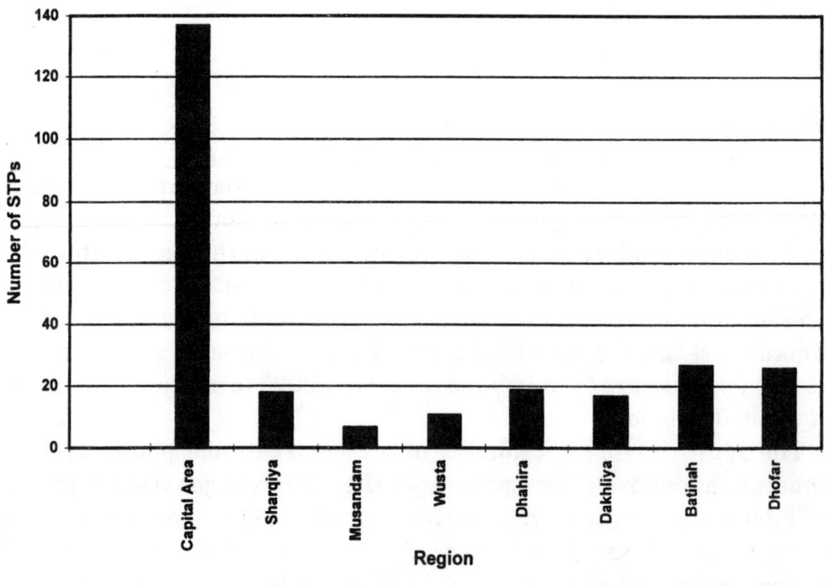

Figure 6. Number of sewage treatment plants for different regions of Oman.

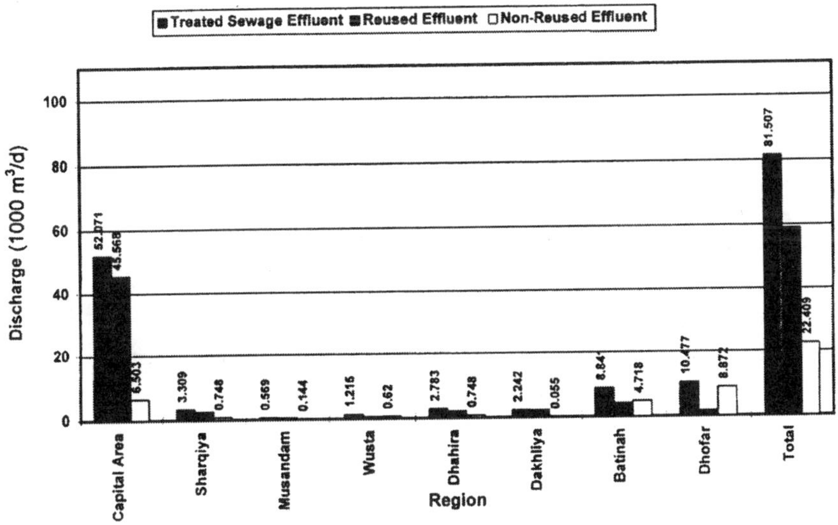

Figure 7. Treated, reused, and non-reused flows in Oman.

Figure 7 shows the distribution of the capacity of the STPs over different regions of Oman, as well as the effluent reuse (Al-Sabahi, 1997). The total treated sewage effluent all over Oman is 81,507 m³/d with a maximum treated sewage of 52,071 m³/d in the capital area. The percentages of effluent reuse, nonreuse, and the treated sewage (regional to total) are given in Figure 8. The capital area shows the maximum percentages of reuse and treated sewage amounting to 87.5% and 63.8%, respectively. The lowest percentage of treated sewage is in Musandam (0.7%), while the highest nonreuse percentage is in Dhofar (84.7%).

Based on the curves plotted in Figure 5, the domestic and industrial water demand in the year 1997 was about 352,500 m³/d. Assuming that 60% is available for treatment, the water available for treatment may reach 211,500 m³/d. The actual treated water as given in Figure 6 was 81,500 m³/day, amounting to 38% of the water available for treatment. Moreover, the total reused effluent was 59,100 m³/d, or 27% of the available water, indicating a high percentage of nonreused water (73%).

At the present time, a number of sewage treatment plants are in operation in the capital area. One in Darsait produces about 11,500 m³/day with input from the Greater Muttrah area and another in Al-

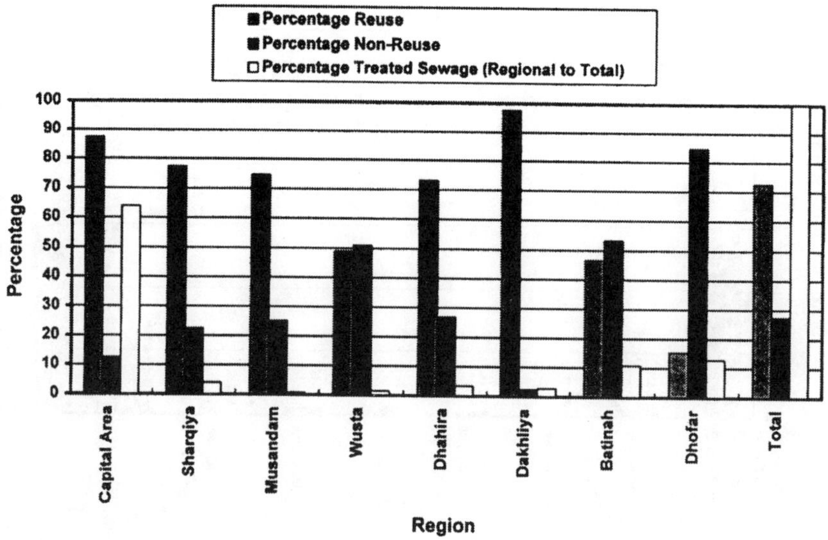

Figure 8. Percentage of treated, reused, and non-reused effluents (regional to total) in Oman.

Ansab is producing 5400 m³/day to be expanded in the future. The effluent of these plants is used for irrigating tree plantations in the capital area from Darsait to Assahwah Tower. A third treatment plant is at Sultan Qaboos University producing around 3000 m³/day and is used to irrigate the university tree plantation. Five other small treatment works exist in Maabaila, Al Khod, Murtafaa, Amrat, and Diwan in Al-Seeb. Figure 9 gives an account of the existing sewage treatment works, their location, design capacity, flow rate of the effluent, and the amount of wastewater utilized for irrigation purposes. Some of the averaged wastewater quality parameters of the plants are summarized with the standards in Figure 10 (Abdel-Magid and El-Zawahry, 1992).

SEPTIC TANKS

The widely used wastewater treatment system in Oman is a septic tank connected to either a soakaway or a holding tank. The partially treated water from the septic tank is allowed to seep into the ground through the soakaway or is stored in a holding tank to be collected by tankers for ultimate disposal. In the Muscat area, about 16,000 dwellings are connected to the sewerage system. The rest are connected to septic tanks. Other connections are directly discharging to a crude

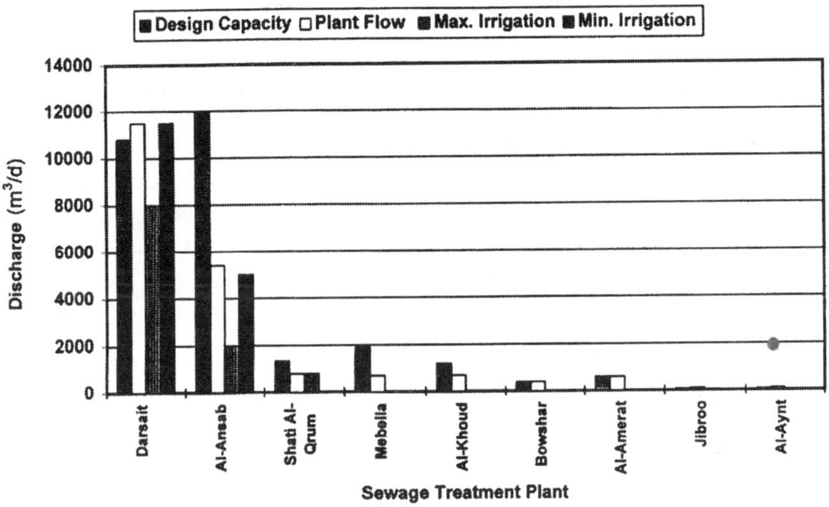

Figure 9. Sewage treatment plants' capacity and irrigation reuse in Muscat, Oman.

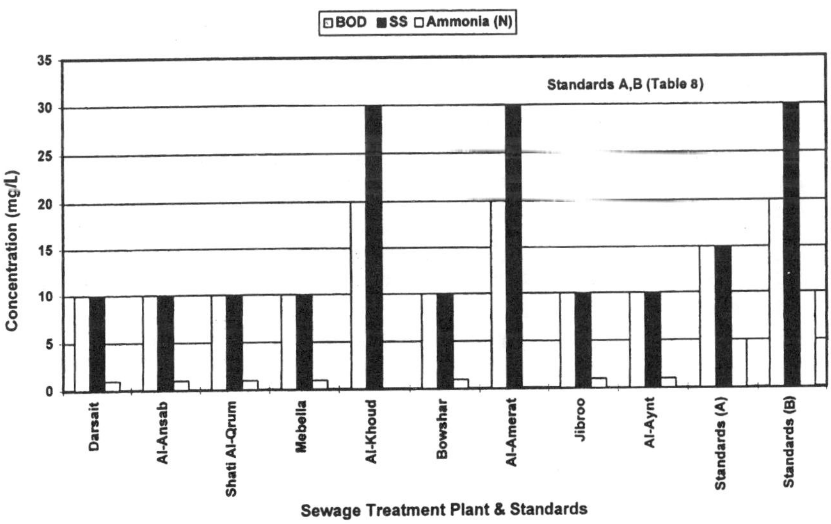

Figure 10. Sewage treatment plants' quality parameters in Muscat, Oman.

soakaway without any treatment. The percentages of sewered sewage, tankered septage, and sewage to soakaway are 24%, 37%, and 39%, respectively (Al-Sabahi, 1997). Muscat is the only city with significant sewerage while other towns are using septic tanks and/or holding tanks. The total infiltration to groundwater through soakaways, cess pools, and wastewater disposal is approximately 70,000 m^3/d (25 million $m^3/year$) (Al-Sabahi, 1997; McDonalds and Hawksley, 1993). The quality of the septic tank effluent is far above the maximum allowed for irrigation (Al-Sabahi, 1997).

Irrigation Systems

The reuse of treated wastewater is currently restricted to irrigation of ornamental plants in the sultanate. Table 9 gives the size of the green areas and beautification in the capital area. About 80% of the water requirements is covered by reused treated wastewater, and the remaining (20%) is obtained from artesian wells and potable water network (Al-Qasmi, 1997).

The construction of the existing irrigation network in the capital area started in 1985 for the purpose of using treated wastewater to irrigate trees and green areas. The existing network consists of three networks. The main network starts from Darsait STP (the biggest in the country) and ends at Al-Sahaw roundabout with a length of 50 km and a diameter of 300 mm. The second network goes from Al-Sahwa roundabout to the Al-Khod roundabout with a length of 9 km. The source of water is the Ministry of Defense STPs in al-Murtafaa Camp. The third network starts from Al-Seeb roundabout to Bait Al-Barka roundabout with a length of 15 km, with Al-Mebella and Al-Khod STPs being the source of the water. The three networks consist of pipe lines, lift stations, automatic control valves, and storage tanks. A sprinkler irrigation system is used to irrigate the green areas, while a trickle system is used to irrigate the trees. To guarantee a good quality of

Table 9. Green areas in the capital area.

Item	Total	Item	Total
No. of big trees	100,000	Palm trees	15,000
No. of small trees	300,000	Flower boxes	800
Length of planted fences (m)	100,000	Grassed areas (m^2)	50,000
Green areas (m^2)	1,539,940		

water, the STPs peak effluent is mixed with the rest of the effluent. The water is stored to improve the water quality parameters such as nitrogen and suspended solids.

Case Study Sewage Treatment Plant (SQU)

A study case from Oman was selected to demonstrate the importance and the value of wastewater reuse in irrigation. Sultan Qaboos University (SQU) irrigation system is an ideal case to be studied since the entire campus is sewered and the treated effluent is fully utilized for irrigation. The sewage treatment plant (STP) is located in the southeast part of SQU. The untreated wastewater coming from the residential areas and buildings flows through the sewage pipe lines and is propelled around the network systems by sewage lifting stations, reaching the sewage treatment plant where the treatment processes take place. The plant is designed to accommodate a population of 10,000 and has a potential capacity of 2,200 m^3/day. It is operated by a specialist operation and maintenance contractor and is closely supervised by qualified engineers from the Department of Projects and Maintenance at SQU to ensure that the standard control parameters required by the Ministry of Regional Municipalities and Environment are strictly followed. The SQU Radiation Protection Officer of the Health and Safety Committee is involved in ensuring the safe maintenance level of radiation in the STP. The entire area of the university is covered by trees, grass and beautification irrigated by the treated wastewater in addition to potable water from the university water supply. The irrigation system combines both sprinkler and trickle irrigation systems.

The main components of the STP of SQU include a sewage inlet pumping station, flow measurement, biological aeration and settlement, tertiary treatment, chlorination, and sludge drying beds. The two biological treatment sections operate on an extended aeration-activated sludge principle with a long period of retention of the mixed liquor of sewage and activated sludge. The sludge pumping station delivers sludge back to the inlet works where it is returned to the distribution chamber for recirculation. A valved connection in the pump chamber enables the flow to be diverted and surplus sludge within the system is removed by this means to the thickening tanks. After thickening, pumps adjacent to the tanks transfer the sludge to beds for drying and subsequent removal.

Untreated wastewater carries contaminated substances that can create serious environmental problems. Hence, recycling or treatment of

Table 10. Typical analysis of wastewater (Bahwan Engineering Company, 1996).

Parameter	Untreated Wastewater Concentration (ppm)	Treated Wastewater Concentration (ppm)	Control Limits
pH	6.5	7.61	6.5–8.5
Biochemical oxygen demand (BOD_5)	300	3.4	10 ppm
Suspended solids	218	12.3	10 ppm
Ammoniacal nitrogen	31	1.0	1 ppm (max. 5)
Chemical oxygen Demand (COD)	428	< 10	100 ppm
Residual chlorine	—		> 0.5 ppm

wastewater is of paramount importance for its reuse for irrigation purposes. Typical analyses of untreated and treated wastewater of the SQU STP are given in Table 10.

As shown in the results above, all samples were within the control limits with the exception of the suspended solids concentration, which was slightly higher by 2.26 ppm. However, this value is acceptable according to the quality standards. The other parameters that were investigated include: total water consumption, STP-produced water, irrigation demand, and surplus/shortage at SQU. Figure 11 gives the water consumption for the last five years. It reveals an annual increase from 1991 to 1994, while the consumption in 1995 was relatively the same as in 1994. The water consumption has risen from 851,900 m^3 in 1991 to 1,387,200 m^3 in 1994. During May to October (summer months) the consumption is higher than the rest of the year.

A study of the actual recorded total monthly irrigation demand and treated water during 1995 as well as the percentages of treated water and irrigation water compared to the consumption, as outlined in Figures 12 and 13 reveal the following. First, there appears to be a general increase in the irrigation demand during the period and the demand is higher in the summer months (i.e., June to September). Second, the minimum STP production is during the month of July when the students and staff are on holidays. Finally, the percentage of treated water compared to consumption varies between a minimum of 32% and a maximum of 72%. The percentage of irrigation water compared to consumption is higher most of the time, with a minimum of 45% and a maximum of 68%. The treated water satisfies about 67% of the irriga-

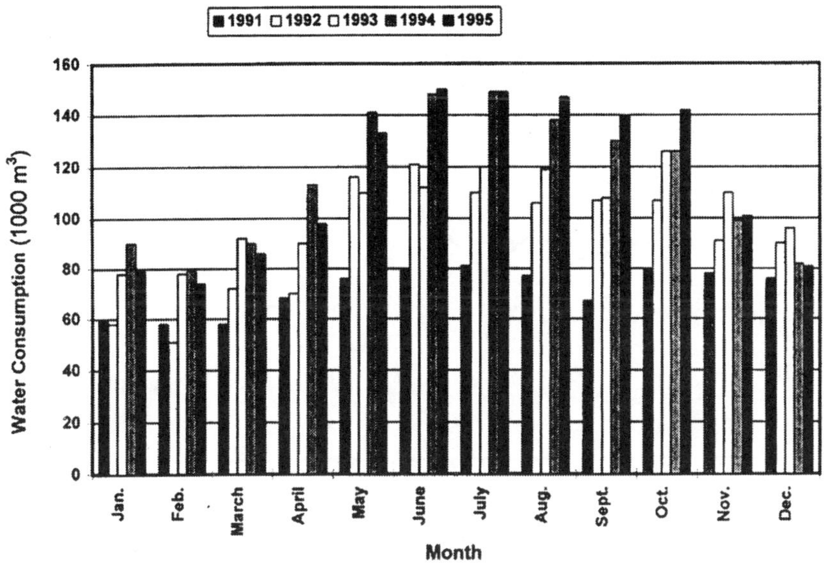

Figure 11. Sultan Qaboos University water consumption.

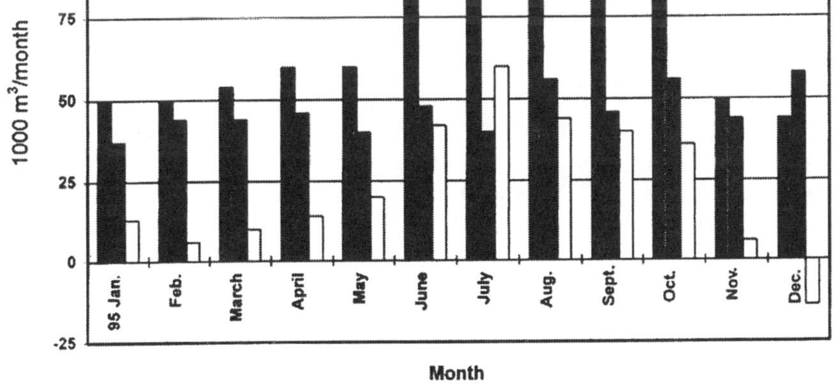

Figure 12. Treated water, irrigation reuse, and shortages at Sultan Qaboos University.

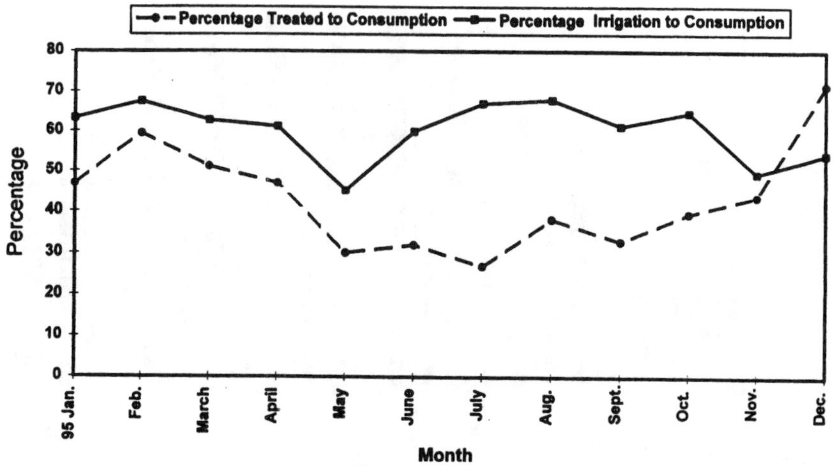

Figure 13. Percentage treated wastewater and irrigation reuse to water consumption (Sultan Qaboos University).

tion requirements, the rest is directly withdrawn from the potable water network.

FUTURE OF WASTEWATER REUSE IN OMAN

Figure 14 shows the treated effluent during the peak summer months for the period of 1987 to 1993 for six sewage treatment plants (Al Ansab, Jibroo, Bausher, Mebilla, Shati Al Qurm, and Darsait). The treated effluent has increased by a factor of about 3 over the six years studied (1987–1993). If the same strategy is adopted in all the sultanate regions, despite the financial and health aspects, the treated effluent consumption might reach 249,000 and 445,000 m^3/d for the years 2000 and 2010, respectively, as shown in Figure 15 (McDonalds and Hawksley, 1993). Such an amount will be sufficient to irrigate about 3113 and 5563 ha, respectively. This area represents 10% of the current cultivated area in the sultanate as shown in Table 5, or about 45 times the green areas in the capital (Table 9).

CONCLUSION

This study focused on the importance of using treated wastewater for irrigation purposes, especially in arid and semi-arid areas. Many

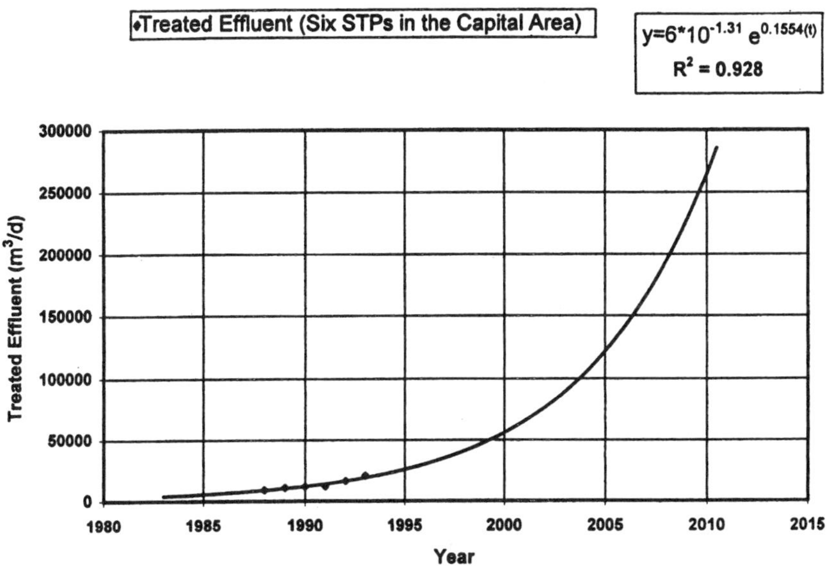

Figure 14. Current and future predicted treated effluents in Muscat, Oman.

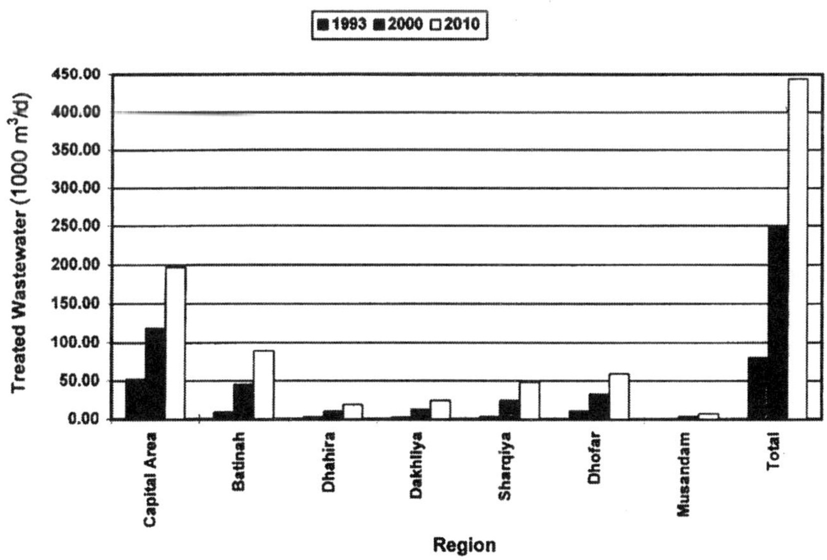

Figure 15. Treated wastewater prediction for different regions of Oman.

quality parameters of importance in the agricultural use of wastewater have to be studied carefully. Appropriate wastewater treatment could provide a high level of health protection. The study indicated a need for using treated wastewater to partially cover the shortage of irrigation of ornamental plants throughout the country, especially in the capital area. The irrigation waste consumption is based on the type of crop and the climatic conditions.

One of the most dangerous sources of pollution to the environment is domestic sewage. In Oman, there is a high percentage of septic and holding tanks in urban housing areas. Septic tanks play a major role in some urban areas and an exclusive role in all rural areas. There are currently about 250 wastewater treatment plants operating in Oman, the majority of them of a localized nature. In total, these plants generate some 28.6 million m^3/d of reclaimed water. Despite the high demand for water for irrigation, the main concern in Oman is the protection of the health of the community and pollution control.

Sultan Qaboos University demonstrates a good example of reuse of treated wastewater. Due to the use of extra treatment units, the water quality of the STP effluent conforms with the Omani standards. At present, the amount of irrigation water used covers the needs of the cultivated land during the winter period and about 50% during the summer months. Considering the ultimate design population, the STP is capable of discharging three times the amount of effluent, which is expected to cover the deficit during the summer period.

The current amount of reused treated wastewater in the Muscat area is about 45,600 m^3/d, sufficient to irrigate about 570 ha. If the strategy of constructing STPs over the period of 1987 to 1993 is followed at the same rate all over the sultanate, an area of 3110 ha and 5560 ha can be irrigated by treated wastewater in the years 2000 and 2010, respectively.

ACKNOWLEDGEMENT

The authors express their thanks to Prof. Dr. Isam Abdel-Magid, Faculty of Engineering, Omdurman Islamic University, Sudan, for his partial participation in the analysis of the results.

REFERENCES

Abdel-Magid, I. M. and A. El-Zawahry. 1992. Establishment of water quality guidelines for the Sultanate of Oman. *Arab Water World J.* 16(5):18–22.

Abdel-Rahman, H. A. and I. M. Abdel-Magid. 1993. Water conservation in Oman. *Water International J.,* 18(2):95–102.

Al-Qasmi, S. M. 1997. Experience of Muscat Municipality in the reuse of treated wastewater. The Third Gulf Water Conference, Muscat, Sultanate of Oman, Arabic 1:67–75.

Al-Sabahi, A. M. 1997. Wastewater treatment and use in the Sultanate of Oman. The Third Gulf Water Conference, Muscat, Sultanate of Oman 3:859–888.

Al-Salem, S. S. 1987. Evaluation of the Al-Asmara waste stabilization pond system and its suitability for unrestricted irrigation. Paper prepared for the Land and Water Development Division, FAO, Rome.

Ayres, R. S. and D. W. Westcot. 1976. Water quality for agriculture. FAO, Rome, Irrigation and Drainage Paper 29.

Bahwan Engineering Company. 1996. Analysis Data. Monthly report, 2844/lab/96, submitted to the Department of Projects and Maintenance (unpublished).

FAO. 1985. Water quality for agriculture. R. S. Ayers and D. W. Westcot. Irrigation and Drainage Paper 29, Rev. 1, FAO, Rome.

Feachem, R. G., D. J. Bradley, H. Garelick, and D. D. Mara. 1983. *Sanitation and Disease: Health Aspects of Excreta and Wastewater Management.* John Wiley, Chichester.

Hillman P. J. 1988. Health aspects of reuse of treated wastewater for irrigation. In: *Treatment and Use of Sewage Effluent for Irrigation,* M. B. Pescod and A. Arar (Eds.). Butterworths, Sevenoaks, Kent.

McDonalds, M. and W. Hawksley. 1993. Wastewater master plans for nine towns of Oman. Ministry of Regional Municipality and Environment, Oman.

MEWR. 1986. Sultanate of Oman law on the conservation of the environment and prevention of pollution: Regulations for waste-water reuse and discharge. Ministerial Decision 5/86 dated 17th May, Muscat, Oman.

MRME. 1993. Regulations for wastewater re-use and discharge. Ministry of Regional Municipalities and Environment, Sultanate of Oman.

MWR. 1990. National Water Resources, Master Plan. Ministry of Water Resources, Sultanate of Oman.

Shuval, H. I., A. Adin, B. Fattal, E. Rawitz, and P. Yekutiel. 1986. Wastewater irrigation in developing countries: Health effects and technical solutions. Technical Paper No. 51. World Bank, Washington, D.C.

Shuval, H. I., P. Yekutiel, and B. Fattal. 1985. Epidemiological evidence for helminths and cholera transmission by vegetables irrigated with wastewater: Jerusalem—case study. *Water Science and Technology* 17 (4/5):433–442.

WHO. 1989. Health guidelines for the use of wastewater in agriculture and aquaculture. Technical Report No. 778. WHO, Geneva.

Index